APPLIED STATISTICAL SCIENCE V

Applied Statistical Science

M. Ahsanullah (Ed.)

Applied Statistical Science V

M. Ahsanullah, J. Kennyon and S.K. Sarkar (Eds.)

ISBN 1-56072-923-6

Applied Statistical Science IV

M. Ahsanullah and F. Yildirim (Eds.)

ISBN 1-56072-682-2

Applied Statistical Science III
Nonparametic Statistics and Related Topics

S.E. Ahmed, M. Ahsanullah, B.K. Sinha (Eds.)

ISBN 1-56072-581-1

Applied Statistical Science II

M. Ahsanullah (Ed.)

ISBN 1-56072-469-2

Applied Statistical Science I

M. Ahsanullah and D.S. Bhoj (Eds.)

ISBN 1-56072-405-6

APPLIED STATISTICAL SCIENCE V

M. AHSANULLAH, J. KENNYON AND S.K. SARKAR
(EDITORS)

Nova Science Publishers, Inc.
Huntington, New York

Library of Congress Cataloging-in-Publication Data
available upon request

ISBN 1-56072-923-6

227 Main Street, Suite 100
Huntington, New York 11743
Tele. 631-424-6682 Fax 631-425-5933
e-mail: Novascience@earthlink.net
Web Site: http://www.nexusworld.com/nova

Printed in the United States of America

CONTENTS

POINT ESTIMATORS AND TESTS OF HYPOTHESIS FOR THE PARAMETERS OF THE CLASSICAL EXTREME VALUE DISTRIBUTION FUNCTIONS

JANOS GALAMBOS, Temple University

Summary

The paper discusses a number of applied models in which the underlying distribution is justified to approximate the classical extreme value distribution functions which have three parameters. The estimation of the parameters from data is a very sensitive task since the estimated value of the shape parameter has critical implications which implications frequently contradict biological, medical, or engineering laws. Therefore, the recommended course of action is to follow point estimation with a test of hypothesis to validate the model stemming from the accepted value for the shape parameter. The other two parameters are related to location and scale, and therefore standard methods can be applied for their estimation.

KEY WORDS: Extremes, classical model, parameter estimation, threshold method, curvature method, wind speed, human life span is unlimited.

1 The Classical Extreme Value Distributions

The present section describes the classical extreme value distributions without references to the model leading to such distribution functions. In section 2 we deal with models as well as with practical problems.

We say that the random variable Z is distributed with a classical extreme value distribution function for the maximum if

$$P(Z \le x) = H_c(x; A, B) = \exp\left\{ - \left(1 + c\frac{x-A}{B}\right)^{-1/c} \right\}, \tag{1.1.1}$$

where $c, A,$ and $B > 0$ are parameters, and

$$1 + c\frac{x - A}{B} \geq 0. \tag{1.1.2}$$

The value $c = 0$ is included in the value set of c by a limiting procedure:

$$H_0(x; A, B) = \lim_{c \to 0} H_c(x : A, B) = \exp\left\{-\exp\left(-\frac{x - A}{B}\right)\right\}, \tag{1.1.3}$$

where the limit is with $|c| > 0$.

We shall write $H_c(x)$ for $H_c(x; 0, 1)$, and call it the standard form for the family $H_c(x; A, B)$. Note that by writing the parameters in the form of (1.1), A and $B > 0$ are location and scale parameters, respectively. The parameter c, called the shape parameter, has a very special role due to (1.2) and (1.3):

$$\text{if } c = 0 \text{ or if } c > 0, \; Z \text{ is unbounded} \tag{1.1.4a}$$

but, for $c < 0$, Z or x in the domain (1.2), satisfies

$$x \leq -(1/c)B + A. \tag{1.1.4b}$$

Remark 1. An implication of (1.4a) and (1.4b) is that a slight error in c (assumed or estimated) may result in a qualitative change for Z: if $c = 0$ is mistakenly estimated as $c = -0.08$, say, an unbounded Z (the case $c = 0$) becomes bounded ($c = -0.08$ means, by (1.4b), that $Z \leq 12.5B + A$). We shall see in a subsequent section the amount of damage that can come from such an error in medical research or engineering standards. On the other hand, for c close to zero, a small error in c results in very small error when computing probabilities. This follows from easily established mathematical estimates on $H_{c_1}(x; A, B) - H_{c_2}(x; A, B)$ in terms of $|c_1 - c_2|$.

It is sometime convenient to separate the three cases of $H_c(x)$ as follows: we write

$$H_{1,c}(x) \text{ for } H_c(x) \text{ with } c > 0, H_{2,c}(x) \text{ for } H_c(x) \text{ with } c < 0, \tag{1.1.5a}$$

and

$$H_{3,0}(x) = H_0(x) = \exp(-e^{-x}), \text{ all } x \tag{1.1.5b}$$

If (1.1) holds for Z, then the distribution function of $W = -Z$ is called a classical extreme value distribution for the minimum. Its form is

$$L_c(x) = P(W \leq x) = P(Z \geq -x) = 1 - H_c(-x). \tag{1.1.6}$$

The particular case $L_c(x) = L_{2,c}(x) = 1 - H_{2,c}(-x)$ has the standard form (we write $d = -1/c > 0$)

$$L_{2,c}(x) = 1 - \exp(-x^d), \; x \geq 0, \; d > 0, \tag{1.1.7}$$

which is widely used in engineering applications, independently of extreme value theory, and known as the Weibull distribution.

2 Mathematical and Applied Models Leading to Classical Extreme Value Distributions

We start with the classical model. Let $X_1, X_2, \ldots, X_n$ be independent copies of the random variable X whose distribution function is $F(x)$. Set $Z_n = \max(X_1, X_2, \ldots, X_n)$. Clearly, $P(Z_n \leq z) = F^n(z)$.

Now, since $F^n(z) \longrightarrow 0$ as $n \longrightarrow +\infty$ for all z for which $F(z) < 1$, one has to change z with n in order to obtain a limiting distribution for Z_n. The classical choice of such $z = z_n$ is to seek constants a_n and $b_n > 0$ such that $F^n(a_n + b_n x)$ converges to a nondegenerate distribution function. It turns out that the only possible limiting distribution functions are the members of the family $H_c(x; A, B)$. The existence of a limiting distribution of $F^n(a_n + b_n x)$, of course, imposes restrictive conditions on $F(x)$, but the conditions are weak enough to be satisfied by the popular distribution functions: normal, lognormal, exponential, Weibull, gamma, uniform, logistic, Cauchy, and many others. The popular discrete distributions such as Poisson or geometric are not included. When one knows that the limiting distribution of the (normalized) Z_n exists, an easy rule for deciding whether it is $H_{3,0}(x)$ of (1.5b) is this: $F^n(a_n + b_n x) < 1$ for all x and, upon denoting by Z_n^+ the positive part of $(Z_n - a_n)/b_n$, if all moments $(Z_n^+)^k$, $k > 0$, are finite, then the limiting distribution is $H_{3,0}(x)$. The theory for this model actually is well developed; see Chapter 2 is Galambos (1987) for details.

We want to use the preceding model not for getting information on the largest observation Z_n in a sample X_j, $1 \leq j \leq n$, on X, but rather to determine the underlying distribution function of a quantity Z which can be approximated by the maximum of certain random variables. In this sense the above classical model is too restrictive because the independence of the associated variables is rarely satisfied. This difficulty is overcome in an extension of the classical model known as the graph-dependent model. It allows quite general dependence among the X_j, and yet in this model, too, the only possible limiting distributions of the normalized Z_n are members of the family $H_c(x; A, B)$. A vague description of the graph-dependent model is this: If X_i and X_j are not (asymptotically) independent, call the pair (i, j) exceptional. Collect all exceptional pairs in a set E_n. An evident assumption is that E_n must not contain a positive percentage of all pairs (u, v), $1 \leq u \leq n$, $1 \leq v \leq n$. Now, we assume that the X_j with subscript outside of E_n (that is, no pairs of their subscripts belong to E_n) are approximately independent, but absolutely no assumption is made on their dependence if at least two pairs of their subscripts belong to E_n. Then, with a simple assumption on interdependence of a subset that has exactly one pair of subscripts from E_n, the previously stated limiting behavior of Z_n follows. For details, see Galambos (1972) or Section 3.9 in Galambos (1987) (where the term E_n^*-sequence is used in place of graph-dependence).

Let us look at some applied models in which the previous two mathematical models can be utilized to justify the application of $H_c(x; A, B)$ as the underlying distribution.

The same description applies to the highest wind speed at a given location or the highest flood level of a river at a given location. The time span over which the highest level is evaluated is immaterial, but we specify it to one year. Now, if X_j denotes the highest level (wind or flood) on day j of the year, then Z_{365} is the highest level in a year. Here, the X_j are acceptable to be identically distributed but not independent. However, clearly, the dependence of X_j and X_{j+k} weakens with increasing k, and thus the graph-dependence structure applies, entailing that Z_{365} is asymptotically distributed as $H_c(x; A, B)$.

For the longest living person in a large society, once again the graph-dependent model, and thus $H_c(x; A, B)$, apply. Here, the assumptions are as follows. Every individual at birth has the same random life length X_j. Small groups (family or small community) have interdependence for economical, social, or medical reasons, but non-interacting individuals have independent life lengths.

One model for fatigue failure of materials, a sheet of metal, say, is that strength depends on the size of the sheet only, and if the sheet is cut hypothetically into equal smaller pieces, then those pieces which are 'close' to each other, may be dependent but those at a larger distance are practically independent. Then the strength of the sheet is the minimum of the strengths of the pieces, hence, by the graph-dependent model, strength determined by fatigue failure is $L_c(x; A, B)$. It should be remarked that the Weibull distribution (the special L_c distribution at (1.6a)) is the preferred choice of the 'life distributions' in an engineering setting for purely statistical reasons. Namely, due to the shape parameter d (recall the relation of d and c at (1.6a)), $L_d(x; A, B)$ covers a large variety of densities. The current set-up gives a mathematical justification to assuming Weibull distributions for strength.

3 Estimation of Parameters

When we settled on a model for a random quantity Z and concluded that Z is (asymptotically) H_c - distributed, then from a set $Z_1, Z_2, \ldots, Z_N$ of observations on Z (such as annual flood levels, annual highest wind speeds, longest living individual in a large society), we want to estimate the parameters A, B and c. From the form in which we wrote up $H_c(x; A, B)$ at (1.1), A and B are location and scale parameters, respectively, and therefore their estimation can be done by standard methods. The case of the shape parameter c, however, is different. It may even be impossible to be estimated by any form of point estimators, and it must certainly not be done by the method known as 'peaks over the threshold' method. We shall return to this last method in the next section and explain why it must not be used. Here we concentrate on the difficulties concerning c, and formulate a course of action one can take.

Because in the sequel we frequently refer to order statistics, let us introduce the following notation. Let us rearrange the sample $Z_1, Z_2, \ldots, Z_N$ in nondecreasing order $Z_{1:N} \le Z_{2:N} \le \ldots \le Z_{N:N}$. Then $Z_{r:N}$ is called the r-th order statistic of the sample, and the terms $Z_{N-k:N}$ with k bounded as N increases are called the upper extremes. When k is allowed to increase moderately relative to N, then $Z_{N-k:N}$ is called an order statistic, or an observation, in the upper tail of H_c. Now, since c is sensitive to upper extremes but not to 'middle terms' among the order statistics (such as p – quantiles for any fixed $p < 1$), any estimator of c must involve order statistics from the upper tail only. Those who are opposed to using only some of the order statistics in an estimator on the ground of losing information will be assured by the result of Janssen (1989) which states that in any (upper) extremal problem all information of the sample is contained in the order statistics in the upper tail. In regard to the statement that quantiles cannot contribute to c we mention that the quantiles of a sample from H_c (from any differentiable distribution function – see p. 255 in David (1981)) are asymptotically normally distributed (this is why the 'middle' of an empirical distribution function is always a straight line on a probability paper).

Now, let $S_k = S(Z_{N:N}, Z_{N-1:N}, \ldots, Z_{N-k:N})$ be an estimator of c. Even if all mathematical niceties in relation to c have been established for large N and k such that $k/N \to 0$, we rarely know that how large N must be, and then, for a given N, which range of k will guarantee that S_k is 'close' to c. The problem is further involved by the fact that the Z_j are only asymptotically H_c – distributed, that is, the underlying distribution deviates from H_c, although it is 'close' to H_c with some specific c. On the other hand, H_{c_1} and H_{c_2} are also close to each other if c_1 is close to c_2 (recall Remark 1 in Section 1), so S_k may estimate c_1 rather than the correct value c_2, say. Error also comes from the fact that S_k is a random value, therefore for different k values, one gets different estimated values for c. However, we need quite an accurate value for c in view of (1.4b). Indeed, if the highest wind is to be estimated in an area where an atomic reactor is to be built, and where A and B are estimated from data as $A = 60$ miles/hour, and $B = 5$ m/h, then one estimated value for $c = -0.25$ gives the highest possible value $Z = 80$ m/h, while another estimate $c = -0.1$ leads to the highest value $Z = 110$ m/h. Such large differences in estimated values of c do occur, and did appear in publication (with smaller value for B)(see Simiu and Heckert (1996) and for a better approach, Galambos and Macri (1999)). As another example, let us look at the case of estimating c for the longest living person in a society. If we misjudge c, and thus provide an upper limit on life length (such as $c = -0.08$ and $Z \leq 122$ in Reiss and Thomas (1997)), then the estimate implies that, for some biological reason, everyone must die by a certain age regardless of how close a person is to that age. An ethical problem would then be that no medical help would be provided to a person close to 'the limit' (122 in the quoted example). But that no such upper limit exists is advocated by more and more medical professionals, and the same statistical data set from which $c = -0.08$ and thus $Z \leq 122$, was concluded in the citation above clearly shows that $c = 0$, that is, Z is unbounded (see Abernethy (1998) and Galambos and Macri (2000), and their references). It should be noted that 'no upper limit' does not mean that death can be avoided. Life span will always be finite, but if we have observed someone with age 122, say, then it is sure that there will be someone with age over 123 at death, and then 124, etc. Furthermore, the increased age of the longest living will not come from medical advances; the longest living might be someone who never received medical care. It is the average life span that is lifted up by medical advances by helping more and more people to avoid death at a young age.

Since a point estimator for c currently does not exist (and cannot be expected to ever exist) which would be reliable enough to conclude that c is negative if it is estimated by a negative value, other approaches are necessary for making decisions on c.

Method 1: The curvature method. Step 1. Plot the empirical distribution function of the observations Z_j, $1 \leq j \leq N$, on a Gumbel probability paper, which means that the probabilities are plotted in the scale of $H_{3,0}$ of (1.5b)(which distribution is called Gumbel distribution): $y = -\ln(-\ln p)$. Because the empirical distribution function jumps to $p = r/N$ at the order statistic $Z_{r:N}$, several authors replace r/N by $(r - 1/2)/N$ in order to avoid a blow-up of the graph at $Z_{N:N}$. However, we do not adopt this modification; rather we stop at $Z_{N-1:N}$. Step 2. Choose two integers r_1 and r_2 with $0 < r_1 < r_2 < \sqrt{N}$. There is no general rule for the magnitude of these integers. Step 3. Fit a straight line by the method of least squares to the points $(Z_j, -\ln[-\ln(j/N)])$, $N - r_2 \leq j \leq N - r_1$, and another one to the succeeding points with $N - r_1 \leq j \leq N - 1$. Step 4. If the two fitted lines in Step 3 are so close to each other that it can be concluded by visual inspection that the graph of the empirical distribution function is a straight line, then accept that $c = 0$. On the other hand, if the slopes of the lines in Step 3 indicate that the upper tail is concave up (convex) then accept $c < 0$, while for indication of the upper tail being concave down (concave), accept $c > 0$. If visual inspection is not conclusive, use statistical decision on whether the curvature of the upper tail is a straight line, concave up, or concave down by utilizing the table on p. 225 in the book by Castillo (1988). Another alternative to the visual method is Method 2 below which actually identifies c at an acceptable level. The underlying theory for Method 1 is the result of Castillo, Galambos and Sarabia (1989).

Method 2: The best least square acceptance for c. This is a modification of the method of Filliben (1975), adapted to our extreme value setting. Choose a value $0 < r < \sqrt{N}$. Generate a sequence of c – values $c_1, c_2, \ldots, c_t$, where the c_j should always include 0, then the gap between consecutive c_j is advisable to be small around zero, which can widen as the absolute value of c_j moves away from zero. Compute $h_{j,k} = H_{c_k}(z_j)$ for $N - r \leq j \leq N$ and $1 \leq k \leq t$.

For every fixed k, compute the distance

$$d_k = \sum_{j=N-r}^{N} [F_N(Z_j) - h_{j,k}]^2 = \sum_{j=N-r}^{N} \left(\frac{j}{N} - h_{j,k}\right)^2,$$

where $F_N(x)$ is the empirical distribution function, that is, $F_N(Z_j) = j/N$. Find $M_t = \min\{d_k, 1 \leq k \leq t\}$. Accept that c_j as the value of c for which $M_t = d_j$.

4 The Threshold Method and Why It Should Not Be Used for Estimatingc

The threshold method for estimating c was introduced by de Haan (1994), and it became quite popular among scientists working on applied problems. de Haan himself presented applied problems and the present author indirectly questioned the conclusions about human life length (Galambos (1994), p. 10). We carried the paper because the mathematical theory in it is very nice as a limit theorem. On the other hand, in applications the very restrictive conditions for the validity of the limit theorem are ignored.

The method starts with choosing a k that is very large but at the same time it should be very small compared with the sample size N. The sample size N itself must be very large. It would still be acceptable, but then for the choice of k a condition is introduced which condition is dependent on the population distribution: for different population distributions different limitations on k are imposed: k must be chosen by the combination of the limit theorem in Remark 2 on p. 102 and by formula (4.2). The limit theorem in Remark 2 assumes that the population distribution is known very accurately. It is then added on p.107 that if k is not chosen by the required rule at (4.2), then the recommended estimator becomes biassed, that is, no longer c is the estimated value. The actual bias again depends on the population distribution that is usually unknown. Finally, the procedure breaks down for normal populations. In other words, if the population distribution of the age at death is normally distributed, then this particular estimation is not valid, and if we do not know whether the population distribution is normal or not, then we do not know whether the estimation is valid or not.

I intentionally did not formulate the actual estimator because the discussion in the preceding paragraph clearly indicates that one cannot rely on this estimation procedure.

REFERENCES

Abernethy, J. (1998). Gompertzian mortality originates in the winding-down of the mitotic clock.
Journal of Theoretical Biology 192, 419–435.

Castillo, E.,(1988). Extreme Value Theory in Engineering, Academic Press Inc. Boston.

Castillo, E., Galambos J. and Sarabia, J. M.(1989). The selection of the domain of attraction of an extreme value distribution from a set of data. Lecture Notes in Statistics 51, Editors J. Hüsler and R.-D. Reiss, pp. 181–190, Springer, New York.

David, H. A. (1981). Order Statistics, 2nd ed., Wiley, New York.

De Haan, L. (1994). Extreme value statistics. In: Extreme Value Theory and Applications
(J. Galambos, J. Lechner and E. Simiu eds.), pp. 93–122, Kluwer, Dordrecht.

Filliben, J.J. (1975) The probability plot correlation plot test for normality. Technometrics 17, 111-117.

Galambos, J. (1972). On the distribution of the maximum of random variables. Ann. Math. Statistics 43, 516–521.

Galambos, J. (1987). The Asymptotic Theory of Extreme Order Statistics. 2nd ed., Krieger, Melbourne, Florida.

Galambos, J. (1994). Extreme value theory for applications. In: Extreme Value Theory and Applications (J. Galambos, J. Lechner and E. Simiu, eds.) pp. 1–14, Kluwer, Dordrecht.

Galambos, J. and Macri N. (1999). Classical extreme value model and prediction of extreme
winds. Journal of Structural Engineering. 125, No. 7, 792 – 794.

Galambos, J. and Macri, N. (2000). The life length of humans does not have a limit. Journal of Applied Statistical Science, 10, to appear.

Janssen, A. (1989). The role of extreme order statistics for exponential families. In: Lecture Notes in Statistics 51 (J. Hüsler and R.-D. Reiss, eds.), pp.204–221. Springer, New York.

Reiss R.-D. and Thomas M. (1997) Statistical Analysis of Extreme Values. Birkhauser, Basel.

Simiu, E. and Heckert, N. A.(1996). Extreme wind distribution tails: A Peaks over Threshold approach. Journal of Structural Engineering, 122, No. 5, May, 539–547.

Distributions Between Uniform and Exponential

M. Ahsanullah
Deprtment of Management Sciences
Rider University
Lawrenceville, NJ 08648-3099, USA
V.B. Nevzorov*
Department of Mathematics and Mechanics
St.-Petersburg State University,
St.-Petersburg, Russia

Abstract

. Some relations between classical order statistics, records and generalized order statistics are given . We explain why some properties of order statistics can characterize uniform distributions while the same properties of records give characterizations of exponential distributions. Based on properties of generalized order statistics new characterizations of a distribution family which includes both the uniform and exponential distributions are obtained.

1. Introduction

Let $X_1, X_2, \ldots$, be independent random variables (r.v.'s) with a common continuous distribution function (d. f.) F and $X_{1,n} \leq X_{2,n} \leq \ldots \leq X_{n,n}$, n=1,2,...., be the corresponding order statistics (o.s.). Very close objects to order statistics are record times L(n) and record values X(n), which are defined as follows:

$$L(1)=1,\ L(n+1)=\min\{j: X_j > X_{L(n)}\},\ n=1,2,\ldots$$
$$X(n)=X_{L(n)},\ n=1,2,\ldots.$$

Detailed theory of order statistics and records see Nevzorov (1987), David (1981), Arnold, Balakrishnan and Nagaraja (1992) and Ahsanullah (1995). There are some properties which are mutual for order statistics and for record values. This fact is not surprising because record values $X(n)=X_{L(n),L(n)}$ can be regarded as extreme order statistics for samples with the random sizes L(n). Let us list some important properties of order statistics and records.

Example 1. Let F(x)=max {0,1-exp(-(x-a)/b)}. It is known (see Sukhatme (1937), Tata (1969)) that in this case r.v.'s

$$X_{1,n},\ X_{2,n}-X_{1,n},\ \ldots \qquad (1)$$

as well as differences

$$X(1),\ X(2)-X(1),\ldots \qquad (2)$$

are independent. Moreover, the independence of r.v.'s (1) and (2) under some mild restrictions on d.f. F characterizes the exponential distribution.

Example 2. Let F(x)=max {0,1-exp(-x/b)}. Then (Rossberg (1972))

$$(n-r)(X_{r+1,n}-X_{r,n}) \overset{d}{=} nX_{1,n} \tag{3}$$

where $X \overset{d}{=} Y$ denotes that X and Y have the same distribution, and relation (3) characterizes the distribution F(x)=max {0,1-exp(-x/b)}. Analogously, the following equality

$$X(r+1)-X(r) \overset{d}{=} X(1) \tag{4}$$

characterizes the same distribution.

It appears that such analogy between characterizations fails in some other situations.

Example 3. Lin (1988) has proved that if $E|X_1|^p<\infty$ for some p>2, then

$$(EX(r))^2 \le (r/(r-1))EX^2(r-1)\ , r\ge 2, \tag{5}$$

and the equality in (5) holds for the exponential distribution F(x)=max {0,1-exp(-x/b)} only. Meanwhile Lin has obtained that if $EX_1^2<\infty$ then an analogous inequality is attained for order statistics as well:

$$(EX_{r,n})^2 \le (rn/(r-1)(n+1))EX^2_{r-1,n-1}\ ,2 \le r \le n, \tag{6}$$

but now the equality in (6) is attained for degenerate and uniform U(0,b), b>0, distributions.

Example 4. Compare the following results for order statistics and records.

Theorem 1 (*Szekely, Mori* (1985)). Let $0<Var(X_{r,n})<\infty$ for some r=i,j, where $1\le i<j \le n$. Then the correlation coefficient $\rho(X_{i,n},X_{j,n})$ satisfies the following inequality:

$$\rho(X_{i,n},X_{j,n}) \le (i(n+1-j)/j(n-i+1))^{1/2}, \tag{7}$$

and the equality in (7) holds if and only if X_1 has the uniform distribution.

Theorem 2 (*Nevzorov (1992)*). Let $0<Var(X_{r,n})<\infty$ for some r=i,j, where $1\le i<j$. Then

$$\rho(X(i),X(j)) \le (i/j)^{1/2}, \tag{8}$$

and the equality in (8) holds if and only if

F(x)=max {0,1-exp(-x/b)}.

Note that in all the above examples properties of record values characterize the exponential distribution , meanwhile in two situations the analogous property of order statistics corresponds to exponential distributions as it can be expected and in other two examples we unexpectedly have got characterizations of uniform distributions. Two questions appear in this context. What is the distribution family which connects exponential and uniform types of distributions? Can we predict what distribution is characterized by the same property of records if we know a characterization of a distribution based on a property of order statistics ?

To answer the first question one has to take the so-called generalized uniform distribution (Proctor (1987))

$$F(x)=1-(1-(x-b)/a)^c, \ b<x<b+a, \ -\infty<b<\infty, \ a>0, \ c>0. \quad (9)$$

If c=1, then

$$F(x)=(x-b)/a, \ b<x<b+a,$$

and one gets the uniform type of distributions.

Taking *ac* instead of *a* and letting *c* to infinity one has the exponential distribution function

$$F(x)=1-\exp\{-(x-a)/b\}, \ x>a.$$

We can see that distribution (9) is a bridge between exponential and uniform distributions. Kamps (1995) has considered example 3 and constructed a family of characterizations , based on inequalities of the types (5) and (6), for the generalized uniform distributions. This family of characterizing theorems includes both the results of Lin given in example 3. Below we will get the corresponding characterization of distribution family (9) by the maximal value of correlation coefficient between two generalized order statistics. This new result will tie the theorem of Szekely and Mori and Nevzorov's theorem cited in example 4.

Both the results connected with generalizations of the situations given in examples 3 and 4 show that a characterization of the uniform distribution by a property of order statistics can imply the corresponding characterization of the exponential distribution based on records property. We will try to describe situations when the same distribution can be characterized, as in examples 1 and 2, by identical properties of records and order statistics. We also will give an explanation why "exponential" record characterizations are generated by different ("exponential" and "uniform") characterizations based on properties of order statistics .

It appears that there is one more distribution family which includes exponential and uniform distributions. Both of these types of distributions belong to the domain of attraction of the exponential distribution which is one of extreme (minimal) value distribution. A distribution function F is said to belong to the domain of minimal attraction of a nondegenerate d.f. L if there exist sequences $\{a_n\}$ and $\{b_n>0\}$ such that

$$1-(1-F(a_n+b_nx))^n \to L(x), \ n\to\infty. \quad (10)$$

If (10) holds , we will write $F\in D(L)$.

As it is known from the theory of extremes there are only three types of asymptotic distributions L for minimal values. They are

$L_1(x)=L_{1,\gamma}(x)=1-\exp(-(-x)^{-\gamma})$, if $x<0$, and $L_1(x)=1$, if $x>0$, $\gamma>0$;

$L_2(x)=L_{2,\gamma}(x)=1-\exp(-x^{\gamma})$, if $x>0$, and $L_2(x)=0$, if $x<0$, $\gamma<0$;

$L_3(x)=1-\exp(-\exp x)$, $-\infty < x < \infty$.

For both the types of distribution

$$F(x)=(x-b)/a, \ b<x<b+a,$$

and

$$F(x)=1-\exp\{-(x-a)/b\}, \ x>a,$$

one can write that $F\in D(L_{2,1})$, where $L_{2,1}(x)=1-\exp(-x)$. It will be shown that for any pair (F,L) such that $F\in D(L)$, one can expect that a characterization of distribution function F by a property of order statistics will imply the characterization of d.f. L based on the analogous property of record values. We will consider the case $F(x)=\exp(x)$, $x < 0$, when $F\in D(L_3)$, and discuss the properties of spacings $X_{2,n}-X_{1,n}$, $X_{3,n}-X_{2,n}$,...for d.f. F and the analogous properties of differences X(2)-X(1), X(3)-X(2),... for d.f. $L_3(x)=1-\exp(-\exp x)$.

In the next section we give definitions of generalized order statistics and discuss their relationships with classical order statistics, order statistics with non-integral sample size and record values.

2. Generalized Order Statistics.

The classical order statistics and record values defined in Section 1 are the partial cases of the so-called generalized order statistics (g.o.s.). The detailed theory of g.o.s. is given in Kamps (1995).

Definition 1. Let n=1,2,... , $k\geq 1$, $m_1,\dots,m_{n-1}\in R$ be parameters , $\mathfrak{m}=(m_1,\dots,m_{n-1})$ and $\gamma_r= k+ n-r+ \sum_{j=r}^{n-1} m_j$, r=1,...,n-1. Random variables $U(r,n,\ \mathfrak{m},k)$, r=1,...,n, are said to be uniform generalized order statistics if their joint density function is of the form

$$h_{1,2,\dots,n}(u_1,u_2,\dots,u_n)=k\Big(\prod_{j=1}^{n-1}\gamma_j\Big)\Big(\prod_{j=1}^{n-1}(1-u_j)^{m_j}\Big)(1-u_n)^{k-1},\ 0<u_1\leq u_2\dots\leq u_n<1. \tag{11}$$

Corollary 1. If F is absolutely continuous with density function f, then (11) implies that the joint density function of g.o.s. $X(1,n,\mathfrak{m},k),\dots,X(n,n,\mathfrak{m},k)$ has the following form:

$$f_{1,2,\dots,n}(x_1,x_2,\dots,x_n)= k\Big(\prod_{j=1}^{n-1}\gamma_j\Big)\Big(\prod_{j=1}^{n-1}(1-F(x_j))^{m_j}f(x_j)\Big)(1-F(x_n))^{k-1}f(x_n),$$

$$F^{-1}(0)<u_1\leq u_2\dots\leq u_n<F^{-1}(1), \tag{12}$$

where $F^{-1}(0)=\inf\{x:F(x)>0\}$ and $F^{-1}(1)=\sup\{x:F(x)<1\}$.

In the sequel we will consider the "stationary" case, when $m_1= m_2 =\dots= m_{n-1}$. For the sake of convenience we will denote $m_1=m_2=\dots=m_{n-1}=m-1$ .Let now X(1,n,m,k),...,X(n,n,m,k) be g.o.s. with the joint density function

$$f_{1,2,\dots,n}(x_1,x_2,\dots,x_n)= k\Big(\prod_{j=1}^{n-1}\gamma_j\Big)\Big(\prod_{j=1}^{n-1}(1-F(x_j))^{m}f(x_j)\Big)(1-F(x_n))^{k-1}f(x_n), \tag{13}$$

$$F^{-1}(0)<u_1\leq u_2\dots\leq u_n<F^{-1}(1),$$

where $\gamma_r= k+ (n-r)m$, r =1,2,...,n-1.

If we need to show that g.o.s. correspond to the certain d.f. F we use the following notation $\{X(r,n,m,k),\ 1\leq r\leq n;\ F\}$, as well as notations $\{X_{r,n},\ 1\leq r\leq n;\ F\}$ and $\{X(r)\ 1\leq r\leq n;\ F\}$ are used for the classical order statistics and records from the distribution F. We will denote exponential (when $F(x)=\max\{0,1-\exp(-x)\}$) g.o.s., classical o.s. and record values as Z(r,n,m,k), $Z_{r,n}$ and Z(r) respectively. Evidently,

$$\{X(r,n,1,1),\ 1\le r\le n;\ F\} \overset{d}{=} \{X_{r,n},\ 1\le r\le n;\ F\}$$

and

$$\{X(r,n,0,1),\ 1\le r\le n;\ F\} \overset{d}{=} \{X(r)\ 1\le r\le n;\ F\}. \qquad (14)$$

that is g.o.s. coincide with order statistics, if m=1 and k=1, and g.o.s. coincide with record values , if m=0 and k=1. If m=0 and k=1,2,... then g.o.s. transform to the so-called k-th record values. Since generalized order statistics are interesting for us as random variables that connect records with classical order statistics. We will consider values m only in the interval [0,1].

3. Representations for the Exponential Generalized Order Statistics.

Ordered random variables generally are dependent but there are some representations which help to express these dependent r.v.'s via independent r.v.'s. Let ξ_1 , ξ_2 ,... be independent exponential E(1) r.v.'s. There are classical representations for the exponential order statistics and the exponential record values.

Representation 1. $\{ Z_{r,n},\ 1\le r\le n\} \overset{d}{=} \{ \sum_{j=1}^{r} \frac{\xi_j}{n-j+1},\ 1\le r\le n\}.$ (15)

Representation 2. $\{Z(\ r),\ 1\le r\le n\} \overset{d}{=} \{ \sum_{j=1}^{r} \xi_j\ ,\ 1\le r\le n\}.$ (16)

For g.o.s. Z(r,n,m,k) with the joint density function (14) the following result holds (see Kamps (1995)).

Representation 3. $\{Z(r,n,m,k\},\ 1\le r\le n\} \overset{d}{=} \{ \sum_{j=1}^{r} \frac{\xi_j}{k+(n-j)m},\ 1\le r\le n\}.$ (17)

Then, denote $H(x)=F^{-1}(1-e^{-x})$., F^{-1} being the inverse function of the d.f. F. Equalities (15)-(17) imply the following set of representations for any d.f. F.

Representation 4. $\{X_{r,n},\ 1\le r\le n;\ F\} \overset{d}{=} \{H(\sum_{j=1}^{r} \frac{\xi_j}{n-j+1}),\ 1\le r\le n\}.$ (18)

Representation 5. $\{X(r), 1\le r\le n;\ F\} \overset{d}{=} \{H(\sum_{j=1}^{r} \xi_j\ ),\ 1\le r\le n\}.$ (19)

Representation 6. $\{X(r,n,m,k), 1\le r\le n;\ F\} \overset{d}{=} \{H(\sum_{j=1}^{r} \frac{\xi_j}{k+(n-j)m})),\ 1\le r\le n\}.$ (20)

Rohtagi and Saleh (1988) have brought in the so-called order statistics with non integral sample size. These r.v.'s have no practical interpretations, but they can be interpreted (see, Kamps (1995)) as order statistics or records in some non classical models. One more aim to work with o.s. with non- integral values of the sample size α coinsides with generalized order statistics $X(r,n,m,\ \alpha\text{-}n+1)$ and representations 1 and 4 stay valid if an integer n is substituted for any value $\alpha > 0$.

Representation 1′. $\{ Z_{r,\alpha}, 1\le r\le\alpha\} \overset{d}{=} \{ \sum_{j=1}^{r} \frac{\xi_j}{\alpha - j + 1}, 1\le r\le\alpha\}$. (21)

Representation 4'. $\{ X_{r,\alpha}, 1\le r\le\alpha; F\} \overset{d}{=} \{H(\sum_{j=1}^{r} \frac{\xi_j}{\alpha - j + 1}) 1\le r \le\alpha\}$. (22)

In the next section we consider relationships between g.o.s. , classical o.s. and records which one can get from the given relationships.

4. Relationships Between Generalized Order Statistics, Order Statistics and Record Values

The representations given above clear the correspondence between records and order statistics. We will give some corollaries of these representations. The first of them evidently follows from (17).

Corollary 2. For any continuous d.f. F ,

$$\{X(r), 1\le r\le n; F\} \overset{d}{=} \lim \{ X(r,n,m,1), 1\le r\le n; F\}, m\to 0. \quad (23)$$

Corollary 3. For any continuous d.f. F , $k\ge 1$ and $0<m\le 1$,

$$\{ X(r,n,m,k), 1\le r\le n; F\} \overset{d}{=} \{X_{r,n+k/m-1}, 1\le r\le n; F_m\}, \quad (24)$$

where

$$F_m(x)=1-(1-F(x))^m. \quad (25)$$

Equality (24) follows from the fact that

$$H(\sum_{j=1}^{r} \frac{\xi_j}{k+(n-j)m})= H(\frac{1}{m} \sum_{j=1}^{r} \frac{\xi_j}{k/m+(n-j)})=H_m(\sum_{j=1}^{r} \frac{\xi_j}{\alpha - j + 1}),$$

where

$H_m(x)=H(x/m)= F^{-1}(1-\exp(-x/m))$, and $\alpha=n+k/m-1$.

Now one can check that

$H_m(x)=F_m^{-1}(1-\exp(-x))$, if F_m^{-1} is the inverse function for d.f. $F_m(x)=1-(1-F(x))^m$.

Equalities (23) and (24) immediately imply the following result.

Theorem 3. As $m\to 0$,

$$\{X(r), 1\le r\le n; F\} \overset{d}{=} \lim \{X_{r,n+1/m-1}, 1\le r\le n; F_m\},$$

(26)

where

$F_m(x)=1-(1-F(x))^m$.

In the RHS of (26) we have order statistics with non-integer sample sizes. Choosing a sequence $m_N=1/(N+1)$, N=1,2,..., one can express records via the classical order statistics.

Theorem 3′. As $N\to\infty$,

$$\{X(r), 1\le r\le n; F\} \overset{d}{=} \lim \{X_{r,n+N}, 1\le r\le n; F_N\}, \quad (27)$$

where

$F_N(x)=1-(1-F(x))^{1/(N+1)}$.

Comment 1. Representation (27) and examples 1 and 2 explain why for some distributions a property of order statistics implies the corresponding property of records. Let $F(x)=1-\exp\{-(x-a)/b\}$, then for any N=1,2,..., $F_N(x)=1-\exp\{-(x-a)/b_N\}$, where $b_N = b(N+1)$, also is the exponential d.f. From the well-known property of the spacings of the exponential order statistics one has that for any n=2,3,... and N=1,2,... the random variables $\{X_{1,n+N}, X_{2,n+N}-X_{1,n+N},...,X_{n,n+N}-X_{n,n}\ ;\ F_N\}$ are independent for any exponential d.f. It follows now from (26) that the corresponding transform of the components of the vector in the LHS of (27) gives the independence of r.v.'s X(1), X(2)-X(1),..., X(n)-X(n-1). We cannot get from (27) that the independence of X(1), X(2)-X(1),..., X(n)-X(n-1) characterizes the exponential type of distributions, because the components of the vector $\{X_{1,n+N}, X_{2,n+N}-X_{1,n+N}, ...,X_{n,n+N}-X_{n,n}\ ;\ F_N\}$ are dependent for any F_N in the case when the d.f F is not of the exponential type. For example 2 when $F(x)=1-\exp(-x/b)$ and $F_N(x)=1-\exp\{-x/b(N+1)\}$. In this case

$$\{X_{1,n+N};\ F_N\} \overset{d}{=} \{(n+N-r)/(n+N)(X_{r+1,n+N}-X_{r,n+N});F_N\}, \qquad (28)$$

for any r=1,2,...,n and N=1,2,.... From (27) and (28) one gets that

$$X(1) \overset{d}{=} X(r+1)-X(r),\ r=1,2,...,$$

if $F(x)=1-\exp(-x/b)$ and this property characterizes the exponential distributions.

Of course, the latter facts are known but these examples help to illustrate the method.

Other distribution families for which the inheritance of some properties of order statistics by record values become evident if one applies Theorem 3′ are the following:

$F(x)=1-\exp(-\exp(x-a))$, for which $F_N(x)=F(x-\log(N+1))$;

$F(x)=1-\exp(-(x/b)^{\alpha})$, $x>0$, $b>0$, $\alpha>0$, with $F_N(x)=F(x/(N+1)^{1/\alpha})$ (which includes the exponential distribution, if $\alpha=1$);

$F(x)=1-\exp(-(-x/b)^{-\alpha})$, $x<0, b>0$, $\alpha>0$, for which $F_N(x)=F(x(N+1)^{1/\alpha})$;

$F(x)=F^{(\alpha)}(x)=1-((x-a)/b)^{-\alpha}$, $x>a$, $b>0$, $\alpha>0$, with $F_N(x)=F^{(\alpha/(N+1))}(x)$.

5. Asymptotic Distributions for Minimal Values

There is a correspondence between record values and order statistics which can be explained if one applies the theory of extremes. As it is well - known (see, for example, Galambos (1987)) there are three types of limit distributions L(x) for a suitably centering and normalizing minimal o.s. $(X_{1,n} -a_n)/b_n$, as $n\to\infty$. They are

$$L_1(x)=L_{1,\gamma}(x)=1-\exp(-(-x)^{-\gamma}),\ \text{if } x<0, \text{ and } L_1(x)=1,\ \text{if } x>0,\ \gamma>0; \qquad (29)$$

$$L_2(x)=L_{2,\gamma}(x)=1-\exp(-x^{\gamma}),\ \text{if } x>0, \text{ and } L_2(x)=0,\ \text{if } x<0,\ \gamma<0; \qquad (30)$$

$$L_3(x)=1-\exp(-\exp x). \qquad (31)$$

Definition 3. A distribution function F is said to belong to the domain of minimal attraction of a nondegenerate d.f. L ($F\in D(L)$) if there exist sequences $\{ a_n \}$ and $\{b_n>0\}$ such that

$$1-(1-F(a_n+b_nx))^n \to L(x),\ n\to\infty.$$

Example 5. If $F(x)=1-\exp(-x)$, $x>0$, then $1-(1-F(x/n))^n = F(x) = L_{2,1}(x)$ and $F\in D(L_{2,1}))$

Example 6. If $F(x)=x$, $0<x<1$, then

$1-(1-F(x/n))^n = 1-(1-x/n)^n \to 1-\exp(-x) = L_{2,1}(x)$, $n\to\infty$

and $F\in D(L_{2,1}))$.

Example 7. If $F(x)=\exp(x)$, $x<0$, then

$1-(1-F(x-\log n))^n = 1-(1-\exp(x)/n)^n \to 1-\exp(-\exp(x)) = L_3(x)$, $n\to\infty$,

and $F\in D(L_3))$.

The following result connects order statistics for d.f. F and record values for d.f. L, when F belongs to the minimal attraction of a non-degenerate d.f. L.

Theorem 4. Let F be a continuous d.f. and $F\in D(L)$,with centering and normalizing constants $\{a_n\}$ and $\{b_n\}$, where L is one of the d.f.'s $L_{1,\gamma}(x)$, $L_{2,\gamma}(x)$ and $L_3(x)$. Then for any fixed j=1,2,... ,

$$\{X(j),\ 1\le j\le r;\ L\} \overset{d}{=} \lim \{(X_{j,n}-a_n)/b_n,\ 1\le j\le r;\ F\}, \tag{32}$$

as $n\to\infty$.

Proof of Theorem 4. The joint distribution function of record values X(1), X(2),..., X(r), corresponding to the d.f. L, can be expressed via r.v.'s ξ_1, ξ_2 ,...from representation 5, in the following form:

$$T(x_1,x_2,...,x_r)=P\{X(1)\le x_1, X(2)\le x_2,..., X(r)\le x_r\}=$$
$$P\{\xi_1\le R(x_1), \xi_1+\xi_2\le R(x_2),..., \xi_1+\xi_2+...+\xi_r\le R(x_r)\}, \tag{33}$$

where $R(x)=-\log(1-L(x))$.

Contrariwise , representation 4 implies that

$$T_n(x_1,x_2,...,x_r)=P\{X_{1,n}-a_n\le x_1b_n, X_{2,n}-a_n\le x_2b_n,..., X_{r,n}-a_n\le x_rb_n\}=$$

$$P\{\xi_1\le R_n(x_1), \xi_1+\xi_2\frac{n}{n-1}\le R_n(x_2),..., \xi_1+\xi_2\frac{n}{n-1}+...+\xi_r\frac{n}{n-r+1}\le R_n(x_r)\}, \tag{34}$$

where $R_n(x)= -n\log(1-F(a_n+b_n x))$.

By comparing (33) and (34) one gets that

$$|T(x_1,x_2,...,x_r)-T_n(x_1,x_2,...,x_r)|\le S_1+S_2,$$

where

$$S_1=\sup_{x_1,...,x_r} |P\{\xi_1\le R_n(x_1), \xi_1+\xi_2\frac{n}{n-1}\le R_n(x_2),..., \xi_1+\xi_2\frac{n}{n-1}+...+\xi_r\frac{n}{n-r+1}\le R_n(x_r)\} -$$
$$P\{\xi_1\le R_n(x_1), \xi_1+\xi_2\le R_n(x_2),..., \xi_1+\xi_2+...+\xi_r\le R_n(x_r)\}|,$$

and

$$S_2=\sup_{x_1,...,x_r} |P\{\xi_1\le R_n(x_1), \xi_1+\xi_2\le R_n(x_2),..., \xi_1+\xi_2+...+\xi_r\le R_n(x_r)\}-$$
$$P\{\xi_1\le R(x_1), \xi_1+\xi_2\le R(x_2),..., \xi_1+\xi_2+...+\xi_r\le R(x_r)\}|.$$

Since ξ_1, ξ_2,..., ξ_r are independent exponentially E(1) distributed r.v.'s it is easy to be convinced that for any fixed r the vector

$$(\xi_1, \xi_1+\xi_2\frac{n}{n-1},..., \xi_1+\xi_2\frac{n}{n-1}+...+\xi_r\frac{n}{n-r+1})$$

converges in distribution as $n\to\infty$ to the vector $(\xi_1 , \xi_1+\xi_2 ,..., \xi_1+\xi_2 +...+\xi_r)$. Hence $S_1\to 0$, if $n\to\infty$. Now let us recall that $1-(1-F(a_n+b_nx))^n\to L(x)$, $n\to\infty$. It implies that

$$R_n(x)=-\log((1-F(a_n+b_n x))^n) \to -\log(1-L(x))=R(x),\ n\to\infty,$$

and then $S_2\to 0$. It completes the proof of the theorem.

Comment 2. Theorem 4 explains the reason why we can have "exponential" record characterizations generated by different ("exponential" and "uniform") characterizations based on properties of order statistics . Both the distributions, exponential and uniform, belong to the domain of minimal attraction of the exponential distribution. Hence, for wide class of functionals Φ , (32) implies the convergence

$$\Phi(\bar{X}_n) \to \Phi(\bar{X}), \tag{35}$$

as $n\to\infty$, where

$$\bar{X}_n=\{(X_{j,n} -a_n)/b_n ,1\le j\le r; F\} ,\ \bar{X}=\{X(j), 1\le j\le r; L\},$$

F is the exponential or uniform d.f. and L is the exponential d.f. . Relation (35) immediately holds for continuous bounded functionals Φ, but in other situations we need to check this relation because it can fail if one has no additional restrictions on the initial distributions. In any case, now it seems rather natural that some property is valid for exponential records if it is true for exponential or uniform normalized order statistics. Below we give one more example of such agreement between the analogous properties of order statistics and record values for another pair (F,L) of d.f.'s.

Example 8. Let $F(x)=\exp(x)$, $x<0$. From example 7 we know that $F\in D(L_3)$. It is known that components of the vector

$$(X_{2,n}-X_{1,n}, 2(X_{3,n} -X_{2,n}),...,r(X_{r+1,n}-X_{r,n}); F)$$

are independent and have the same exponential E(1) distribution for any r=1,2,..., and n>r. In this situation the r.v. $X_{1,n}$, unlike the case $F(x)=1-\exp(-x)$, $x>0$, is not independent with the given spacings of order statistics. The statement of theorem 4 immediately implies that for any r=2,3,... r.v.'s

$$X(2)-X(1), 2(X(3)-X(2)),..., r(X(r+1)-X (r))$$

are also independent and identically E(1) distributed for d.f. $L_3(x)=1-\exp(-\exp(x))$ (Houchens (1984)).

6. Characterizations by Properties of Generalized Order Statistics.

Corollary 3 gives the tools to investigate generalized order statistics. We can rewrite (24) in the following way:

$$\{ X(r,n,m,k), 1\le r\le n; F_{1/m}\} \overset{d}{=} \{X_{r,n+k/m-1}, 1\le r\le n; F\}, \tag{36}$$

where

$$F_{1/m}(x)=1-(1-F(x))^{1/m}.$$

Relation (36) helps us to extend the known results of inter valued sample sizes to non-integer valued sample sizes to obtain new results for g.o.s.

Now there are a lot of results for the classical o.s. rather than for o.s. with non-integral sample size, but very often it is enough only to analyze the proof of a result for o.s. $X_{r,n}$ to make sure that it can be reformulated for o.s. $X_{r,\alpha}$ with any $\alpha \geq r$. For example, looking through the proof of Theorem 1 one can check that the parameter n of a sample size doesn't need to be an integer. Hence Theorem 1 can be given in the following more general form.

Theorem 1′. Let $0<Var(X_{r,\alpha})<\infty$ for some r=i,j, where $1\leq i<j \leq \alpha$. Then the correlation coefficient $\rho(X_{i,\alpha},X_{j,\alpha})$ satisfies the following inequality:

$$\rho(X_{i,\alpha},X_{j,\alpha}) \leq (i(\alpha+1-j)/j(\alpha-i+1))^{1/2}, \qquad (37)$$

and the equality in (7) holds if and only if X_1 has the uniform distribution.

For the uniform d.f. F(x)=(x-a)/b, a<x<a+b, d.f.'s $F_{1/m}(x)=1-(1-F(x))^{1/m}$, $0<m\leq 1$, are of the form

$$F(x,m)=F_{1/m}(x)=1-(1-(x-a)/b)^{1/m}, \quad a<x<a+b. \qquad (38)$$

From (36) and (37) , with α= n+k/m-1, one immediately gets a new generalizations of Theorem 1.

Theorem 5. Let $0<Var(X(r,n,m,k))<\infty$ for some r=i,j, where $1\leq i<j \leq n$. Then for any $0<m\leq 1$ and $k\geq 1$, the correlation coefficient $\rho(X(i,n,m,k),X(j,n,m,k))$ satisfies the following inequality:

$$\rho(X(i,n,m,k),X(j,n,m,k)) \leq (i(n+ k/m -j)/j(n+ k/m -i))^{1/2}, \qquad (39)$$

and the equality in (39) holds if and only if X_1 has the d.f. F(x,m).

Comment 3. If m in Theorem 5 tends to 0, then the RHS of (39) transforms into $(i/j)^{1/2}$, d.f.'s F(x,m) , b being written as b/m, converge to the exponential d.f. $F_0(x)=1-\exp(-(x-a)/b\}$, and we get the statement of Theorem 2 as the limit form of Theorem 5 (under k=1). Taking m=1 in Theorem 5 one comes to Theorem 1.

In Example 2 we discussed Rossberg's (1972) equality (3)

$$(1-r/n)(X_{r+1,n}-X_{r,n}) \overset{d}{=} X_{1,n}$$

which characterizes the distribution F(x)=max {0,1-exp(-x/b)}.

There are some modifications of Rossberg' s theorem (see, Ahsanullah (1976, 1977) and some theorems for record values in the spirit of this result. For example, Ahsanullah (1977) has proved that if X_1 has a density function f and a monotone failure rate h(x)=f(x)/(1-F(x)) then the next two statements are equivalent:

(1) X_1 has d.f. F(x)= max {0,1-exp(-x/b)};

(2) for any $1\leq m<n$, $X(n)-X(m) \overset{d}{=} X(n-m)$. (40)

It appears, that the parameter n in (3) can take non-integral values. The following generalization of Rossberg' s result is valid.

Theorem 6. Suppose X_1 has a non-latticed distribution. Then the statements:

(1) X_1 has d.f. F(x)= max {0,1-exp(-x/b)};

(2) For any r=1,2,... and $\alpha\geq r+1$, $X_{1,\alpha} \overset{d}{=} \frac{\alpha-r}{\alpha}(X_{r+1,\alpha}-X_{r,\alpha})$; (41)

are equivalent.

For $F(x)=\max\{0,1-\exp(-x/b)\}$ the corresponding d.f.'s $F(x,m)=F_{1/m}(x)$ also have the same form for any $0<m\leq 1$. Hence, combining (36), which has been formulated for continuous d.f.'s F and (41), with $\alpha= n+k/m-1$, implies the following result for g.o.s.

Theorem 7. Suppose X_1 has a continuous distribution. Then the next two statements are equivalent:

(1) X_1 has d.f. $F(x)=\max\{0,1-\exp(-x/b)\}$;

(2) For any r=1,2,..., n-1 and $0<m\leq 1$,

$$X(1,n,m,k) \overset{d}{=} \frac{k+m(n-r+1)}{k+m(n-1)}(X(r+1,n,m,k)-X(r,n,m.k)) . \qquad (42)$$

Corollary 4. It follows from relation (23) and Theorem 7 that the equality

$$X(1) \overset{d}{=} X(r+1)-X(r), \qquad (43)$$

which presents a partial case of (40), holds for d.f. $F(x)=\max\{0,1-\exp(-x/b)\}$. However for any n the relation (42) is not valid for all distributions, except for the exponential one, we can not state that (43) fails for non-exponential distributions, because the situation is possible when for some distributions (42) fails for any $0<m\leq 1$ but it holds asymptotically as $m\to 0$. Therefore using Theorem 7 one can predict that (43) characterizes the exponential distributions but this fact needs additional arguments and may be some additional restrictions on the initial set of d.f. have to be introduced. For example, the corresponding result of Ahsanullah sited in Comment 3 is formulated under additional condition that r.v. X_1 has a monotone failure rate.

We have obtained here generalizations of the results given in Examples 2 and 4. The properties of g.o.s. which are analogous to the properties of the classical o.s. and record values from Examples 1 and 3 were discussed in Kamps (1995, p.81,146).

One more theorem for g.o.s. which we consider in the paper is connected with example 8.

Let $F(x)=\exp(x)$, $x<0$. Then $F(x,m)=F_{1/m}(x)=1-(1-\exp(x))^{1/m}$. One knows from Example 8 that components of the vector

$$(X_{2,n}-X_{1,n},\ 2(X_{3,n}-X_{2,n}),\ldots,r(X_{r+1,n}-X_{r,n});\ F)$$

are independent and have the same exponential E(1) distribution for any r =1,2,..., and n>r. Combining this result with relation (36) immediately implies the following theorem.

Theorem 8. Let $F(x,m)=1-(1-\exp(x))^{1/m}$. Then for any r=2,3,...,n-1 and $0<m\leq 1$ elements of the vector

$$(X(2,n,m,k)-X(1,n,m,k),\ 2(X(3,n,m,k)-X(4,n,m,k),\ldots,\ r(X(r+1,n,m,k)-X(r,n,m,k)))$$

are independent and have the common exponential E(1) distribution.

REFERENCES

Ahsanullah, M (1976). On a characterization of the exponential distribution by order statistics. J. Appl. Prob,13 , 818-822.

Ahsanullah, M. (1977). A characteristic property of the exponential distribution. Ann.Statist., 5 (3), 580-582.

Ahsanullah, M (1995). Record statistics. Nova Science Publishers, Inc. Commack,NY

Arnold,B.C., N. Balakrishnan and H.N. Nagaraja (1992). A first course in order statistics. Wiley, New York.

David, H.A. (1981), Order Statistics. Second Edition, Wiley, New York.

Galambos, J. (1987)). The asymptotic theory of extreme order statistics. Kreiger, FL
Houchens, R.L. (1984)).Record value theory and inference. Ph.D. Thesis, University of California, Riverside.
Kamps, U. (1995). A concept of generalized order statistics. B.G. Teubner Stuttgart.
Lin, G.D. (1988). Characterizations of uniform distributions and of exponential distributions. Sankhya 50 A, 64-69.
Nevzorov, V. (1987). Records. Theory Probab. and Appl., v.32, 201-228.
Nevzorov, V. (1992). A characterization of exponential distributions by correlation between records. Mathematical methods of statistics,1,49-54.
Proctor, J.W. (1987). Estimation of two generalized curves covering the Pearson system. Proceedings of ASA Section on Statistical Computing. 287-292.
Rohatgi, V.K.and A,K.Saleh (1988).A class of distributions connected to order statistics with nonintegral sample size. Commun.Statist.-Theory Meth., 17, 2005-2012.
Rossberg, H.J. (1972). Characterization of the exponential and the Pareto distributions by means of some properties of the distributions which the differences and quotients of order statistics are subject to. Mathematish Operationsforschung und Statistik, Series Statistics, 3, 207-215.
Tata, M.N. (1969). On outstanding values in a sequence of random variables. Z. Wahrscheinlichkeitsth. verw. Geb. 12, 9-20.
Sukhatme, P.V. (1937). Tests of significance for samples of the χ^2 population with two degrees of freedom. Ann. Eugenics 8, 52-56.
Szekely, G.J. and T.F. Mori (1985)). An extremal property of rectangular distributions. Stat.& Probab. Lett., 3, 107-109.

STABILITY ANALYSIS OF PHARMACEUTICAL PRODUCTS IN MULTIPLE BATCHES: PRELIMINARY TEST AND SHRINKAGE APPROACH

by

A.K.Md.E. Saleh
Carleton University, Ottawa
and
K.M. Hassanein and Edward F. Brown
Kansas University Medical Centre, Kansas City

ABSTRACT

Pharmaceutical products are routinely monitored for their stability over time. Stability studies generally consist of a random sample of dosage units from a batch or several batches placed in a storage room under controlled temperature and humidity, then assayed on the drug content. The degradation of the drug product is modeled by linear regression functions, and according to FDA, the shelf-life is calculated as the timepoint at which the lower 95% confidence limit below the fitted regression line crosses the lowest acceptable limit of the drug content (usually 90% of labeled unit).

When multiple batches are concerned, preliminary test procedure is applied for testing the equality of slopes of the fitted lines to decide to pool or not to pool the slopes. The shelf-life is then computed batch by batch following FDA regulations as stated above. In this paper, we apply the appropriate preliminary test and shrinkage (empirical Bayes) estimation procedure for the slopes and determine the common shelf-life based on confidence sets using the estimators. Some display of data analysis is also presented to illustrate the procedure.

1. INTRODUCTION

One aspect of the property of pharmaceutical products is the stability of the drug contents over time. In the pharmaceutical industry stability studies are routinely carried to measure degradation rates of new drug products or old drug repackaged in a new way. Stability studies consist of a random sample of dosage units (tablets, capsules, vials etc.) from a manufactured batch stored under controlled conditions of temperature and humidity. Individual samples are then taken at predetermined time points and essayed for drug contents. Based on these studies, the rate of decrease of drug contents (degradation rate) is modelled by regression lines with time as the independent variable. Shelf-life is then calculated according to FDA guidelines, as the time point where the lower 95% confidence interval below the fitted line intersects with the lower acceptable limit for the drug content which is usually 90% of the labeled amount. The shelf-life is used to determine the expiration date which is printed on the package label of the drug.

Usually, for the multibatch problem, the slopes and intercepts of simple regression lines are pooled to determine one shelf-life for all batches. This pooling method is questionable, hence a preliminary test is carried out on the equality of slopes and according to the outcomes they are pooled or never pooled and shelf-life is calculated batch by batch and minimum of the shelf-life is chosen as one value.

2. STATISTICAL METHODOLOGY

2.1 Statistical Model and Estimation of Parameters

Consider the p batches with simple linear regression models:

$$Y_{ij} = \theta_i + \beta_i t_{ij} + e_{ij}; \quad j = 1, \ldots, n_i, \ i = 1, \ldots, p \tag{2.1}$$

where θ_i and β_i are the **intercept** and **slope** (degradation rate) of the ith batch, e_{ij} is the error which is distributed as $\mathcal{N}(0, \sigma^2)$, Y_{ij} is the jth response in the ith batch at time t_{ij}. For the stability studies of the p batches, statistical problem relates to the combined estimation of the predictive vector parameter $\boldsymbol{\theta} + \boldsymbol{\beta} t$ at time t [where $\boldsymbol{\theta} = (\theta_1, \ldots, \theta_p)'$ and $\boldsymbol{\beta} = (\beta_1, \ldots, \beta_p)'$ are the **intercept** and the **slope** vector parameters respectively] when it is suspected that the slope parameters may be equal, i.e. $\boldsymbol{\beta} = \beta_0 \mathbf{1}_p$, $\mathbf{1}_p = (1, \ldots, 1)'$ – a p-tuple of ones and β_0 is a scalar.

It is known from the least-square principle that the least-squares estimators of $\boldsymbol{\theta}$ and $\boldsymbol{\beta}$ are given by

$$\widetilde{\boldsymbol{\theta}} = (\widetilde{\theta}_1, \ldots, \widetilde{\theta}_p)' \quad \text{and} \quad \widetilde{\boldsymbol{\beta}} = (\widetilde{\beta}_1, \ldots, \widetilde{\beta}_p)' \tag{2.2}$$

respectively, where

$$\widetilde{\theta}_i = \bar{Y}_i - \widetilde{\beta}_i \, \bar{t}_i, \quad \widetilde{\beta}_i = \sum_{j=1}^{n_i} (t_{ij} - \bar{t}_i) y_{ij} / n_i Q_i, \tag{2.3a}$$

$$\bar{Y}_i = \frac{1}{n_i}\sum_{j=1}^{n_i} y_{ij}, \quad \bar{t}_i = \frac{1}{n_i}\sum_{j=1}^{n_i} t_{ij} \text{ and } n_i Q_i = \sum_{j=1}^{n_i}(t_{ij} - \bar{t}_i)^2. \tag{2.3b}$$

Let us define $\bar{\mathbf{Y}}$ as

$$\bar{\mathbf{Y}} = (\bar{Y}_1, \ldots, \bar{Y}_p)' \tag{2.4a}$$

and the diagonal matrix $\mathbf{T}_p$ as

$$\mathbf{T}_p = \text{Diag}(\bar{t}_1, \ldots, \bar{t}_p). \tag{2.4b}$$

Then, the estimators of $\boldsymbol{\theta}$ and $\boldsymbol{\eta} = \boldsymbol{\theta} + \boldsymbol{\beta} t$ may be written as

$$\widetilde{\boldsymbol{\theta}}_n = \bar{\mathbf{Y}} - \mathbf{T}_p \widetilde{\boldsymbol{\beta}}_n, \tag{2.5a}$$

$$\widetilde{\boldsymbol{\eta}}_n = \widetilde{\boldsymbol{\theta}}_n + \widetilde{\boldsymbol{\beta}}_n t. \tag{2.5b}$$

All these estimators with "tilde" will be referred to as **unrestricted estimators** (UE).

If the hypothesis $H_0 : \boldsymbol{\beta} = \beta_0 \mathbf{1}_p$ is true, then the pooled estimator of $\boldsymbol{\beta}$ is given by

$$\widehat{\beta}_n \mathbf{1}_p = (\widehat{\beta}_n, \ldots, \widehat{\beta}_n)' \tag{2.6}$$

where

$$\widehat{\beta}_n = \frac{n_1 Q_1 \widetilde{\beta}_1 + \cdots + n_p Q_p \widetilde{\beta}_p}{nQ}, \quad nQ = n_1 Q_1 + \cdots + n_p Q_p. \tag{2.7}$$

Then, the **restricted estimator** (RE) of $\boldsymbol{\theta}$ and $\boldsymbol{\eta} = \boldsymbol{\theta} + \boldsymbol{\beta} t$ are given by

$$\widehat{\boldsymbol{\theta}}_n = \bar{\mathbf{Y}} - \mathbf{T}_p \widehat{\beta}_n \mathbf{1}_p = \widetilde{\boldsymbol{\theta}}_n + \mathbf{T}_p \mathbf{J} \widetilde{\boldsymbol{\beta}}_n \tag{2.8}$$

and

$$\widehat{\boldsymbol{\eta}}_n = \widetilde{\boldsymbol{\theta}}_n + \mathbf{T}_p\, \mathbf{J}\, \widetilde{\boldsymbol{\beta}}_n + \widehat{\beta}_n \mathbf{1}_p t, \tag{2.9}$$

where

$$\mathbf{J} = I_p - \frac{\mathbf{1}_p\, \mathbf{1}_p'}{nQ}\mathbf{D}_{22}^{-1}, \text{ and } \mathbf{D}_{22} = \text{diag}\left(\frac{1}{n_1 Q_1}, \ldots, \frac{1}{n_p Q_p}\right). \tag{2.10}$$

2.2 Testing of Hypothesis.

For the testing of the null-hypothesis $H_0 : \boldsymbol{\beta} = \beta_0 \mathbf{1}_p$ Vs. $H_A : \boldsymbol{\beta} \neq \beta_0 \mathbf{1}_p$, we consider the likelihood ratio statistic given by

$$\mathcal{L}_n = \frac{(\widetilde{\boldsymbol{\beta}}_n - \widehat{\beta}_n \mathbf{1}_p)' \mathbf{D}_{22}^{-1} (\widetilde{\boldsymbol{\beta}}_n - \widehat{\beta}_n \mathbf{1}_p)}{(p-1) S_e^2} = \frac{\widetilde{\boldsymbol{\beta}}_n' \mathbf{J}' \mathbf{D}_{22}^{-1} \mathbf{J} \widetilde{\boldsymbol{\beta}}_n}{(p-1) S_e^2} \tag{2.11}$$

with

$$S_e^2 = (n - 2p)^{-1} \sum_{i=1}^{p} \sum_{j=1}^{n_i} \left\{ (Y_{ij} - \bar{y}_i) - \widetilde{\beta}_i (t_{ij} - \bar{t}_i) \right\}^2. \tag{2.12}$$

Under H_0, $\mathcal{L}_n$ follows an F-distribution with $[(p-1),\ (n-2p)]$ degrees of freedom (D.F.) while under H_A, it follows a non-central F-distribution with $[(p-1),\ (n-2p)]$ D.F. and non-centrality parameter Δ given by

$$\Delta = \sigma^{-2}\boldsymbol{\delta}'\mathbf{D}_{22}^{-1}\boldsymbol{\delta}, \quad \boldsymbol{\delta} = \mathbf{J}\widetilde{\boldsymbol{\beta}}_n. \tag{2.13}$$

See for example ANCOVA tables 2 and 4.

2.3 Preliminary Test Estimator of θ and $\eta = \theta + \beta t$.

If the null-hypothesis $H_0 : \boldsymbol{\beta} = \beta_0 \mathbf{1}_p$ is **uncertain**, one may remove this uncertainty by testing it out. Then, we may define the preliminary test estimator (PTE) of $\boldsymbol{\theta}$ and $\boldsymbol{\eta} = \boldsymbol{\theta} + \boldsymbol{\beta} t$ as follows:

$$\widehat{\boldsymbol{\theta}}_n^{PT} = \widehat{\mathbf{Y}} - \mathbf{T}_p\,\widehat{\boldsymbol{\beta}}_n^{PT} = \boldsymbol{\theta}_n + \mathbf{T}_p\,\mathbf{J}\widetilde{\boldsymbol{\beta}}_n\, I(\mathcal{L}_n < F_\alpha) \tag{2.14}$$

and

$$\widehat{\boldsymbol{\beta}}_n^{PT} = \widetilde{\boldsymbol{\beta}}_n - (\widetilde{\boldsymbol{\beta}}_n - \widehat{\beta}_n \mathbf{1}_p) I(\mathcal{L}_n < F_\alpha) \tag{2.15}$$

where $I(A)$ is the indicator function of the set A and F_α is the upper $\alpha\%$–point of the central F-distribution with $[(p-1),\ (n-2p)]$ D.F. Then,

$$\widehat{\eta}^{PT} = \widetilde{\boldsymbol{\theta}}_n + \mathbf{T}_p\,\mathbf{J}\,\widetilde{\boldsymbol{\beta}}_n I(\mathcal{L}_n < F_\alpha) + \left\{\widetilde{\boldsymbol{\beta}}_n - (\widetilde{\boldsymbol{\beta}}_n - \widehat{\beta}_n\,\mathbf{1}_p) I(\mathcal{L}_n < F_\alpha)\right\} t. \tag{2.16}$$

2.4 Shrinkage Estimators of θ and β.

One may notice from section 2.3 that the PTE depends on the choice of α, the level of significance and it is not a smooth function. That is to say, the transition between the restricted and unrestricted estimators is not smoooth to obtain intermediate values. Thus, we attempt to obtain the smooth versions of the PTE as given below. First, we modify (2.15) to obtain the shrinkage estimator (SE) of $\boldsymbol{\beta}$ as,

$$\widehat{\boldsymbol{\beta}}_n^{s} = \widetilde{\boldsymbol{\beta}}_n - k(\widetilde{\boldsymbol{\beta}}_n - \widehat{\beta}_n\,\mathbf{1}_p)\mathcal{L}_n^{-1} \tag{2.17}$$

where we have modified $I(\mathcal{L}_n < F_\alpha)$ by $k\mathcal{L}_n^{-1}$ and k is the shrinkage constant such that $0 \leqslant k \leqslant \frac{2(p-3)m}{(p-1)(m+2)}$, $(m = n-2p)$. For convenience, we take

$$k = \frac{(p-3)m}{(p-1)(m+2)}. \tag{2.18}$$

As $\mathcal{L}_n \to \infty$, $\mathcal{L}_n^{-1} \to 0$, hence $\widehat{\boldsymbol{\beta}}_n^{s} \to \widetilde{\boldsymbol{\beta}}_n$ while as $\mathcal{L}_n \to k$, $\widehat{\boldsymbol{\beta}}_n^{s} \to \widehat{\beta}_n \mathbf{1}_p$. Thus, $\widehat{\boldsymbol{\beta}}_n^{s}$ behaves like a PTE at these limits. However, as $\mathcal{L}_n \to 0$, $\widehat{\boldsymbol{\beta}}_n^{s}$ is undefined and unreasonable. This leads to the consideration of the *positive-rule shrinkage estimator* (PRSE)

$$\widehat{\boldsymbol{\beta}}_n^{s+} = \widehat{\beta}_n \mathbf{1}_p + \{1 - k\mathcal{L}_n^{-1}\} I(\mathcal{L}_n > k)\mathbf{J}\widetilde{\boldsymbol{\beta}}_n \tag{2.19}$$

which is a PTE based on $\widehat{\beta}_n \mathbf{1}_p$ and $\widehat{\beta}_n^s$ with $F_\alpha = k$, that is,

$$\begin{aligned}\widehat{\beta}_n^{s+} &= \widehat{\beta}_n \ \mathbf{1}_p I(\mathcal{L}_n \leqslant k) + \widehat{\beta}_n^s I(\mathcal{L}_n > k) \\ &= \widehat{\beta}_n \ \mathbf{1}_p + \{1 - k\mathcal{L}_n^{-1}\} I(\mathcal{L}_n > k) \mathbf{J} \ \widetilde{\beta}_n. \end{aligned} \tag{2.20}$$

Based on these two estimators of $\boldsymbol{\beta}$, we define the two shrinkage estimators of $\boldsymbol{\theta}$ as

$$\widehat{\boldsymbol{\theta}}_n^s = \widetilde{\boldsymbol{\theta}}_n + k\mathbf{T}_p \mathbf{J} \widetilde{\boldsymbol{\beta}}_n \mathcal{L}_n^{-1},$$

and

$$\widehat{\boldsymbol{\theta}}_n^{s+} = \widetilde{\boldsymbol{\theta}}_n + \{1 - k\mathcal{L}_n^{-1}) I(\mathcal{L}_n > k)\} \mathbf{T}_p \mathbf{J} \widetilde{\boldsymbol{\beta}}_n, \quad \text{respectively.} \tag{2.21}$$

3. SHELF-LIFE BASED ON SINGLE BATCH.

First, note that the distribution of $\begin{pmatrix} \widetilde{\boldsymbol{\theta}}_n \\ \widetilde{\boldsymbol{\beta}}_n \end{pmatrix}$ on the assumption of normality of errors with fixed variance, σ^2 is given by $\mathcal{N}_{2p} \left\{ \begin{pmatrix} \boldsymbol{\theta} \\ \boldsymbol{\beta} \end{pmatrix}, \sigma^2 \mathbf{D} \right\}$, where

$$\mathbf{D} = \begin{pmatrix} \mathbf{D}_{11} & \mathbf{D}_{12} \\ \mathbf{D}_{21} & \mathbf{D}_{22} \end{pmatrix} \tag{3.1}$$

$$\mathbf{D}_{11} = \Lambda_0 + \mathbf{T}_p \ \mathbf{D}_{22} \mathbf{T}_p, \ \Lambda_0 = \text{Diag}\left(\frac{1}{n_1}, \ldots, \frac{1}{n_p}\right) \tag{3.2}$$

$$\mathbf{D}_{22} = \text{Diag}\left(\frac{1}{n_1 Q_1}, \ldots, \frac{1}{n_p Q_p}\right) \tag{3.3}$$

$$\mathbf{D}_{12} = -\mathbf{D}_{22} \ \mathbf{T}_p. \tag{3.4}$$

Further,

$$\mathbf{J} \ \mathbf{D}_{22} \mathbf{J}' = \mathbf{D}_{22} \ \mathbf{J}' \ \text{ and } \ \mathbf{D}_{22} \ \mathbf{J} \ \mathbf{T}_p = -\mathbf{D}_{22} \mathbf{T}_p + \frac{\bar{\mathbf{t}} \ \bar{\mathbf{t}}'}{nQ} \tag{3.5}$$

with $\bar{\mathbf{t}} = (\bar{t}_1, \ldots, \bar{t}_p)'$.

From the above (3.1) - (3.5), we conclude that

$$\widetilde{\boldsymbol{\theta}}_n + \widetilde{\boldsymbol{\beta}}_n t \sim \mathcal{N}_p\{\boldsymbol{\theta} + \widetilde{\boldsymbol{\beta}} t, \ \sigma^2 [\Lambda_0 + \mathbf{D}^*]\} \tag{3.6}$$

where

$$\mathbf{D}^* = \text{Diag}\left\{ \frac{(t - t_1)^2}{n_1 Q_1}, \ldots, \frac{t - t_p)^2}{n_p Q_p} \right\}. \tag{3.7}$$

Thus, for the ith batch

$$\widetilde{\theta}_i + \widetilde{\beta}_i t \sim \mathcal{N} \left\{ \theta_i + \beta_i t, \ \sigma^2 \left[\frac{1}{n_i} + \frac{(t - \bar{t}_i)^2}{n_i Q_i} \right] \right\} \tag{3.8}$$

and the 95% confidence interval for $\theta_i + \beta_i t$ is given by

$$\widetilde{\theta}_i + \widetilde{\beta}_i t \pm \; t(.025, n_i - 1)s_i\sqrt{\frac{1}{n_i} + \frac{(t - \bar{t}_i)^2}{n_i Q_i}} \tag{3.9}$$

where s_i^2 is the error-variance from the ith batch, i.e.

$$s_i^2 = \frac{1}{n_i - 1}\sum_{j=1}^{n_i}\left\{(y_{ij} - \bar{y}_i) - \widetilde{\beta}_i(t_{ij} - \bar{t}_i)\right\}^2. \tag{3.10}$$

Thus, the shelf-life of the ith batch is the minimum of the two solutions of the quadratic equation

$$(\eta_0 - \widetilde{\theta}_i - \widetilde{\beta}_i t)^2 = F_{1,n_i-2}(.05)s_i^2\left\{\frac{1}{n_i} + \frac{(t - \bar{t}_i)^2}{n_i Q_i}\right\} \tag{3.11}$$

$i = 1, 2, \ldots, p$ and η_0 is the 90% of the labeled value. Here $F_{1,n_i-2}(.05)$ is the upper 5% F-value of an F-distribution with $(1,\ n_i - 2)$ D.F.

The following Figure gives the graphical explanation of the procedure

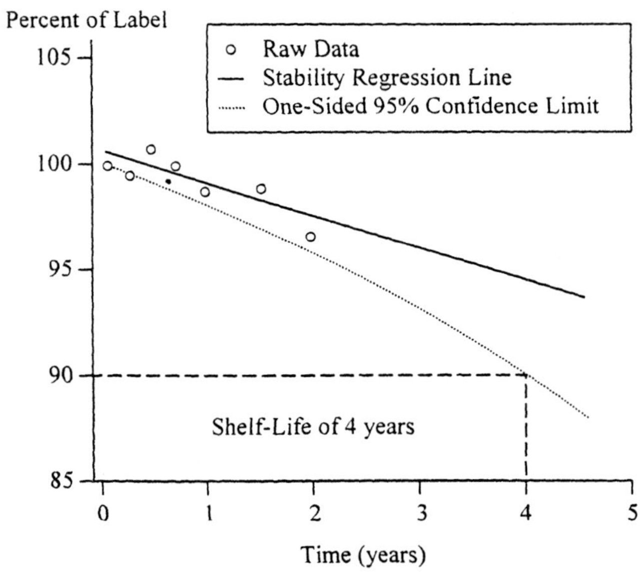

Figure 1. Shelf-life calculation for a single batch.

Now, we consider the data-set 1 of 6 batches discussed by Ruberg and Stegress (1991) in Table 1 and see the shelf-life of the six batches in the next page. The Table 2 gives the ANCOVA as well as the shelf-life results. ANCOVA table gives the value of $\mathcal{L}_n$ = Time × Batch Source F-value = 1.64. The P-value for this $F = 1.64$ is $P = .186$ which is less than .25, hence the null-hypothesis $H_0 : \boldsymbol{\beta} = \beta_0 \mathbf{1}_p$ is rejected in favour of $H_A : \boldsymbol{\beta} \neq \beta_0 \mathbf{1}_p$.

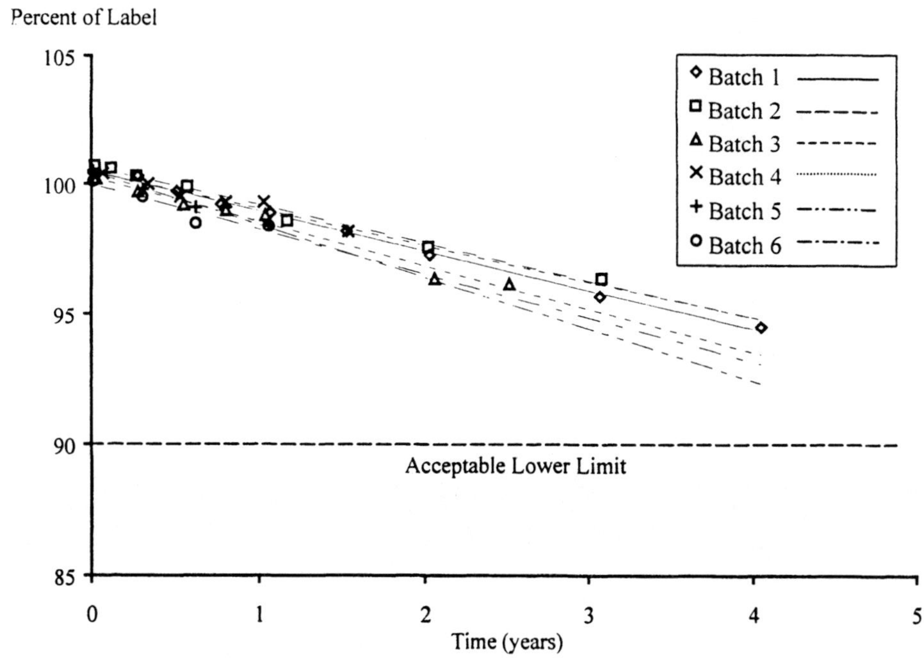

Figure 2. Stability data and individual regression lines for six batches (data set 1).

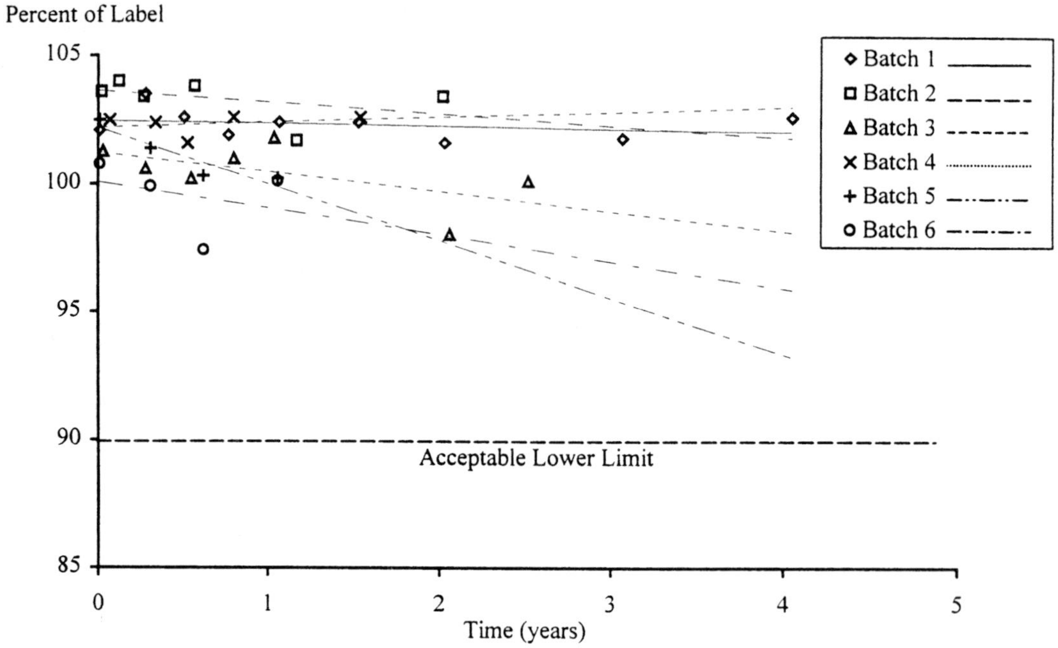

Figure 3. Graphs of regression lines of the six batches (data set 2).

We note that individual batches have shelf-life varying from 3.5 to 6.8 years. The calculations have been done by using equation (3.8). The last line gives shelf-life of 6.2 years after *pooling both the intercepts* and the *slopes*. But this value is not legitimate in view of the fact that both the equality of intercepts as well as the slopes are rejected at .25-level of significance.

Similarly, we consider the second set of data with six batches in table 3. We notice from the ANCOVA table that the null-hypothesis $H_0 : \boldsymbol{\beta} = \beta_0 \mathbf{1}_p$ is *accepted* at .25 level of significance since $\alpha < .359$ which is the P-value of the $\mathcal{L}_n$-statistic and $\mathcal{L}_n = 1.16$. The shelf-life of the six batches is computed using formula (3.8) along with the *pooled* intercepts and slopes with a shelf-life of 18.2 years. The shelf-life for six batches varies from 1.7 to 32.3 years. The data-set, the corresponding graphs with the ANCOVA table together with the shelf-life calculation are given in the sequel.

Table 1. Listing of data set 1

Batch	Time (years)	Concentration (%)	Batch	Time (years)	Concentration (%)
1	.014	100.4	4	.066	100.4
	.280	100.2		.343	100.0
	.514	99.7		.533	99.3
	.769	99.2		.802	99.3
	1.074	98.9		1.033	99.3
	1.533	98.2		1.533	98.2
	2.030	97.3			
	3.071	95.7	5	.011	100.5
	4.049	94.5		.310	99.8
				.624	99.1
2	.022	100.7		1.063	98.4
	.110	100.6			
	.272	100.3	6	.011	100.1
	.566	99.9		.310	99.5
	1.165	98.6		.624	98.5
	2.022	97.6		1.063	98.4
	3.077	96.4			
3	.025	100.2			
	.275	99.7			
	.547	99.2			
	.797	99.0			
	1.041	98.8			
	2.053	96.4			
	2.519	96.2			

Table 2. ANCOVA, regression results and shelf-life calculations for data set 1

ANCOVA results for testing batch slopes

Source	df	Sum of squares	F-value	P-value
Batch	5	.60	2.85	.038
Time	1	21.52	503.24	$< .001$
Time × Batch	5	.35	1.64	.186
Error	25	1.07		

Individual batch regression

Batch	Intercept (%)	Slope (%/year)	MSE	Weight n_iQ_i	Shelf-life (years)
1	100.49	- 1.515	.019	14.63	6.7
2	100.66	- 1.419	.043	7.83	6.8
3	100.25	- 1.612	.062	5.14	5.5
4	100.45	- 1.393	.035	1.21	6.2
5	100.45	- 1.999	.011	.61	4.4
6	99.98	- 1.701	.124	.61	3.5
Pooled batch regression with equivalent slopes					
6	99.90	-1.534	.047		6.2

Table 3. Listing of data set 2

Batch	Time (years)	Concentration (%)	Batch	Time (years)	Concentration (%)
1	.014	102.1	4	.066	102.5
	.280	103.5		.343	102.4
	.514	102.6		.533	101.6
	.769	101.9		.802	102.6
	1.074	102.4		1.538	102.6
	1.533	102.4			
	2.030	101.6			
	3.071	101.7	5	.011	102.5
	4.049	102.6		.310	101.4
				.624	100.3
2	.022	103.6		1.063	100.2
	.110	104.0			
	.272	103.4			
	.566	103.8	6	.011	100.8
	1.165	101.7		.310	99.9
	2.022	103.4		.624	97.4
	3.077	96.4		1.063	100.1
3	.025	101.3			
	.275	100.6			
	.547	100.2			
	.797	101.0			
	1.041	101.8			
	2.053	98.0			
	2.519	100.1			

Table 4. ANCOVA, regression results and shelf-life calculations for data set 2

ANCOVA results for testing batch slopes

Source	df	Sum of squares	F-value	P-value
Batch	5	15.93	4.01	.009
Time	1	4.14	5.21	.002
Time × Batch	5	4.60	1.16	.359
Error	23	18.28		

Individual batch regression

Batch	Intercept (%)	Slope (%/year)	MSE	Weight n_iQ_i	Shelf-life (years)
1	102.48	- .109	.343	14.63	32.3
2	103.63	- .449	.702	2.96	9.6
3	101.24	- .778	1.189	5.14	7.0
4	102.21	.194	.221	1.26	16.2
5	100.21	- 2.218	.256	.61	3.2
6	100.07	- 1.045	2.973	.61	1.7
Pooled batch regression with equivalent slopes					
6	100.41	- .330	.628		18.2

4. SHELF-LIFE BASED ON MULTIPLE BATCHES; RECENTERED CONFIDENCE SETS METHOD (RCSM).

For the determination of shelf-life from a single batch, FDA recommends using lower 95% confidence interval for the fitted regression line. The method is applied batch by batch for multiple batch problem, then choosing the **minimum** of the shelf-lives of the p shelf-lives. We note from data-set 1, it is 3.5 years and for the data-set 2, it is 1.7 years. Both of these solutions are the extremes of shelf-lives in the two data-sets. Thus, an attempt is made to obtain reasonable **interpolated** shelf-life for the p batches. Thus, we propose recentered confidence set method (RCSM) with the five estimators we discussed, as center of the confidence sets.

4.1 Unrestricted confidence set method (UCSM)

The p batches yield to p independent data sets. Thus, we consider the **unrestricted** estimator of prediction parameter $\boldsymbol{\eta} = \boldsymbol{\theta} + \boldsymbol{\beta} t$. The $(1-\alpha)\%$ confidence set for $\boldsymbol{\eta} = \boldsymbol{\theta} + \boldsymbol{\beta} t$ for fixed t is defined by

$$C^{(U)}_{F_{p,n-2p}(1-\alpha)} = \left\{ \boldsymbol{\eta} : \frac{1}{pS_e^2}(\boldsymbol{\eta} - \widetilde{\boldsymbol{\theta}}_n - \widetilde{\boldsymbol{\beta}}_n t)' \mathbf{D}_{11}^{-1} (\boldsymbol{\eta} - \widetilde{\boldsymbol{\theta}}_n - \widetilde{\boldsymbol{\beta}}_n t) \leqslant F_{p,n-2p}(1-\alpha) \right\}. \tag{4.1}$$

Thus, for the choice of shelf-life, we solve for t such that

$$\frac{1}{pS_e^2}(\eta_0 \mathbf{1}_p - \widetilde{\boldsymbol{\theta}}_n - \widetilde{\boldsymbol{\beta}}_n t)' \mathbf{D}_{11}^{-1} (\eta_0 \mathbf{1}_p - \widetilde{\boldsymbol{\theta}}_n - \widetilde{\boldsymbol{\beta}}_n t) \leqslant F_{p,n-2p}(1-\alpha). \tag{4.2}$$

The solution $t^{(U)}$ is the shelf-life for the combined shelf-lives of the p batches using .95 as confidence coefficient. For the data set 1 it is 6.72 years and for the data set 2, it is 15.44 years.

4.2 Restricted Confidence Set Method (RCSM)

If the null-hypothesis $H_0 : \boldsymbol{\beta} = \beta_0 \mathbf{1}_p$ is true then the estimate of $\boldsymbol{\eta}$ is given by $\widetilde{\boldsymbol{\theta}}_n + \widehat{\beta}_n \mathbf{1}_p t$ for fixed t. Then, the recentered $(1-\alpha)\%$ confidence set is defined by

$$C^{(R)}_{F_{p,n-2p}(1-\alpha)} = \left\{ \boldsymbol{\eta} : \frac{1}{pS_e^2}(\boldsymbol{\eta} - \widehat{\boldsymbol{\theta}}_n - \widehat{\beta}_n \mathbf{1}_p t)' \mathbf{D}_{11}^{-1} (\boldsymbol{\eta} - \widehat{\boldsymbol{\theta}}_n - \widehat{\beta}_n \mathbf{1}_p t) \leqslant F_{p,n-2p}(1-\alpha) \right\}. \tag{4.3}$$

Thus, for the choice of t, we solve for t such that

$$\frac{1}{pS_e^2}(\eta_0 \mathbf{1}_p - \widehat{\theta}_n - \widehat{\boldsymbol{\beta}}_n t)' \mathbf{D}_{11}^{-1} (\eta_0 \mathbf{1}_p - \widehat{\theta}_n - \widehat{\boldsymbol{\beta}}_n \mathbf{1}_p t) \leqslant F_{p,n-2p}(1-\alpha). \tag{4.4}$$

The solution $t^{(R)}$ is the shelf-life for the combined shelf-lives of the p batches using .95 as confidence coefficient. For the data set 1 it is $t^{(R)} = 6.34$ years and for the data set 2, it is 12.69 years.

Note that for the data set 1, the hypothesis is rejected at .25 level of significance. However, the above represents the methodology to illustrate the present restricted method. For the dataset 2 it is valid as we accept the hypothesis and the combined shelf-life is 12.69 years.

4.3 Preliminary Test Confidence Set Method (PTCSM)

From section 2.3 we know that the PTE of $\boldsymbol{\eta} = \boldsymbol{\theta} + \boldsymbol{\beta} t$ is

$$\widehat{\boldsymbol{\eta}}^{PT}(t) = \widetilde{\boldsymbol{\theta}}_n^{PT} + \widetilde{\boldsymbol{\beta}}_n^{PT} t \tag{4.5}$$

for α_0-level of significance for testing H_0. Hence, the $(1-\alpha)\%$ confidence set for $\boldsymbol{\eta}$ is defined by

$$C^{PT}_{F_{p,n-2p}}(1-\alpha) = \left\{\boldsymbol{\eta}; (\boldsymbol{\eta} - \widehat{\boldsymbol{\eta}}^{PT})'\mathbf{D}_{11}^{-1}(\boldsymbol{\eta} - \widehat{\boldsymbol{\eta}}^{PT}) \leqslant F_{p,n-2p}(1-\alpha)\right\}. \tag{4.6}$$

To obtain the shelf-life t we solve for t the following equation

$$\frac{1}{pS_e^2}\left(\eta_0 \mathbf{1}_p - \widehat{\boldsymbol{\theta}}_n^{PT} - \widehat{\boldsymbol{\beta}}_n^{PT} t\right)' \mathbf{D}_{11}^{-1}\left(\eta_0 \mathbf{1}_p - \widehat{\boldsymbol{\theta}}_n^{PT} - \widehat{\boldsymbol{\beta}}_n^{PT} t\right) \leqslant F_{p,n-2p}(1-\alpha). \tag{4.7}$$

Let $t^{(PT)}$ be the solution for t, then $t^{(PT)}$ is the combined shelf-life for the p batches using .95 as confidence coefficient. For the data set 1, $t^{(PT)} = 6.34$ years since H_0 is rejected at $\alpha_0 = .25$ level of significance. For the data set 2, it is 14.23 years since H_0 is accepted.

4.4 Shrinkage Confidence Set Method (SCSM)

From Section 2.4 we know that the shrinkage estimator of $\boldsymbol{\eta}$ is $\widehat{\boldsymbol{\theta}}_n^s + \widetilde{\boldsymbol{\beta}}_n^s t$. Hence, $(1-\alpha)\%$ confidence set for $\boldsymbol{\eta}$ is for $p \geqslant 4$,

$$\begin{aligned} &C^{s}_{F_{p,n-2p}}(1-\alpha) \\ &= \left\{\boldsymbol{\eta} : \frac{1}{pS_e^2}\left(\eta_0 \mathbf{1}_p - \widehat{\boldsymbol{\theta}}_n^s - \widehat{\boldsymbol{\beta}}_n^s t\right)' \mathbf{D}_{11}^{-1}\left(\eta_0 \mathbf{1}_p - \widehat{\boldsymbol{\theta}}_n^s - \widehat{\boldsymbol{\beta}}_n^s t\right) \leqslant F_{p,n-2p}(1-\alpha)\right\}. \end{aligned} \tag{4.8}$$

Thus, the shelf-life $t^{(s)}$ is solution of the equation

$$\frac{1}{pS_e^2}(\eta_0 \mathbf{1}_p - \widehat{\boldsymbol{\theta}}_n^s - \widehat{\boldsymbol{\beta}}_n^s t)'\mathbf{D}_{11}^{-1}(\eta_0 \mathbf{1}_p - \widehat{\boldsymbol{\theta}}_n^s - \widehat{\boldsymbol{\beta}}_n^s t) \leqslant F_{p,n-2p}(1-\alpha) \tag{4.9}$$

For the data set 1, $t^{(s)}$ is equal to 6.34 years and for the data set 2, it is equal to 12.91 years using .95 as confidence coefficient.

Similarly, the $(1-\alpha)\%$ confidence for $\boldsymbol{\eta}$ based on the positive-rule shrinkage estimator, we have for $p \geqslant 4$

$$\begin{aligned} &C^{S+}_{F_{p,n-2p}}(1-\alpha) \\ &= \left\{\boldsymbol{\eta} : \frac{1}{pS_e^2}(\boldsymbol{\eta} - \widehat{\boldsymbol{\theta}}_n^{S+} - \widehat{\boldsymbol{\beta}}_n^{S+} t)'\mathbf{D}_{11}^{-1}(\boldsymbol{\eta} - \widehat{\boldsymbol{\theta}}_n^{S+} - \widehat{\boldsymbol{\beta}}_n^{S+} t) \leqslant F_{p,n-2p}(1-\alpha)\right\}, \end{aligned} \tag{4.10}$$

Then, the solution t^{s+} of the equation

$$\frac{1}{pS_e^2}(\eta_0 \mathbf{1}_p - \widehat{\boldsymbol{\theta}}_n^{S+} - \widehat{\boldsymbol{\beta}}_n^{S+}t)'\mathbf{D}_{11}^{-1}(\eta_0 \mathbf{1}_p - \widehat{\boldsymbol{\theta}}_n^{S+} - \widehat{\boldsymbol{\beta}}_n^{S+}t) \leqslant F_{p,n-2p}(1-\alpha). \tag{4.11}$$

For the data set 1 it is 6.49 years, and for the data set 2 it is 13.51 years using .95 as confidence coefficient.

5.0 DISCUSSION ON RECENTERED CONFIDENCE SETS

The theory of point estimation in multivariate set-up has been fully developed. Beginning with preliminary test estimation (PTE) theory by Bancroft (1944), to shrinkage estimates by Stein (1961), it is well-known that shrinkage estimators dominate uniformly the usual estimators. Further improvement towards positive rule shrinkage estimators due to Baranchick (1940) or empirical Bayes by Lindley and Smith (1972), Efron and Morris (1973) and Morris (1983) are well-known. The theory of these are extended to nonparametric setup by Saleh and Sen (1978-86) and Sen and Saleh (1979-1989). For some review and discussion the reader may see Judge and Bock (1978) and Berger (1985).

On the issue of confidence sets, Brown (1966) and Joshi (1967) showed that the usual confidence sets may be dominated by other sets with the same coverage probability and smaller volume or same volume and larger probability. Hwang and Casella (1982) proved that confidence sets with center positive-rule Stein estimator dominates the usual estimator have some constant volume for $p \geqslant 4$. Casella and Hwang (1987) extended this to the class of Stein-type estimators shrinking towards linear subspaces. Hwang and Chen (1986) and Hwang and Casella (1984) extended this idea to spherically symmetric distributions.

The difficulty arises when variance is unknown. In this case, Robert and Casella (1987) established dominator results for a class of spherically symmetric distributions including multivariate t distributions with unknown scale parameter. Kim (1987) dealt with the case of normal distributions, Robert and Saleh (1989) studied the problem of confidence sets with PTE as center of the set and proved non-domination of the usual sets.

Thus, considering all of it, we consider the result with confidence set based on the positive part. Some simulation results are being studied at this point.

6.0 SHELF-LIFE FROM DATASET 1 AND DATASET 2.

In this section, we present the total picture of the shelf-lives from dataset 1 and dataset 2 by the use of single batch methodology and the combined methodology that have been proposed. Table 5 and Table 6 present the results. The results are the solution for t of the equation

$$(\eta_0 - \theta_i^* - \beta_i^* t)^2 = F_{1,n_i-2}(.05)s_i^2 \left\{ \frac{1}{n_i} + \frac{(t-\bar{t}_i)^2}{n_i Q_i} \right\} \tag{6.1}$$

$i = 1, \ldots, p$ and θ_i^* and β_i^* are the point estimates designated as unrestricted, restricted, PTE, shrinkage and positive-rule shrinkage estimates available from Section 2 and 4. The details are given below.

A. Unrestricted shelf-life is obtained by solving for t in the equation

$$(\eta_0 - \widetilde{\theta}_i - \widetilde{\beta}_i t)^2 = F_{1,n_i-2}(.05)s_i^2 \left\{ \frac{1}{n_i} + \frac{(t - \bar{t}_i)^2}{n_i Q_i} \right\},$$

$i =, 2, \ldots, p$. The solutions for datasets 1 and 2 are given in Tables 5 and 6 under $t^{(\mathrm{U})}$.

B. Restricted shelf-life is obtained by solving for t in the equation

$$(\eta_0 - \widetilde{\theta}_i - \widehat{\beta}_n t)^2 = F_{1,n_i-2}(.05)s_i^2 \left\{ \frac{1}{n_i} + \frac{(t - \bar{t}_i)^2}{n_i Q_i} \right\},$$

$i =, 2, \ldots, p$. The solutions for datasets 1 and 2 are given in Tables 5 and 6 under $t^{(\mathrm{R})}$.

C. Preliminary test shelf-life is obtained by solving for t in the equation

$$(\eta_0 - \widehat{\theta}_i^{\mathrm{PT}} - \widehat{\beta}_n^{\mathrm{PT}} t)^2 = F_{1,n_i-2}(.05)s_i^2 \left\{ \frac{1}{n_i} + \frac{(t - \bar{t}_i)^2}{n_i Q_i} \right\},$$

$i =, 2, \ldots, p$. The solutions for datasets 1 and 2 are given in Tables 5 and 6 under $t^{(\mathrm{PT})}$.

D. Shrinkage shelf-life is obtained by solving for t in the equation

$$(\eta_0 - \widehat{\theta}_i^{\mathrm{S}} - \widehat{\beta}_n^{\mathrm{S}} t)^2 = F_{1,n_i-2}(.05)s_i^2 \left\{ \frac{1}{n_i} + \frac{(t - \bar{t}_i)^2}{n_i Q_i} \right\},$$

$i =, 2, \ldots, p$. The solutions for datasets 1 and 2 are given in Tables 5 and 6 under $t^{(\mathrm{S})}$.

E. Positive rule shrinkage shelf-life is obtained by solving for t in the equation

$$(\eta_0 - \widehat{\theta}_i^{\mathrm{S+}} - \widehat{\beta}_n^{\mathrm{S+}} t^2 = F_{1,n_i-2}(.05)s_i^2 \left\{ \frac{1}{n_i} + \frac{(t - \bar{t}_i)^2}{n_i Q_i} \right\},$$

$i =, 2, \ldots, p$. The solutions for datasets 1 and 2 are given in Tables 5 and 6 under $t^{(\mathrm{S+})}$.

7.0. CONCLUSION

In this paper, we have presented methodology to determineshelf-life of several batches using confidence set involving recentered estimates. Five different estimates of the prediction parameter is used and applied to two data-sets. The tables 5 and 6 give the results on common shelf-life which are different than the traditional results and are useful. We think that this methodology is the appropriate one to follow in future analysis of stability data.

Data-Set 1

Table 5. Shelf-lives by different methods

batch	$t^{(U)}$	$t^{(RE)}$	$t^{(PT)}$	$t^{(S)}$	$t^{(S+)}$
1	6.68	6.60	6.63	6.61	6.67
2	6.75	6.40	6.61	6.51	6.80
3	5.49	5.96	5.35	5.46	5.10
4	6.16	5.70	5.86	5.72	6.10
5	4.45	5.62	4.16	4.39	3.68
6	3.51	3.70	2.99	3.03	2.90
Combined (All)	6.72	6.34	6.34	6.34	6.49

Data-Set 2

Table 6. Shelf-lives by different methods

batch	$t^{(U)}$	$t^{(RE)}$	$t^{(PT)}$	$t^{(S)}$	$t^{(S+)}$
1	32.32	20.35	31.50	23.04	28.70
2	9.63	10.37	8.53	6.28	7.94
3	6.97	9.29	7.87	6.84	5.61
4	16.23	9.66	8.23	9.53	11.13
5	34.80	5.87	4.48	3.31	1.86
6	1.71	1.90	1.40	1.35	1.26
Combined (All)	15.44	12.69	14.24	12.91	13.51

REFERENCES

Bancroft, T.A. (1964), Analysis and inference for incompletely specified models involving the use of preliminary sums of significance. *Biometrics.* **29**, 427–442.

Brown, L.D. (1966), On the admissibility of invariant estimators of one or more location parameters. *Ann. Math. Stat.* **37**, 1087–1136.

Casella, G., Hwang, J.T. (1983), Empirical Bayes confidence sets for the mean of a multivariate normal distribution, *J.A.S.A.* **78**, 688–698.

Casella, G., Hwang, J.T. (1986), Confidence sets and the Stein-effect, *Comm. Stat.-Theo. Meth.* **15** 2043–2063.

Casella, G., Hwang, J.T., Robert, C. (1989), Loss functions for set estimation. Mathematical Sciences Institute, Cornell University, Tech. Report '89–45.

Chen, J., Hwang, J.T. (1988), Improved set estimators for the coefficients of a linear model when the error distribution is spherically symmetric with unknown variances. *Can. J. Stat.* **16**, 293–299.

Faith, A.E. (1976), Minimax Bayes point and set estimators of a multivariate normal mean. tech. Report #66, Dept. of Statistics, University of Michigan.

Food and Drug Administration (1987), *Guildeline for Submitting Documentation for Stability Studies of Human Drugs and Biologics.* Rockville, Maryland: Center for Drugs and Biologics.

Hwang, J.T., Casella, G. (1982), Minimax Confidence Sets for the Mean of a Multivariate Normal Distribution. *Ann. Stat.* **10**, 868–881.

Hwang, J.T., Casella, G. (1984), Improved Set Estimators for a Multivariate Normal Mean. *Stat. Dec.* Supplement Issue **1**, 3–16.

Joshi, V.M. (1967), Inadmissibility of the usual confidence set for the mean of a multivariate normal population. *Ann. Math. Stat.* **38**, 1868–1875.

Joshi, V.M. (1969), Admissibility of the usual confidence set for the mean of a multivariate normal population. *Ann. Math. Stat.* **40**, 1042–1067.

Judge, G., Bock, M.E. (1978), *The statistical implications of pre-test and Stein-rule estimators in Econometrics.* North–Holland.

Kim, P.T. (1987), Recentered Confidence Sets for the Mean of a Multivariate distribution when the Scale Parameter is Unknown. Ph.D. thesis, Department of Mathematics, University of California at San Diego.

Robert, C., Casella, G. (1987), Improved confidence sets in spherically symmetric distributions, Tech. Report #87-51, Dept. of Statistics, Purdue University. To appear in *J. Mult. Anal.*

Robert, C. and Saleh, A.K.Md.E. (1989), Recentered confidence Sets: A review. Tech. Report No. 89, LSTA.

Ruberg, S.J. and Stegman, J.W. (1991). Pooling Data for Stability Studies: Testing Equality of Batch Degradation Slopes. *Biometrics* **47**, 1059–1069.

Saleh, A.K.Md.E. and Han, C.P. (1990). Shrinkage estimations on regression analysis. *Estadistica.* **42**, 40-63.

Saleh, A.K.Md.E. and Sen, P.K. (1978). Nonparametric estimation of location parameter after a preliminary test on regression, *Annals of Statistics.* **6**, 154-168.

Saleh, A.K.Md.E. and Sen, P.K. (1983). Nonparametric tests of location after a preliminary test on regression in the multivariate case. *Communications in Statistics, Theory and Method.* **12**(16), 1855-1872.

Saleh, A.K.Md.E. and Sen, P.K. (1984a). Least squares and rank order preliminary test estimation in general multivariate linear models. *Proceedings of Indian Statistical Institute, Golden Jubilee Conference in Statistical Applications and New Directions, J.K. Ghosh and Roy, J.* (eds.), 237-253.

Saleh, A.K.Md.E. and Sen, P.K. (1984b). Nonparametric preliminary test inference. *Handbook of Statistics, 4, Krishnaiah, P.R. and Sen, P.K. (eds.), North-Holland.* Amsterdam, 275-297.

Saleh, A.K.Md.E. and Sen, P.K. (1985a). Nonparametric shrinkage estimation in a parallelism problem. *Sankhya.* **47**A, 156-165.

Saleh, A.K.Md.E. and Sen, P.K. (1985b). On shrinkage M-estimators of location parameters. *Communications in Statistics, Theory and Method.* **14**(10), 2313-2329.

Saleh, A.K.Md.E. and Sen, (1985c). Preliminary test predicted in general multivariate liner models, Proceedings of the Pacific Area Statistical Conference, North-Holland Publishers, 619-638.

Saleh, A.K.Md.E. and Sen, P.K. (1985d). Shrinkage least squares estimation in a general multivariate linear model. *Proceedings of the Fifth Pannonian Symposium on Mathematical Statistics*, 307-325.

Saleh, A.K.Md.E. and Sen, P.K. (1986a). On Shrinkage least squares estimation in a parallelism problem. *Communications in Statistics, Theory and Method.* **15**(5), 1451-1466.

Saleh, A.K.Md.E. and Sen, P.K. (1986b). On Shrinkage R-estimation in a multiple regression model. *Communications in Statistics, Theory and Method.* **15**(7), 2229-2244.

Saleh, A.K.Md.E. and Kibria, B.M.G. (1993). Performance of some new preliminary test ridge regresion estimators and their properties. *Communications in Statistics, Theory and Method.* **22**, (10), 2747-2764.

CAN RESPONDENT-GENERATED INTERVAL ESTIMATION IN SAMPLE SURVEYS REDUCE ITEM-NONRESPONSE?[1]

S. JAMES PRESS

University of California, Riverside

JUDITH M. TANUR

State University of New York, Stony Brook

ABSTRACT

Press (1999) proposed a method of questioning in surveys called Respondent-Generated Intervals (RGI) expected to increase the accuracy of estimators of population means obtained by recall from respondents. The method involves requesting that a respondent supply not only a point assessment of a quantity, but also an assessment of the smallest and largest values the quantity could be. Point and interval estimates of the population mean are generated from these assessments. Press and Tanur (2000) provide empirical evidence that these estimates are quite accurate. Here we examine nonresponse using RGI. Across items, up to 14% of respondents supplied an interval, but did not answer the direct question. Perhaps these respondents would have skipped the question had they not been permitted to offer an interval. If that is the case, we find savings in item non-response achieved by offering such intervals. For some cases in which only intervals were offered we have verification data so we can examine the accuracy achieved by using the midpoint of the RGI to estimate the quantity of interest. We find that the intervals formed by the averages of the endpoints of the RGIs cover the true values of all the quantities examined.

Key Words and Phrases: sample survey, respondent generated intervals, nonresponse

INTRODUCTION

Survey researchers do not need to be reminded of the importance of the problem of nonresponse. Methods for avoiding it, minimizing it, measuring it, and compensating for it have generated a large literature (e.g. Panel on Incomplete Data, 1983; Groves,1989). This paper presents a simple method which seems to have some promise for reducing item nonresponse, a method that presented itself to us as a by-product of research we have been conducting in an attempt

---

1. This paper was presented at the International Conference on Applied Statistics held at Ryder University May 22, 23, 1999.

to improve recall accuracy. That method asks respondents to give an interval that they believe to bracket the quantitative item they are being asked to recall.

Of course, intervals have long been used in survey questionnaires, notably in questions about income, where it is believed that respondents are more willing to report on the sensitive issue of income within a preassigned range rather than with an exact number. But the intervals used in the method presented here are generated by the respondent rather than by the survey researcher, which is why we call them Respondent-Generated Intervals, RGI. In an experiment we ran on two college campuses to study the characteristics of the RGI protocol, we noticed that a sizable number of respondents supplied an interval in which they believed the quantity was almost certain to fall, but seemed not to be sure enough of the value of the quantity to answer the direct question. Perhaps these respondents would have simply skipped the question if they had not been permitted to offer an interval. On the assumption that such respondents would have been item nonrespondents, in this paper we explore the savings in item non-response achievable by the offering of such intervals.

DATA AND METHODS

At our respective universities we administered surveys designed to test a method, proposed by Press (see Press, 1999), expected to increase the accuracy of quantitative data recalled by respondents. The method involves requesting that a respondent supply not only a point assessment of the quantity s/he is asked to recall (what we have called the "usage quantity"), but also an interval assessment of the smallest and largest values s/he believes the quantity could possibly be (RGI). Point and interval estimates of the population mean were then generated based on these assessments. Results of these analyses appear in Press and Tanur (2000), and seem very promising.

These data were generated in the spring of 1997 when approximately 1000 students at the University of California at Riverside (UCR) and about 750 at the State University of New York at Stony Brook (SUSB) responded to a paper-and-pencil, self-administered questionnaire asking them to recall factual information about matters relating to their life on campus. At both campuses students were asked for the number of credits they had earned (CRED), the number of grades of C or less they had received (C's), their grade point average (GPA), their Scholastic Aptitude Scores on the math (SATM) and verbal (SATV) tests[2], and the number of traffic tickets they had received on campus during that academic year (TICK). At UCR students were also asked to recall the amount of the registration fee (REG) and the recreation center fee (REC) they paid at the beginning of the quarter. The corresponding questions at Stony Brook asked for the amounts of the health fee

---

2. The SAT verbal and math scores are not used in this analysis because we were unable to separate data missing because the respondent did not take the SAT tests from data missing for reasons that the interval procedure could hope to remedy.

(HFEE) and the student activities fee (SAFEE). Stony Brook students were also asked to recall the amount spent on the food plan (FOOD) and the number of library fines (FINES) they had been assessed. As stated above, we call the answers to these recall questions "usage quantities."

For each usage quantity, respondents were also asked to give an interval in which they were almost certain that the quantity would lie. (In fact, because one of our aims was to test two possible ways of asking this "uncertainty" question, we used a split ballot experimental design, so only half of each sample was offered the interval question as here described for most recall items. For the fee data, however, all respondents were presented with the interval question.) For those respondents who consented to have their data verified and who gave their student identification numbers for that purpose, we were able to get "true" values for the usage quantities the students were recalling from appropriate campus offices. (No permission for verification or identification numbers were needed to ascertain the true values for the fee data, as the fees are standard for all full-time students. We limited our analysis to full-time students.) While we are cognizant of the possible errors in the administrative data we are using for verification, we shall use those data as a "gold standard" in what follows.

FINDINGS

Table 1 shows the number of responses for each question classified as full responses (both a usage quantity and an interval were given, column 3), usage but no interval (column 1), interval but no usage (column 2), and neither usage nor interval (column 4). Column 5 shows the total number of potential respondents, column 6 presents the interval-only responses as a percentage of the total potential respondents to the question (i.e., column 2/column 5), and column 7 presents the interval-only respondents as a percentage of nonresponders to the usage quantity (i.e., column 2/[column 2 + column 4]). We see from column 6 that those who chose to give an interval-only response constitute only a few percent of the total potential responders for all questions except for the fee questions at SUSB where those percents are 9.7% and 13.4%. But we see from column 7 that the interval-only responders constitute an appreciable percentage of potential nonresponders to each item. Indeed, those percentages are never less than 4% and twice are over 40%. We can interpret these results as estimated conditional probabilities of giving an interval among those who did not give a usage quantity. Figure 1 presents the results shown in column 7 graphically.

Of course, having ascertained that respectable proportions of respondents who do not answer the direct question do provide intervals, we must inquire about the accuracy of such intervals if they should be used to produce estimators of the usage quantities. Press (1999) proposed (among other estimators, including a Bayesian estimator) that the average midpoint of the interval be used as a point estimate of the usage quantity and that the interval formed by the average of the lower bounds supplied by the respondents and the average of their upper bounds (termed the Average Respondent-Generated Interval, ARGI) be used as an interval estimator. (In an effort to account for respondent errors that reflect bias errors as well as errors in variance, in Press, 1999, it was suggested that a weighted average of the average upper and lower bounds be used, but until further research provides guidance as to how to choose the weights, we plan to use an equal-weighted, or "midpoint estimator".) We shall evaluate the accuracy of these estimators for those respondents who offered interval but no usage quantity responses. We note that the midpoint of the RGI estimator used in this paper requires no model assumptions, so using this estimator does not require that any such assumptions be checked. (In fact, the alternative Bayesian estimator suggested in Press and Tanur, 2000, does indeed require some model assumptions, and it appears to be robust with respect to moderate violations of assumptions.)

Unfortunately, at UCR, no verification data were collected for respondents who omitted a usage quantity response, and at SUSB the number of respondents for whom such validation data are available is tiny (*n* ranges from 0 to 5). Fortunately, however, verification data are not needed for the fee questions, for which the values are constant for all full-time students, so we can use the fee data to evaluate the accuracy of the estimators. From Table 2 we can see that the average midpoints of the RGIs overestimate the true usage quantities in three out of four cases. The ARGIs, however, cover that true value in all four cases.

The results so far, however, lump together two kinds of respondents. Some students participate in funding their college education, and hence the task of reporting on the amount paid for fees is almost purely one of recall from autobiographical memory. Other students, however, do not so participate; for them, the reporting task may well include elements of proxy reporting for amounts mentioned to them by their parents or noticed on financial aid forms. It seemed sensible to look at the results shown in Table 2 separately for those students who participated in funding their education and those who did not. These results are shown in Table 3.

Table 3 shows that the midpoint of the RGI is always closer to truth for the self-payers than it is for the non-self payers. The ARGI, however, covers truth only for two of the four items for the self-payers and three of the four for the non-self-payers. The final two columns of Table 3 present another interval estimator proposed by Press (1999), the ARGI (1σ), which adds one estimated standard error of the upper bounds to the average upper bound and subtracts one estimated standard error of the lower bounds from the average lower bound. Examining the ARGI (1σ), we find that the extended interval covers truth for all items except the UCR registration fee for both self-payers and non-self-payers.

We predicate our thinking about respondents who gave an interval but no usage quantity on the premise that they are trying their utmost to be cooperative. Indeed, we visualize individuals sufficiently unsure of the correct answer that they do not wish to risk giving a point estimate, but taking the survey task sufficiently seriously that they are willing to codify their partial knowledge into an interval. The fee data, for which there are an appreciable number of RGI-only respondents, afford us the opportunity for a partial test of this premise. In addition, the fee data recommend themselves for this analysis because these questions seem to embody few elements of threat or of social desirability. Hence they should not pose special problems to challenge or to abet the cooperation of respondents.

To test the premise that "interval only" respondents are cooperators we ran logistic regressions to predict the probability of giving an interval when a usage quantity was not given. We asked whether those who we can conceptualize as more prone to cooperation are more likely to offer an interval than those who we cannot conceptualize as prone to cooperation. We operationalize "proneness to cooperation" with three independent variables: whether a respondent gave a social security number, thus granting permission for verification of his/her data (SSNIND), year in school (as a rough measure of maturity -- YEAR), and GPA (as a rough indicator of "playing by the rules" at the university). In the logistic regression we used whether the respondent supplied an interval as the dependent variable; we also introduce an indicator of whether the respondent participated in funding his/her education (SELFPAY), reasoning as above that those who do so participate will be more cognizant of the amount of fees paid.

The results of the logistic regressions are shown in Table 4. We see that the analyses for both fees at UCR do not capture a dependence on the explanatory variables selected. We speculate that this is a result of the relatively small sample sizes for the UCR fee data (relative to the SUSB fee data) and the fact that 20% or fewer of the UCR no-usage-quantity respondents offered an interval. The smaller sample sizes may have resulted in insufficient power of the Wald test to detect significance of the relevant coefficients. At SUSB, however, in both cases over 40% of the no-usage-quantity respondents supplied an interval, and the results are more revealing. Both logistic regression models at SUSB exhibit statistically significant fits, and the coefficients of all the independent variables are positive, indicating that those respondents whom we would expect to be more cooperative (or more knowledgeable) are indeed more likely to supply an interval. The social security number indicator (indicating the respondent's willingness to have his/her data verified from academic records) attains statistical significance at the .044 level for the student activities fee, and year in school (YEAR) and SELFPAY are statistically significant for the health fee at p=.076 and p=.066 respectively. Thus those respondents we would expect to be more prone to cooperate do seem to cooperate more often.

CONCLUSION

We have shown that, in a substantial proportion of cases, respondents who do not supply an estimate of usage quantities do indeed supply intervals, thus reducing the amount of item nonresponse appreciably. Further, in all cases we could examine (the four fee items), the ARGI covered the true value of the usage quantity, thus supplying useful data for respondents who would otherwise have been coded as missing. Those who participated in funding their own education gave slightly more accurate intervals when they did not give a usage quantity, and we have some evidence that those respondents we would expect to cooperate are more likely to give an interval when they do not give a usage quantity than are those who we would not expect to cooperate. Of course, the reader should note that these respondents are already in some sense uncooperative, having refused to follow instructions to give both a usage quantity and an interval. Thus, although whether asking a respondent to supply only an interval will be effective as a device to reduce item nonresponse cannot be tested with these data, the results are suggestive and merit further research.

ACKNOWLEDGEMENT

We are very grateful to Dr. Robert Creecy for his suggestions and computer programming help; to Dr. Jacob Pozharny, Ms. Lucinda Manolakes, Mr. Andrew Catalano, Mr. Nephtali O'Connor, Mr. Alex Trillo, Ms., Waragkana Sangchan, and Mr. Jeffery Dennis for their help in collecting, cleaning and analyzing the data; to Dr. Kent Marquis for helpful discussions and suggestions; to the U.S. Census Bureau for its financial support; and to the campus administrators who provided verification.

REFERENCES

Groves, Robert M. 1989. *Survey Errors and Survey Costs*. New York: Wiley

Panel on Incomplete Data 1983 *Incomplete Data in Sample Surveys*. New York: Academic Press.

Press, S. James and Tanur, Judith (2000) "Experimenting with Respondent-Generated Intervals in Sample Surveys", with Discussions, in *Survey Research at the Intersection of Statistics and Cogintive Psychology,* Working Paper Series #28, Monroe G. Sirken, Editor, National Center for Health Statistics, Jan. 2000, U.S. Dept. of Health and Human Services, Center for Disease Control and Prevention, pp. 1-18.

Press, S. James (1999). "Respondent-Generated Intervals For Recall in Sample Surveys", manuscript submitted for publication, and available for viewing and downloading at the website http://cnas.ucr.edu/~stat/press.htm

Table 1

Responses Classified by Completeness/Incompleteness

Item	1 Usage, no Interval	2 Interval, no Usage	3 Interval and Usage	4 Neither Interval nor Usage	5 Total Respondents	6 % of Total giving Interval but Not Usage	7 % "Non Responders" Giving Interval but Not Usage
SUSB							
Credits	39	4	328	11	382	1.0%	26.7%
C's	41	11	309	16	377	2.9%	40.7%
Fines	59	2	282	40	383	0.5%	4.8%
Food	7	7	158	35	207	3.4%	16.7%
GPA	26	2	340	15	383	0.5%	11.8%
Health Fee	27	74	509	150	760	9.7%	33.0%
SA Fee	29	102	483	146	760	13.4%	41.1%
Tickets	50	6	294	27	377	1.6%	18.2%
UCR							
Credits	38	8	437	25	508	1.6%	24.2%
C's	41	6	447	24	518	1.2%	20.0%
GPA	35	4	444	25	508	0.8%	13.8%
RecFee	52	26	827	121	1026	2.5%	17.7%
RegFee	49	24	779	174	1026	2.3%	12.1%
Tickets	31	2	462	23	518	0.4%	

Table 2
Accuracy of Respondent-Generated Intervals for RGI-Only Respondents

Item	*N*	Truth	Average Midpoint	Av. Lower Bound	Av. Upper Bound
SUSB Health Fee	74	$70.00	$123.98	$41.92	$206.04
SUSB Activity Fee	102	$78.75	$87.14	$33.06	$141.23
UCR Rec Fee	26	$59.00	$121.71	$47.73	$195.69
UCR Reg Fee	24	$1,371.25	$1,343.65	$922.33	$1,764.96

Table 3
Accuracy of Respondent-Generated Intervals for RGI-Only Respondents by Payment Status: Fee Data

				AVERAGE RGI		AVERAGE RGI (1σ)	
ITEM	TRUTH	N	AVERAGE MIDPOINT	AVERAGE LOWER BOUND	AVERAGE UPPER BOUND	AVERAGE LOWER BOUND −1σ	AVERAGE UPPER BOUND +1σ
SELF-PAYERS							
SUSB Health Fee	**$70.00**	**39**	**$120.14**	**$47.31**	**$192.97**	**$39.26**	**$224.77**
SUSB Activity Fee	**$78.75**	**48**	**$98.90**	**$36.65**	**$161.15**	**$30.69**	**$188.37**
UCR Rec Fee	**$59.00**	**6**	**$33.08**	**$20.17**	**$46.00**	**$12.06**	**$61.32**
UCR Reg Fee	**$1371.25**	**8**	**$895.35**	**$857.38**	**$933.33**	**$729.01**	**$1123.15**
NON-SELF-PAYERS							
SUSB Health Fee	**$70.00**	**35**	**$128.26**	**$35.91**	**$220.60**	**$29.34**	**$265.73**
SUSB Activity Fee	**$78.75**	**49**	**$117.38**	**$48.44**	**$186.33**	**$28.11**	**$248.52**
UCR Rec Fee	**$59.00**	**18**	**$155.53**	**$54.72**	**$256.33**	**$38.36**	**$316.84**
UCR Reg Fee	**$1371.25**	**18**	**$734.36**	**$613.70**	**$855.56**	**$465.30**	**$1098.18**

Table 4
Coefficients of Logistic Regresions: Fee Data
(p values given for results statistically significant at .10 or below)

	UCR REG	UCR REC	SUSB ACTIV.	SUSB HEALTH
SSNIND	7.9195	--	0.651 *(p=.044)*	0.5472
YEAR	-0.1165	0.2625	0.1292	0.2493 *(p=.076)*
GPA	-0.7029 *(p=.097)*	-0.4326	0.4077	0.2579
SELFPAY	0.4412	0.037	0.4195	0.5994 *(p=.066)*
CONSTANT	0.7624	-0.6907	-2.1302 *(p=.012)*	-2.2191 *(p=.015)*
Model Chi Square (df)	6.502 (4)	2.837 (3)	9.925 (4) *(p=.042)*	11.110 (4) *(p=.025)*
Percent Giving RGI	18.31%	20.19%	45.90%	41.80%
N	142	104	183	172

FIGURE 1:
% of Cases with Missing Usage
Quantities Having Respondent-Generated Intervals

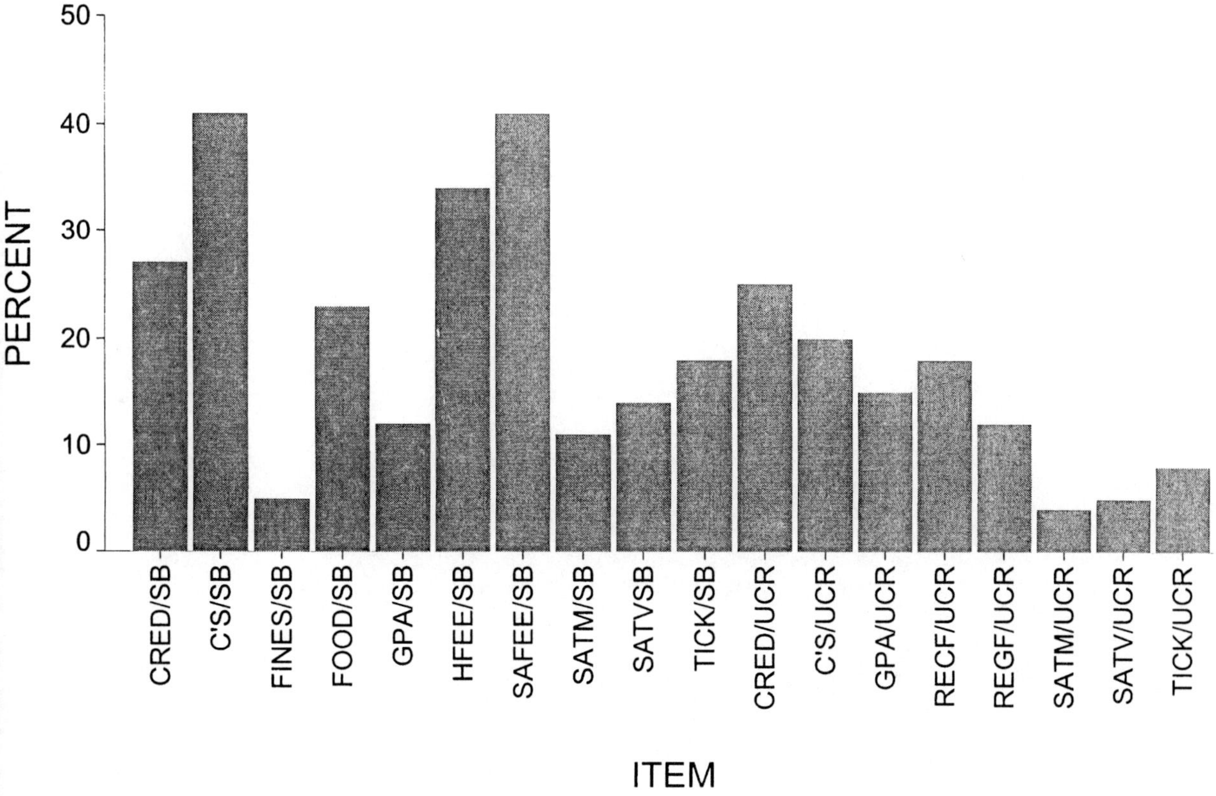

RECURSIVE PREDICTION FOR BIOSTATISTICAL TIME SERIES MODELS

A. Thavaneswaran[1], Jagbir Singh [2] and Nivedita Bhatt[2]

[1] University of Manitoba, Winnipeg, Canada R3T 2N2

[2] Temple University, Phila, PA 19122

SUMMARY

Godambe's (1999) theorem on an optimal estimating function for stochastic processes are applied for nonlinear time series predictions.A recursive estimation procedure based on optimal estimating equations is provided. Examples are considered in the usual classes of non-linear non-Gaussian time series models. An application to non-linear non-Gaussian biostatistical time series is also discussed in some detail.

Keywords: Biostatistical time series, non-linear non-Gaussian time series, optimal prediction, recursive prediction.

1 INTRODUCTION

Much prediction theory, both in the stationary process context and otherwise, is based on ad hoc application of minimum mean squares or minimum dispersion methods. This has the advantage of avoiding specific distributional assumptions. However, the standard results can sometimes be improved and the general estimating function theory which is now available provides a clear focus on particular classes of estimating functions and, within these, leads to optimal predictors whose information associated with the combined estimating function is large. This theory is not yet widely used in prediction.

Recently there has been growing interest in biostatistics using non-linear time series. In order to use non-linear time series models in practice, one must be able to fit the models to the data and estimate the parameters. Computational procedures for determining parameters together with the theoretical properties of the resulting estimates are outlined in Thavaneswaran, and Abraham (1988). The prediction via estimating functions is dicussed in Thavaneswaran and Heyde (1999).

In this paper, we illustrate ideas on combining estimating functions on a nonlinear non-Gaussian prediction problem and provide an information updating formula. In Section 2, using a theorem by Gill and Levit (1995), it is shown that the minimum mean squared error predictor has same efficiency as the posterior mode for a certain class of models. In Section 3, recursive predictors for biostatistical time series models are discussed and applied to polio data.

2 ESTIMATING FUNCTION PREDICTORS

Linear prediction is optimal for Gaussian processes and for linear model time series models. When one leaves this context it becomes important to assess possible improvements in efficiency associated with various non-linear predictors. In this section we also note that the optimal nonlinear predictor has same second order efficiency as the posterior mean.

Let $(\Omega, F, P_\theta : \theta \in \Theta)$ be a family of distributions on some sample space Ω and let μ be the dominating measure. Take the parameter space Θ to be a closed interval on the real line. Let $f(y|\theta)$ be the density of P_θ . Let π be some probability distribution on Θ with density $\lambda(\theta)$ with respect to Lebesque measure. Suppose that λ and $f(y|\theta)$ are both absolutely continuous and that λ converges to zero at the end points of the interval Θ .

Let $\theta = \hat{\theta}(Y)$ be any estimator of θ, $Y \sim P_\theta$. Let the corresponding information associted with the likelihood and the prior be $I(\theta) = E[\frac{\partial \ln f(Y|\theta)}{\partial \theta}]^2$ and $I(\lambda) = E[\frac{\partial ln\lambda(\theta)}{\partial \theta}]^2$ respectively. Then the following lemma on the optimality of the combined estimating function holds under the regularity conditions given in Gill and Levit (1995).

Lemma 2.1:

(a) $E[\hat{\theta}(Y) - \theta]^2 \geq \frac{1}{E[I(\theta)]+I(\lambda)}$.

(b) For an absolutely continuous function ψ of θ ,

$$E[\hat{\psi}(Y) - \psi(\theta)]^2 \geq \frac{(E\frac{\partial \psi(\theta)}{\partial \theta})^2}{E[I(\theta)] + I(\lambda)}.$$

(c) The posterior mode estimate is more informative than the posterior mean in general, and for exponential family distributions, the posterior mode has same efficiency as the posterior mean.

Proof: The proof of parts (a) and (b) follows from Gill and Levit (1995) while (c) follows by the definition of information associated with estimating functions (see Heyde (1997) for a comprehensive discussion).

Note: It follows from Godambe (1999) that the posterior score function attains the lower bound in (c). However, the corresponding estimate need not have the minimum mean square error. Moreover, if, for example, Y has a Cauchy distribution then the MMSE of θ is not defined but the information associated with the corresponding estimating function is well defined.

3 RECURSIVE PREDICTION

Mean squared errors of prediction are usually computed, on the basis of the assumption that the fitted model is in fact the true model for the data. As a result, they do not reflect the variability in the estimation of the model parameters. Consider the case where $Y_1, Y_2, \ldots, Y_n$ are generated from a causal ($|\phi| \leq 1$) $AR(1)$ model:

$$Y_t = \phi Y_{t-1} + a_t$$

where $\{ a_t \}$ is a white noise series having mean zero and variance σ^2. If $\hat{\phi}$ is the maximum likelihood estimate of ϕ, based on $Y_1, Y_2, \ldots, Y_n$, then it can be shown that the one step ahead prediction of Y_{n+1} (i.e. $\hat{\phi} Y_n$) has mean squared error

$$\begin{aligned} E(Y_{n+1} - \hat{\phi} Y_n)^2 &= E((\phi - \hat{\phi}) Y_n)^2 + \sigma^2 \\ &\approx \frac{n+1}{n} \sigma^2. \end{aligned}$$

Thus, the error in parameter estimation contributes the term σ^2/n to the mean squared error of prediction. If the sample size is large, this factor is negligible and so, for the purpose of mean squared error computation, the estimated parameters can be treated as the true model parameters. Similar considerations hold for much more general models. On the other hand, for small sample sizes, ignoring parameter variability can lead to a severe underestimate of the actual mean squared error of the predictor.

In the literature the prediction based on all the available data has been considered by assuming that the model parameters are known. When data come successively in time, it is natural to look for a recursive predictor for future observations which can also incorporate variability due to parameter estimation.

3.1 NONLINEAR MODELS

In this section, we provide a theorem on optimal prediction for discrete time stochastic processes based on estimating functions with applications to non-linear, non-Gaussian time series models.

Let $\{Y_t : t \in I\}$ be a discrete-time stochastic process taking values on the real line and defined on a probability space (Ω, F, P). The index set I is the set of all integers. We assume that the observations from the infinite past $(Y_{-\infty}, \ldots, Y_n)$ are available and that $\hat{Y}_t(1) = E\left[Y_{t+1} | F_t^Y\right]$ is a function of F_t^Y, the σ-field generated by $Y_{-\infty}, \ldots, Y_t$. Then the following theorem gives the form of the optimal one step ahead forecast of Y_{n+1} based on the observed values $Y_{-\infty} \ldots, Y_n$.

Let $Y_{n+1} - \hat{Y}_n(1) = a_{n+1}$, where $\{a_1, \ldots a_t, \ldots a_{n+1}\}$ is a stationary sequence of martingale differences whose one dimensional distribution has probability density function $f(.)$. This choice of a_t 's covers most of the recent popular models in finance such as autoregressive conditional heteroskedastic models having uncorrelated but not independent errors. Assume that $f(.)$ is known and that $E\left[-\frac{d^2}{da^2} f(a)\right] < \infty$ and $\int_{-\infty}^{\infty} f(a)\, da$ is twice differentiable under the integral sign. Let $\mathcal{G}$ be the class of unbiased estimating functions $g(a_{n+1})$ such that $E\; g(a_{n+1}) = 0$. The following theorem in Thavaneswaran and Heyde (1999) gives the recursive form of the predictor based on combined estimating functions.

Theorem 3.1.

(i) In the class $\mathcal{G}$, the optimal predictor of a_{n+1} which minimizes $\dfrac{\text{Var}\; g(a_{n+1})}{(E\frac{dg}{da})^2}$ is given by $m_a(f)$, where $m_a(f)$ is the mode of f, and the optimal predictor of Y_{n+1} is given by $Y_n^{\text{opt.}}(1) = \hat{Y}_n(1) + m_a(f)$.

(ii) The efficiency of the optimal estimating function, g°, $Eff(g^\circ)$ for a_{n+1} is $\frac{\{E(\frac{dg^\circ}{da})\}^2}{E(g^{\circ 2})} = E(g^{\circ 2})$.

(iii) Let $\hat{Y}_t(1) = E\left[Y_{t+1}|F_t^Y\right] = m(\theta, F_t^Y)$ be differentiable with respect to θ .

Then the recursive predictors are given by

$$\hat{\theta}_t = \hat{\theta}_{t-1} + \frac{K_{t-1}b^*_{t-1}}{1 + [\frac{\partial m(\hat{\theta}_{t-1}, F^Y_{t-1})}{\partial \theta}]b^*_{t-1}K_{t-1}} g[Y_t - m(\hat{\theta}_{t-1}, F^Y_{t-1})] \tag{3.1}$$

$$K_t = \frac{K_{t-1}}{1 + [\frac{\partial m(\hat{\theta}_{t-1}, F^Y_{t-1})}{\partial \theta}]b^*_{t-1}K_{t-1}} \tag{3.2}$$

and $\hat{Y}_{t+1}$ is given by

$$\hat{Y}_{t+1} = m(\hat{\theta}_t, F_t^Y) \tag{3.3}$$

where b^*_{t-1} is a function of g, θ and the observations.

If we take $g(x) = x$, then

$$b^*_{t-1} = \frac{[\frac{\partial m(\hat{\theta}_{t-1}, F^Y_{t-1})}{\partial \theta}]}{Var\,(a_t|F^Y_{t-1})},$$

while for any function g (e. g. if we take $g = \dfrac{d \log f(a)}{da}$, then)

$$b^*_{t-1} = \frac{[\frac{\partial m(\hat{\theta}_{t-1}, F^Y_{t-1})}{\partial \theta}][\frac{\partial g(\hat{\theta}_{t-1}, F^Y_{t-1})}{\partial \theta}]}{Var\,(g|F^Y_{t-1})}$$

Note: If $g(x) = x$, then $\dfrac{Var\ g(a_{t+1})}{(E\frac{dg}{da})^2} = \sigma_a^2$. Thus the minimum mean square error predictor is a special case of the Theorem 3.1. If we take $g = \dfrac{d \log f(a)}{da}$, then

$$\frac{\text{Var } g(a_{t+1})}{(E\frac{dg}{da})^2} = \frac{1}{\text{Var}(\frac{d\log f}{da})} = \frac{-1}{E\frac{d^2 \log f}{da^2}}.$$

Hence the maximum likelihood predictor is also a special case of the Theorem 3.1. For autoregressive conditional heterocedastic models with condtional mean and variance depending on the parameter of interest the estimating function approach provides a better predictor than even the maximum likelihood approach. Moreover, it is of interest to note that when the distribution of $\{a_t\}$ is stable, for example Cauchy, the MMSE predictor cannot be defined but the MLE can be defined and it has finite information.

Example 3.1.

We now give an interesting biostatistical example of recursive prediction. Let $X_1, \ldots, X_n$ be i.i.d. with unknown d.f. F For the uncensored case, it can be easily shown that the recursive form of the empirical estimates are given by

$$t\hat{F}_t(x) = (t-1)\hat{F}_{t-1}(x) + I_{(X_t \le x)}, \tag{3.4}$$

or equivalently

$$\hat{F}_t(x) = (1-\omega_t)\hat{F}_{t-1}(x) + \omega_t I_{(X_t \le x)}, \tag{3.5}$$

where $\omega_t = 1/t$. Moreover if we assume a Dirichelt process prior it can be shown that the partial predictive recursion can be given as

$$\hat{F}_t(B) = (1-\omega_t)\hat{F}_{t-1}(B) + \omega_t \frac{\hat{F}_{t-1}(B \bigcap A_t)}{\hat{F}_{t-1}(A_t)}, \tag{3.6}$$

where $\omega_t = 1/(t+\alpha)$ and B is a Borel set .

Example 3.2. We give an interesting biostatistical time series example of prediction with censored autocorrelated data. Censored observations may arise naturally in time series if there is an upper or lower limit of detection. For example when one is monitoring levels of an airborne contaminant or recording daily bioassays of hormone levels in a patient. Regression analysis with autoregressive errors when some observations are censored had been studied in Zeger and Brookmeyer (1986). Let y_t satisfy

$$y_t = \theta y_{t-1} + \epsilon_t$$

where the ϵ_t 's are independent and identically distributed Gaussian variates with mean zero and variance σ^2 . We observe possibly censored observations, (i.e.) $y_t = y_t$ for $y_t \ge L_t$ and $y_t = L_t$ for $y_t < L_t$ where L_t 's are known limits of detection.

If the ith observation is censored,

$$E(y_i|y_i < L_i, y_{i-1}, ..) = \theta y_{i-1} + E(\epsilon_i|\theta y_{i-1} + \epsilon_i < L_i)$$

and then

$$E(y_i|y_i < L_i, y_{i-1}, ...) = \theta y_{i-1} - \frac{\phi(\theta y_{i-1} - L_i)}{1 - \Phi(\theta y_{i-1} - L_i)}$$

where ϕ and Φ are, respectively, the density and distribution functions of the standard normal law.

Similar results can be obtained for right censored observations. Now, by taking $g(h_t) = h_t$ where $h_t = E_\theta(y_t|y_t < L_t) - \theta y_{t-1}$ when $t \in C$ (C corresponds to the time index set for the censored observations) and $h_t = y_t - \theta y_{t-1}$ for uncensored observations we can use the algorithm given in Theorem 3.1 to obtain nonlinear predictors for a linear time series model with censored observations as well.

It is of interest to note that even if the error distribution is normal, the posterior mean (i.e. the mean square predictor) based on censored observations becomes a nonlinear function of the observations. This shows that even for linear models with censored observations nonlinear predictors are more appropriate.

3.2 NUMERICAL EXAMPLE

As an illustration of recursive estimation for a nonlinear mod el of polio incidences is given in this section. Table 1 lists the monthly cases of poliomyelitis reported by the U. S. Center for Disease Control for the years 1970 to 1996. Of interest is to estimate the parameters of the model which will help forecast the future poliomyelitis incidences.

From Table 1, one notices that the polio incidences has the form of count data and there are many months with count zero. Also, the data have some counts which are very large. For example, the month of November in 1972 has a count of 14. Descriptive statistics such as histogram and normal qq-plot show that the data is not normal.

In this data set there is overdispersion i.e. variance = 2.266 is greater than mean = 0.8735. This suggests that the data does not follow a Poisson distribution either.

Table 1: **Monthly number of U.S. cases of poliomyelitis for the years 1970 to 1996**

	Jan.	Feb.	Mar.	Apr.	May.	Jun.	Jul.	Aug.	Sep.	Oct.	Nov.	Dec.
1970	0	1	0	0	1	3	9	2	3	5	3	5
1971	2	2	0	1	0	1	3	3	2	1	1	5
1972	0	3	1	0	1	4	0	0	1	6	14	1
1973	1	0	0	1	1	1	1	0	1	0	1	0
1974	1	0	1	0	1	0	1	0	1	0	0	2
1975	0	1	0	1	0	0	1	2	0	0	1	2
1976	0	3	1	1	0	2	0	4	0	2	1	1
1977	1	1	0	1	1	0	2	1	3	1	2	4
1978	0	0	0	1	0	1	0	2	2	4	2	3
1979	3	0	0	2	7	8	2	4	1	1	2	4
1980	0	1	1	1	3	0	0	0	0	1	0	1
1981	1	0	0	0	0	0	1	2	0	2	0	0
1982	0	1	0	1	0	1	0	2	0	0	1	2
1983	0	1	0	0	0	1	2	1	0	1	3	6
1984	1	2	0	0	0	1	1	0	0	0	0	3
1985	0	1	1	0	0	1	0	2	1	1	0	0
1986	0	0	1	0	0	0	0	0	1	0	0	1
1987	0	0	0	0	0	0	0	0	0	0	0	0
1988	0	0	0	1	2	0	1	2	0	0	3	0
1989	1	0	0	0	0	0	1	0	1	1	0	1
1990	0	1	0	2	0	0	0	1	0	1	2	0
1991	0	0	1	0	0	0	1	1	2	1	0	0
1992	0	0	0	1	0	0	2	0	0	0	1	0
1993	0	0	0	0	0	0	1	1	0	0	0	1
1994	0	0	0	0	0	0	0	0	0	0	0	0
1995	0	2	0	0	0	0	0	0	0	0	0	0
1996	0	0	0	1	0	0	0	0	1	0	3	0

Next, to check whether the observations are independent, conditional probability of observing one of $(0, 1, 2, 3, 4, \geq 5)$, given an observed incidences are calculated. These probabilities are listed in Table 2. If the observed counts are independent there will be homogeneity among the rows. It can be concluded that the polio incidences are not independent. Hence the time series methods are more appropriate to analyse the data.

Table 2: **Transition probability**

	0	1	2	3	4	5
0	0.583	0.314	0.0686	0.0286	0.005	0
1	0.540	0.23	0.137	0.06	0.011	0.023
2	0.516	0.194	0.065	0.065	0.129	0.032
3	0.333	0.2	0.067	0.133	0	0.267
4	0.67	0.17	0.17	0	0	0
5	0.11	0.22	0.33	0.11	0	0.22

With independent observations, log linear models can be used. Under the log linear model, $\mu_t = \exp(x_t'\beta)$ where β is a vector of unknown coefficients to be estimated. If y_t is Poisson, likelihood methods can be used to estimate β. Under the Poisson model, $\text{Var}(y_t) = \mu_t$. In practice, however, it is common that $\text{Var}(y_t) > \mu_t$. In this case, quasilikelihood methods (Heyde (1997)) which allow a variety of variance-mean relations are appropriate. The two most common assumptions are: (i) $\text{Var}(y_t) = \mu_t\phi$; (ii) $\text{Var}(y_t) = \mu_t + \mu_t^2\sigma^2$, where ϕ and σ^2 are unknown scale parameters.

Extensions of log linear models which account for dependence are necessary to obtain valid inferences about the relationship of y_t and x_t.

In parameter-driven models, autocorrelation is introduced through a latent process. Let $\theta_t = \log \mu_t$ be the canonical parameter for the log linear model. Then θ_t is assumed to depend on an unobservable noise process, ϵ_t, that is $\theta_t = \theta(\epsilon_t, y_{t-1}, \ldots, y_1)$. For example, y_t given ϵ_t might be Poisson with $E(y_t|\epsilon_t) = \exp(X_t'\beta)\epsilon_t$. The latent process,$\epsilon_t$ introduces both overdispersion and autocorrelation in y_t

Zeger (1988) has used the parameter-driven model to polio incidence data as an illustration. Of interest in the paper was whether this record provides evidence of a long-term decrease in the rate of U.S. polio infection.

We fit an AR(1) model to the data. The first step in this direction was to generate a time series plot of the polio incidences. From the plot, it can be concluded that the polio incidences are not a linear function of time. For the next step, ARIMA procedure of SAS was used to make preliminary identification of the process underlying in this time series. From the SAS output, it is noticed that the sample autocorrelation function of the time series polio incidences tails off and its sample partial autocorrelation function cuts off at lag 1. Therfore, this time series was identified as an AR(1) model. Also, the t-ratio for coefficient of lag 1 are significant. Therefore, AR(1) model used for LS parameter estimation and residual analysis was carried out to validate the assumptions made for the analysis.

From the residual analysis it can be concluded that the normality assumption is not valid, and $R^2 = 0.1105$. Due to the non-normality of the polio incidences, the standard tests carried out for AR(1) analysis (from output of ARIMA) are no longer valid, which in turn casts doubt about the conclusion that the coefficient is significant. This, led us tox conclude that AR(1) model is not adequate. This is supported by the visual inspection of the times series overlay plot of actual versus the predicted polio incidences.

```
                              Autocorrelations

Lag Covariance Correlation -1 9 8 7 6 5 4 3 2 1 0 1 2 3 4 5 6 7 8 9 1
  0   2.258678    1.00000  |                    |********************|
  1   0.750151    0.33212  |                  . |*******             |
  2   0.508579    0.22517  |                  . |*****               |
  3   0.231533    0.10251  |                  . |**.                 |
  4   0.302862    0.13409  |                  . |***                 |
  5   0.417401    0.18480  |                  . |****                |
  6   0.398442    0.17641  |                  . |****                |
  7   0.259856    0.11505  |                  . |**.                 |
  8   0.265980    0.11776  |                  . |**.                 |
  9   0.166384    0.07366  |                  . |* .                 |
 10   0.263928    0.11685  |                  . |**.                 |
 11   0.377686    0.16722  |                  . |***                 |
 12   0.333255    0.14754  |                  . |***                 |
 13   0.231353    0.10243  |                  . |**.                 |
 14   0.268341    0.11880  |                  . |**.                 |
 15   0.136357    0.06037  |                  . |* .                 |
 16   0.226204    0.10015  |                  . |**.                 |
 17   0.186812    0.08271  |                  . |**.                 |
 18   0.125425    0.05553  |                  . |* .                 |
 19   0.066343    0.02937  |                  . |* .                 |
 20   0.164668    0.07290  |                  . |* .                 |
 21   0.121409    0.05375  |                  . |* .                 |
 22   0.131010    0.05800  |                  . |* .                 |
 23   0.287977    0.12750  |                  . |***                 |
 24   0.178731    0.07913  |                  . |**.                 |
                           "." marks two standard errors
```

```
                          Partial Autocorrelations

           Lag Correlation -1 9 8 7 6 5 4 3 2 1 0 1 2 3 4 5 6 7 8 9 1
             1    0.33212  |                  . |*******             |
             2    0.12910  |                  . |***                 |
             3   -0.00628  |                  . | .                  |
             4    0.08797  |                  . |**                  |
             5    0.12849  |                  . |***                 |
             6    0.06910  |                  . |*.                  |
             7   -0.00094  |                  . | .                  |
             8    0.04874  |                  . |*.                  |
```

```
 9   -0.00566  |                     .  |  .                |
10    0.05243  |                     .  |*.                 |
11    0.09766  |                     .  |**                 |
12    0.03518  |                     .  |*.                 |
13   -0.00621  |                     .  |  .                |
14    0.05576  |                     .  |*.                 |
15   -0.03059  |                     .*|  .                 |
16    0.02466  |                     .  |  .                |
17    0.00556  |                     .  |  .                |
18   -0.02630  |                     .*|  .                 |
19   -0.03086  |                     .*|  .                 |
20    0.05023  |                     .  |*.                 |
21   -0.00614  |                     .  |  .                |
22   -0.01382  |                     .  |  .                |
23    0.10567  |                     .  |**                 |
24   -0.00763  |                     .  |  .                |
```

```
SSE            650.2423     DFE                321
                  MSE              2.025677    Root MSE     1.423263
                  SBC               1154.19    AIC          1146.635
                  Reg Rsq            0.1105    Total Rsq      0.1105
                  Durbin-Watson      2.0800
```

```
 Conditional Least Squares Estimation

                                       Approx.
                 Parameter    Estimate     Std Error    T Ratio   Lag
                 MU            0.87012      0.11811        7.37    0
                 AR1,1         0.33247      0.05259        6.32    1

                 Constant Estimate   = 0.58082424

                 Variance  Estimate = 2.02175364
                 Std Error Estimate = 1.42188383
```

We have fitted a nonlinear time series model using the recursive estimation procedure given in equation (3.1) and (3.2) with $E[y_t|\mathcal{F}_{t-1}^y] = \mu + \theta y_{t-1}$ and $Var[y_t|\mathcal{F}_{t-1}^y] = \sqrt{y_{t-1}}$. The recursive equations for the above model are

$$\begin{aligned}\hat{\theta}_t &= \hat{\theta}_{t-1} + \frac{k_{t-1}\sqrt{y_{t-1}}}{1 + k_{t-1}\sqrt{y_{t-1}}y_{t-1}}[y_t - \hat{\theta}_{t-1}y_{t-1}] \\ k_t &= \frac{k_{t-1}}{1 + k_{t-1}\sqrt{y_{t-1}}y_{t-1}}.\end{aligned}$$

This nonlinear model of counts imply that if an immediate past month had no polio incidences then the variability for the current month is zero, i.e. the incidences to be observed for the current month is a unknown constant with probabilty one. On the other hand if for an immediate past month has a large number of polio incidences then the observed variability in the current month will be large. Based on the residual analysis and the time series overlay plot of actual as well as the predicted polio incidences we conclude that the proposed nonlinear model is adequate.

REFERENCES

Gill, R. and Levit (1995). Applications of the van Trees Inequality: a Bayesian Cramer-Rao bound. *Bernoulli* 1, 59-79.

Godambe, V. P. (1999) Linear Bayes and optimal estimation. *Annals of the Institute of Statistical Mathematics* 51, 201-215.

Heyde, C.C. (1997) Quasi-likelihood and its application. A general approach to optimal parameter estimation. *Springer-Verlag* , New York.

Thavaneswaran, A. and Abraham, B. (1988). Estimation of nonlinear time series models using estimating equations. *J. Time Series Anal.* 9, 99-108.

Thavaneswaran, A. and Heyde, C.C.(1999) Prediction via estimating functions. *Journal of Statistical Planning and Inf.* 77, 89-101.

Zeger, L. and Brookmeyer, R. (1986). Regression Analysis with Censored Autocorrelated Data. *J. Amer. Statist. Assoc.*, 81, 722-729.

Zeger, L.(1988) A regression model for time series of counts. *Biometrika* 75 621-629.

THE MAXIMUM PENALIZED LIKELIHOOD APPROACH USING NON-HOMOGENEOUS POISSON PROCESSES FOR SOFTWARE RELIABILITY AND RELIABILITY GROWTH MODELS

By

KARTHIK DEVARAJAN

Pharmaceutical Research Institute

Bristol-Myers Squibb Company

Princeton, NJ 08543*

and

NADER EBRAHIMI

Division of Statistics

Northern Illinois University

DeKalb, IL 60115

SUMMARY

In this paper we propose a non-parametric estimate for the intensity function of a non-homogenous Poisson Process (NHPP) using penalized likelihood methods. We also develop bootstrap methods for constructing confidence regions for the intensity function. Finally, we illustrate our approach with an actual failure data set.

Key words and phrases: Non-homogenous Poisson process, Software Reliability, Reliability growth models, Penalized likelihood function, Time truncated model, Failure truncated model.

*The work was done when he was at Northern Illinois University.

1 INTRODUCTION

Considerable research has been carried out to study software failure phenomena and to develop and apply software reliability models to predict software performance. Different stochastic models have been proposed to describe the reliability of software. We refer to the book by Musa et al. (1987) and the recent review article by Nayak (1998).

In looking through the literature, we observe that there are broadly two modeling strategies employed. These strategies do not cover all the models that have been proposed, but do cover most of them. One can approach software reliability by either

i. Modeling the (inter-failure) times between successive failures

or

ii. Modeling the number of failures of the software up to a given time.

Let $N(t)$ be the total number of failures of the software that are observed during time $[0, t]$. Nonhomogeneous Poisson Processes (NHPP's), which are special cases of point processes, have been used frequently for describing the behavior of $N(t)$. A nonhomogenous Poisson process on $(0, \infty)$ with intensity function $\lambda(t),\ t > 0$, is defined by the property that the probability of an "event" (in the present context, a failure) in a short time interval $(t,\ t + \delta t)$ is of the form $(\lambda(t))\delta t + o(\delta t)$, the probability of more than one such event being $o(\delta t)$, and such events are independent over different time intervals. For such a process, the number of events observed in any finite time interval, (t_1, t_2) say, has a Poisson distribution with mean $\int_{t_1}^{t_2} \lambda(x)dx$. Many parametric models have been proposed for modeling the intensity function of $N(t), \lambda(t)$, the first of which was by Jelinski and Moranda (1972). Others have also developed models under different assumptions. For more background see review papers by Nayak (1998), Ries and Basu (1998) and Singpurwalla and Wilson (1994). Specific inferences for various models can also be obtained from the literature cited.

Although some of these parametric models are popular due to their simplicity, their statistical methods for predicting software performance are often sensitive to the parametric assumptions. In fact, each model may lead to a different assessment of software reliability. In this paper, our goal is to present a non-parametric estimate of $\lambda(t)$ using penalized likelihood methods. The method of penalized likelihood estimation has become an important tool in non-parametric estimation since it was introduced by Good and Gaskins (1971). An excellent exposition of penalized likelihood estimates is given by Tapia and Thompson (1978). Our method is designed to avoid assumptions and has broader applicability.

The general statistical methodology developed in this paper have different ramifications in different application areas. The software reliability problem mentioned above is one such application.

Another important application is in the estimation of the current reliability for systems that undergo a developmental testing program. During the developmental phase of a production involving complex systems, a test-analyze-and fix process is often undertaken to promote the detection of system faults and incorporate appropriate corrective actions. As a consequence of the redesign efforts contributing to the development of the system, failures constitute the only meaningful source of data available to the engineer. In this context NHPP with an intensity function of the Weibull form was used by Crow (1974) to model these failures. Statistical inference concerning the current reliability of a reliability growth model has an extensive literature. For the latest references we refer you to Sen (1998) and Erkanli et al. (1998).

The outline of the paper is as follows: The mathematical formulations of the method are given in Section 2 for both failure truncated and time truncated models. In Section 3, we describe our bootstrap algorithms and confidence intervals for the intensity function. Finally, in Section 4, our method is applied to a real data set.

2 MODEL AND INFERENCE

Here are the typical scenarios for statistical inference about $\lambda(t)$: Given the time truncated model, testing until time t, the ordered epochs of the observed n failure times are denoted by $x_1, \cdots, x_n$. Therefore, the data set D_t consists of $\{n, x_1, \cdots, x_n; t\}$. Given the failure truncated model (observed until n-th failure), the data set D_{x_n} consists of $\{x_1, x_2, \cdots, x_n\}$.

To draw statistical inference from either D_t or D_{x_n} for the intensity function, the likelihood of a curve $\lambda(x)$ for the time-truncated model is given by

$$L(\lambda|D_t) = (\prod_{i=1}^{n} \lambda(x_i))exp(-m(t)), \tag{2.1}$$

where $m(t) = \int_0^t \lambda(x)dx$. For the failure truncated model a similar expression can be applied with t replaced by x_n.

The likelihood in (2.1) has no finite maximum over the class of all intensity functions and hence it is not possible to use the maximum likelihood method for intensity function estimation. There are, nevertheless, possible approaches to maximum likelihood. One method, introduced by Good and Gaskins (1971), is to incorporate into the likelihood function a term which describes the roughness of the curve under consideration. In this section we use this method to obtain a non-parametric estimator of $\lambda(t)$. To enforce non-negativity of our estimator we first estimate the square root of the intensity function $h(t) = (\lambda(t))^{\frac{1}{2}}$ and then square the result.

Suppose $R(h)$ is a functional which quantifies the roughness of h. Define the penalized log-likelihood function by

$$\ell_{K_0}(h) = 2\sum_{i=1}^{n} log\ h(x_i) - \int_0^t h^2(u)du - K_0 R(h), \tag{2.2}$$

where the parameter K_0 controls the amount by which the data are smoothed. For example, larger values of K_0 will produce smoother function estimates. For our purposes we take $R(h) = \int_I (h''(u))^2 du$, where I is an interval in the positive real axis containing the closed interval $[t_0, t_m], m = n+1, t_i = x_i, i = 1, \cdots, m-1, t_0 = 0$, and $t_m = t$. That is, we penalize the log-likelihood for curvature.

Now the penalized likelihood function reduces to

$$\ell_{K_0}(h) = \sum_{i=1}^{m-1} [2\ log\ h(t_i) - \int_{t_{i-1}}^{t_i} h^2(u)du] - \int_{t_{m-1}}^{t_m} h^2(u)du - K_0 \int_I (h''(u))^2 du. \tag{2.3}$$

It should be noted that $h(u)$ is restricted to the class of continuous functions which have continuous first and second derivatives and have the third derivative that is a step function with jumps at $t_1, \cdots, t_m$.

Using the methods in Calculus of Variations, (see Reisnch (1967) and Anderson and Senthilselvan (1980)), it can be shown that the unique function maximizing (2.3) over the class of all continuous functions which have piecewise continuous third derivative satisfies the following conditions. (For details see also the Appendix.)

i. $h(u), h^{(1)}(u)$ and $h^{(2)}(u)$ are continuous on I

ii. $\lim_{\Delta \to 0} h^{(3)}(\Delta) = \lim_{\Delta \to 0} h^{(3)}(t_m + \Delta) = 0,\ h^{(2)}(0) = 0$

iii. $\lim_{\Delta \to 0} [h^{(3)}(t_i - \Delta) - h^{(3)}(t_i + \Delta)] = \dfrac{-1}{K_0 h(t_i)},\ i = 1, \cdots, m-1$

iv. $h^{(4)}(u) + \frac{1}{K_0} h(u) = 0, u \in (t_i, t_{i+1}),\ i = 0, 1, \cdots, m-1$

v. $h^{(2)}(u) = 0$ for $u \in I - [t_0, t_m]$

vi. $h^{(3)}(u)$ is continuous at t_m

Here $h^{(\ell)}(u)$ means the ℓ-th derivative, $I - [t_0, t_m]$ means the complement of $[t_0, t_m]$ with respect to I.

A solution to the differential equation (iv) is given by the following equation, (see Boyce and Di Primor (1960), page 211),

$$\begin{aligned} \hat{h}(t) &= \exp(w(t-t_i))\,[a_i \cos\ w(t-t_i) + b_i \sin\ w(t-t_i)] \\ &+ \exp(-w(t-t_i))\,[c_i \cos w(t-t_i) + d_i \sin w(t-t_i)] \\ \text{for}\quad t &\in [t_i, t_{i+1}],\ i = 0, 1, \cdots, m-1, \end{aligned} \tag{2.4}$$

where $w = (\frac{1}{\sqrt{2}})(\frac{1}{K_0})^{\frac{1}{4}}$. Further, from (i) - (vi),

$$\begin{aligned}\hat{h}(t) &= \{\exp(wu_{m-1})[a_{m-1}\cos\ wu_{m-1} + b_{m-1}\sin\ wu_{m-1}] + \exp(-wu_{m-1})[c_{m-1}\cos\ wu_{m-1} \\ &+ d_{m-1}\sin\ wu_{m-1}] + e(t - t_m)\},\ t \in I - [t_0, t_m].\end{aligned} \tag{2.5}$$

The equations (i), (ii), (iii) can be solved to give the following recurrence relations for a_i, b_i, c_i, and d_i's and e.

(i)′ $a_i = \exp(wu_{i-1})[a_{i-1}\cos\ wu_{i-1} + b_{i-1}\sin\ wu_{i-1}] - \frac{wE_i}{2}$

(ii)′ $b_i = \exp(wu_{i-1})[b_{i-1}\cos\ wu_{i-1} - a_{i-1}\sin\ wu_{i-1}] + \frac{wE_i}{2}$

(iii)′ $c_i = \exp(-wu_{i-1})[c_{i-1}\cos wu_{i-1} + d_{i-1}\sin\ wu_{i-1}] + \frac{wE_i}{2}$

(iv)′ $d_i = \exp(-wu_{i-1})[d_{i-1}\cos\ wu_{i-1} - c_{i-1}\sin\ wu_{i-1}] + \frac{wE_i}{2}$

(v)′ $b_0 = d_0, b_0 - a_0 + c_0 + d_0 = 0,$

$$\{\exp(wu_{m-1})[(b_{m-1} - a_{m-1})\cos wu_{m-1} - (a_{m-1} + b_{m-1})\sin\ wu_{m-1}]$$
$$+ \exp(-wu_{m-1})[(c_{m-1} + d_{m-1})\cos\ wu_{m-1} + (d_{m-1} - c_{m-1})\sin wu_{m-1}]\} = 0,$$
$$\{\exp(wu_{m-1})[b_{m-1}\cos\ wu_{m-1} - a_{m-1}\sin\ wu_{m-1}]$$
$$+ \exp(-wu_{m-1})[-d_{m-1}\cos\ wu_{m-1} + c_{m-1}\sin wu_{m1}]\} = 0,$$

(vi)′ $\{\exp(wu_{m-1})[(b_{m-1} + a_{m-1})\cos wu_{m-1} + (b_{m-1} - a_{m-1})\sin wu_{m-1}]$

$$+ \exp(-wu_{m-1})[(d_{m-1} - c_{m-1})\cos wu_{m-1} - (d_{m-1} + c_{m-1})\sin wu_{m-1}]\} = \tfrac{e}{w},$$

where $u_i = t_{i+1} - t_i$, $i = 0, 1, \cdots, m-1$ and $E_i = \{\exp(wu_{i-1})(a_{i-1}\cos\ wu_{i-1} + b_{i-1}\sin\ wu_{i-1}] + \exp(-wu_{i-1})[c_{i-1}\cos wu_{i-1} + d_{i-1}\sin wu_{i-1}]\}^{-1}$, $i = 1, \cdots, m$.

It is seen from the equations (i)′ - (iv)′ that if a_0, b_0, c_0 and d_0 are known, other coefficients can be determined for a given value of K_0. However, since $a_{m-1}, b_{m-1}, c_{m-1}$ and d_{m-1} are functions of a_0, b_0, c_0, and d_0 one can get hold of them simply by solving (v)′. The key to this method is to use the equations (i)′ – (v)′ to search for a_0, b_0, c_0 and d_0 and then to obtain a_i, b_i, c_i, d_i and e by using equations (i)′ - (vi)′.

For the failure truncated model, $\ell_{K_0}(h)$ is

$$\ell_{K_0}(h) = \sum_{i=1}^{n}\left[log\ h(x_i) - \int_{x_{i-1}}^{x_i} h^2(u)du\right] - K_0 \int_I (h''(u)^2 du, \tag{2.6}$$

where I is a closed interval containing $[x_0, x_n]$. Similar to the time truncated model case, the unique function maximizing the class of all continuous functions which have piecewise continuous third derivative satisfies the following conditions.

i. $h(u), h^{(1)}(u)$ and $h^{(2)}(u)$ are continuous on I

ii. $\lim_{\Delta \to 0} h^{(3)}(\Delta) = \lim_{\Delta \to 0} h^{(3)}(x_n + \Delta) = 0,\ h^{(2)}(0) = 0$

iii. $\lim_{\Delta \to 0} \left[h^{(3)}(x_i - \Delta) - h^{(3)}(x_i + \Delta)\right] = \dfrac{-1}{K_0 h(x_i)},\ i = 1, \cdots, n$

iv. $h^{(4)}(u) + \frac{1}{K_0} h(u) = 0,\ u \epsilon (x_i, x_{i+1}),\ i = 0, \cdots, n-1$

v. $h^{(2)}(u) = 0$ for $u \in I - [x_0, x_n]$.

Again the solution is

$$\begin{aligned} \hat{h}(x) &= \{\exp(w(x - x_i))\ [a_i \cos\ w(x - x_i) + b_i \sin\ w(x - x_i)] \\ &+ \exp(-w(x - x_i))[c_i \cos\ w(x - x_i) + d_i \sin\ w(x - x_i)]\} \text{ for } x \in [x_i, x_{i+1}],\ i = 0, \cdots, n-1 \end{aligned}$$

and

$$\hat{h}(x) = \{\exp(wu_{n-1})[a_{n-1} \cos\ wu_{n-1} + b_{n-1} \sin\ wu_{n-1})] + \exp(-wu_{n-1})[c_{n-1} \cos\ wu_{n-1} + d_{n-1} \sin\ wu_{n-1})]\} + e(x - x_n), \text{ for } x \in I - [x_0, x_n],$$

where a_i, b_i, c_i and d_i's are obtained by solving the following recursive equations.

(i)′ $a_i = \exp(wu_{i-1})[a_{i-1} \cos\ wu_{i-1} + b_{i-1} \sin\ wu_{i-1}] - \frac{wE_i}{2}$

(ii)′ $b_i = \exp(wu_{i-1})[b_{i-1} \cos\ wu_{i-1} - a_{i-1} \sin\ wu_{i-1}] + \frac{wE_i}{2}$

(iii)′ $c_i = exp(-wu_{i-1})[c_{i-1} \cos\ wu_{i-1} + d_{i-1} \sin\ wu_{i-1}] + \frac{wE_i}{2}$

(iv)′ $d_i = exp(-wu_{i-1})[d_{i-1} \cos\ wu_{i-1} - c_{i-1} \sin\ wu_{i-1}] + \frac{wE_i}{2}$

(v)′ $b_0 - a_0 + c_0 + d_0 = 0, b_0 = d_0$

(vi)′ $\{\exp(wu_{n-1})(b_{n-1} \cos\ wu_{n-1} - a_{n-1} \sin\ wu_{n-1}) + \exp(-wu_{n-1})(-d_{n-1} \cos\ wu_{n-1} + c_{n-1} \sin\ wu_{n-1})\}$
0,
$\{\exp(wu_{n-1})[(b_{n-1} + a_{n-1}) \cos\ wu_{n-1} + (b_{n-1} - a_{n-1}) \sin\ wu_{n-1}] + exp(-wu_{n-1})$
$[(d_{n-1} - c_{n-1}) \cos\ wu_{n-1} - (d_{n-1} + c_{n-1}) \sin wu_{n-1}]\} = \dfrac{e}{w}$

(vii)′ $(E_n^{-1}\{(\exp(wu_{n-1})[(b_{n-1} - a_{n-1}) \cos\ wu_{n-1} - (a_{n-1} + b_{n-1}) \sin\ wu_{n-1}] + \exp(-wu_{n-1})[(c_{n-1} + d_{n-1}) \cos wu_{n-1} + (d_{n-1} - c_{n-1}) \sin wu_{n-1}]\} = -2w$.
Here $u_i = x_{i+1} - x_i,\ i = 0, 1, \cdots, n-1,\ w = (\dfrac{1}{K_0})^{\frac{1}{4}} \dfrac{1}{\sqrt{2}}$, and $E_i = \{\exp(wu_{n-1})[a_{i-1} \cos\ wu_{i-1} + b_{i-1} \sin\ wu_{i-1}] + \exp(-wu_{i-1})[c_{i-1} \cos\ wu_{i-1} + d_{i-1} \sin wu_{i-1}]\},\ i = 1, \cdots, n.$

3 CONFIDENCE BANDS

There are several ways of bootstrapping the process $\{X_i\}$ to obtain a resample $\{X_i^*\}$. One method is simply to generate x_i's as points of a NHPP of intensity $\hat{\lambda}$, where $\hat{\lambda}$ is given by either the equation (2.4) or (2.7). The second method is to draw a sample $X_1^*, \cdots, X_{N^*}^*$ by sampling randomly with N^* times from $\{X_1, \cdots, X_n\}$, where N^* is a random variable having the Poisson distribution with intensity function $\int_0^t \hat{\lambda}(x)dx$. Finally, the third method is to draw $X_1^*, \cdots, X_n^*$ by sampling uniformly, with replacement from $\{X_1, \cdots, X_n\}$ and take this collection to be our resample from $\{X_1, \cdots, X_n\}$. Cowling et al. (1996), using simulations, showed that all three resampling methods produce confidence bands with low coverage errors. Thus in practice, there is no need to do computationally intensive parametric simulation from a Poisson process, as in methods 1 and 2, since the resampling method 3, the one used in the independent and identically distributed case, seems to work as well.

For our purpose we use the method 3, which is much easier to implement. For more details about the Bootstrap method we recommend the excellent book by Efron and Tibshirani (1993).

There are basically two techniques for constructing confidence regions for λ, the percentile approach and the percentile-t approach. In this paper, we take the percentile approach to constructing confidence intervals which is based on percentiles of the bootstrap distribution of a statistic. A bootstrap data set of X^* is generated according to the method 3 and Bootstrap replications $\hat{\lambda}^*(x)$ are computed as follows.

Let $\hat{G}_x(t)$ be the cumulative distribution function corresponding to $\hat{\lambda}(x)$. The $(1-2\alpha)$ percentile interval for the intensity function at a given point x is defined by α and $(1-\alpha)$ percentiles of $\hat{G}_x$.

For each bootstrap data set, we solve the recurrence equations (i)$'$ - (iv)$'$ using the method outlined earlier. Hence, we get a bootstrap replication of a_i, b_i, c_i, d_i, $i = 1, m-1$ and $\hat{\lambda}(x)$. We repeat this procedure for $B = 100$. For each replication we obtain $\hat{\lambda}(x)$ and obtain the 2.5th and 97.5th percentiles of $\hat{\lambda}(x)$. Thus, we obtain a 95% confidence band for $\hat{\lambda}(x)$.

4 EXAMPLE

We illustrate our methods for computation of $\hat{\lambda}(t)$ using the data from the development of the Naval Tactical Data System (NTDS). The NTDS system had several modules, the data are based on one of the larger modules and its three phases, the production phase, the test phase and the user phase. These data were analyzed by Goel-Okumoto (GO)(1979), Singparwalla and Wilson (1994), Kuo and Yang (1995, 1996) and many others. Following Kuo and Yang (1995) we only consider the 26 failures that occured during the production phase. The inter-failure times for these 26 failures are: 9, 12, 11, 4, 7, 2, 5, 8, 5, 7, 1, 6, 1, 9, 4, 1, 3, 3, 6, 1, 11, 33, 7, 91, 2, 1.

In order to reduce the spread of the failure times the data were transformed by dividing by the largest failure time, which is 250. Starting with an initial guess for each of a_0 and c_0 we perform a grid research as follows. For each combination of (a_0, c_0) such that $a_0 + c_0$ satisfies our intial guess, we compute the corresponding values of b_0 and d_0 using (v)$'$. We then compute $E_i, i = 1, \ldots, m$ and hence a_i, b_i, c_i and $d_i, i = 1, \ldots, m-1$ using equations (i)$'$ - (iv)$'$. We check to see whether the equations (vi)$'$ are satisfied. If it is satisfied we stop. Otherwise we repeat the procedure over a grid of values of (a_0, c_0) such that $a_0 + c_0$ satisfies our initial guess. We terminate the procedure when a sufficiently small z is obtained, where $z = \sqrt{q_1^2 + q_2^2}$, q_1 and q_2 are the left hand side of the first equation in (vi)$'$ and q_2 is the difference between the left hand side and right hand side of the second equation in (vi)$'$, respectively. In this example our $z = 10^{-4}$ The estimate of $\lambda(t)$ for $K_0 = .001,\ .01,\ .1,\ 2,\ 4,\ 6,\ 15$ are plotted in Figures 1(a) - 1(g). From these Figures, it is clear that as the smoothing parameter increases the curve becomes smoother and smoother. In fact, beyond $K_0 = 1$, we have noticed that the value of K_0 does not have much effect on the behavior of $\hat{\lambda}(t)$. In this example, to construct confidence interval we take the value $K_0 = 6$.

Assuming $K_0 = 6$, Figure 2 gives 95% bootstrap confidence bands obtained for the $B = 100$ replications.

In likelihood criterion, Goel and Okumoto (1979) found that $\lambda(t) = ab\exp(-bt)$ provides a good model for this data. Thus assuming $\lambda(t) = ab\ exp(-bt)$ for this data, they obtained an estimate for the intensity function $\lambda(t)$ as $\hat{\lambda}(t) = (34)(.00579)exp(-.00579t)$. Figure 3 shows the plot of our estimator with $K_0 = 6$ and Goel and Okumoto (1979)'s estimator.

REFERENCES

Anderson, J.A. and Senthiselvan, A. (1980). McGraw-Hill, New York. Smooth estimates for the hazard function. JRSS-Series B, 42, 322-327.

Boyce, W. and DiPrima, R.C. (1967). Elementary differential equations and Boundary value problems. John Wiley & Sons, Inc., New York.

Cowling, A., Hall, P., Phillips, M.J. (1996) Bootstrap confidence regions for the intensity of a Poisson process. Journal of the American Statistical Association, 91, 1516-1524.

Crow, L.H. (1974). Reliability analysis for complex repairable systems. Reliability and Biometry, edited by Proschan, F. and Serfling, R.J., Philadelphia: Society of industrial and applied mathematics, 379-410.

Efron, B. and Tibshirani, R.J. (1993). An introduction to the Bootstrap, Chapman and Hall, New York.

Erkanli, A., Mazzuchi, T.A. and Soyer, R. (1998). Baysian computations for a class of reliability growth models. Technometrics, 40, 14-24.

Goel, A.L. and Okumoto, K. (1979). Time dependent error detection rate model for Software Reliability and other performance measures. IEEE trans. rel., R-28, 206-211.

Good, I.J. and Gaskins, R. (1971). Non-parametric roughness penalties for probability densities. Biometrika, 58, 255-277.

Hewitt, A. and Stromberg, P. (1975). Real and Abstract Analysis. Springer, New York.

Jelenski, Z. and Moranda, P.B. (1972). Software Reliability Research. In statistical computer performance evaluation, [edited by Freinburger]. Academic Press, New York, 465-497.

Kuo, L. and Yang, T.Y. (1951). Bayesian computation of software reliability. J. Computational and Graphical Statistics, 4, 65-82.

Kuo, L. and Yang, T.Y. (1996). Bayesian computation for NHPP in software reliability. Journaal of the American Statistical Association, 9, 763-773.

Musa, J.D., Iannino, A. and Okumoto, K. (1987) Software reliability measurement, prediction, application. McGraw-Hill, New York.

Nayak, T. (1998) Statistical approaches to modeling and estimating software reliability. Frontiers in reliability, edited by A.P. Basu, World Scientific, 271-281.

Reisnch, C.H. (1967). Smoothing by Spline functions. Numer. Math., 10, 1977-183.

Reis, L.D. and Basu, A.P. (1998) Parameter estimation in software reliability growth models. Frontiers in reliability, edited by A.P. Basu, World Scientific, 303-316.

Sen, A. (1998). Estimation of current reliability in a Duane based reliability growth model. Technometrics, 40, 334-344.

Singpurwalla, N.D. and Wilson, S.P. (1994). Software reliability modeling. International Statistical Review, 62, 289-317.

Tapia, R.A. and Thompson, J.R. (1978). Nonparametric probability density estimation. John Hopkins University Press, Baltimore, MD.

APPENDIX

Mathematical details of the optimization of the functional

The penalized likelihood function in terms of $h(t)$ is

$$J[h]=\sum_{i=1}^{n}2log\ h(x_i)-\int_0^t h^2(u)du-K_0\int_I\{h''(u)\}^2du. \tag{A1}$$

Consider a small increment $\delta\eta(u), \delta>0$, to the function $h(u)$. Now a necessary condition for J to have a maximum is

$$\frac{\partial J[h+\delta\eta]}{\partial\delta}=0 \text{ when } \delta=0.$$

This implies that

$$2\sum_{i=1}^{n}\frac{\eta(x_i)}{h(x_i)}-2\int_0^t\eta(u)h(u)du-2K_0\int_I\eta''(u)h''(u)du=0. \tag{A2}$$

Using the fact that $\int_a^b f(x)dg(x) = f(b-)g(b-)-f(a+)g(a+)-\int_a^b g(x)df(x)$, see Hewitt and Stromberg (1975), page 419,

$$\int_I\eta''(u)h''(u)du\sum_{i=1}^{m}\int_{t_{i-1}}^{t_i}\eta''(u)h''(u)du+\int_{I-[t_0,t_m]J}\eta''(u)h''(u)du,$$

where $t_i=x_i, i=1,\cdots,n, t_m=t, t_0=0$ and $m=n+1$,

$$\begin{aligned}
&=\sum_{i=1}^{m}(\eta'(t_i-)h^{(2)}(t_i-) &-& \eta'(t_{i-1}+)h^{(2)}(t_{i-1}+))-\sum_{i=1}^{m}\int_{t_{i-1}}^{t_i}h^{(3)}(u)\eta'(u)du\\
&&+&\int_{I-[t_0,t_m]}\eta^{(2)}(u)h^{(2)}(u)du\\
&=\sum_{i=1}^{m}(\eta'(t_i-)h^{(2)}(t_i-) &-& \eta'(t_{i-1}+)h^{(2)}(t_{i-1}+))-\sum_{i=1}^{m}(h^{(3)}(t_i-)-h^{(3)}(t_i+))\eta(t_i)\\
&&+&\sum_{i=1}^{m}\int_{t_{i-1}}^{t_i}h^{(4)}(u)\eta(u)du\\
&&+&h^{(3)}(t_0+)\eta(t_0)-h^{(3)}(t_m+)\eta(t_m)\\
&&+&\int_{I-[t_0,t_m]}\eta^{(2)}(u)h^{(2)}(u)du
\end{aligned} \tag{A3}$$

Now combining (A2) and (A3) we get that

$$\sum_{i=1}^{n}[2K_0(h^{(3)}(t_i-)-h^{(3)}(t_i+)) \quad + \quad \frac{2}{h(t_i)}]\eta(t_i)$$

$$
\begin{aligned}
-\sum_{i=1}^{m}\int_{t_{i-1}}^{t_i}(2h(u) &+ 2K_0h^{(4)}(u))\eta(u)du \\
-2K_0(h^{(3)}(t_m-) &- h^{(3)}(t_m+))\eta(t_m) \\
-2K_0[\sum_{i=1}^{m}(\eta'(t_{i-1})\ h^{(2)} &\ (t_{i-1}-) - \eta'(t_{i-1}+)h^{(2)}(t_{i-1}+)) \\
&+ h^{(3)}(t_0+)\eta(t_0+) - h^{(3)}(t_m+)\eta(t_m) \\
&+ \int_{I-[t_0,t_m]}\eta^{(2)}(u)h^{(2)}(u)du]. \qquad \text{(A4)}
\end{aligned}
$$

Now $\eta(\cdot)$ is an arbitrary function and by choosing $\eta(\cdot)$ suitably, it can be easily shown that each term in (A4) which is multiplied by $\eta(u), \eta'(u)$ and $\eta''(u)$ is identically zero. Hence we get the condition (i) to (v) in Section 2. The above solution is a maximum for the functional $J[\cdot]$ since it is concave. To see this consider

$$
\begin{aligned}
\frac{\partial^2 J[h+\delta\eta]\partial\delta^2}{\partial\delta^2} = &- \sum_{i=1}^{n}\left[\frac{\eta(t_i)}{h(t_i)+\delta\eta(t_i)}\right]^2 \\
&- \int_0^t \eta^2(u)du \\
&- 2K_0\int_I(\eta^{(2)}(u))^2du < 0.
\end{aligned}
$$

Hence the functional $J[\cdot]$ is concave.

Figure 1.

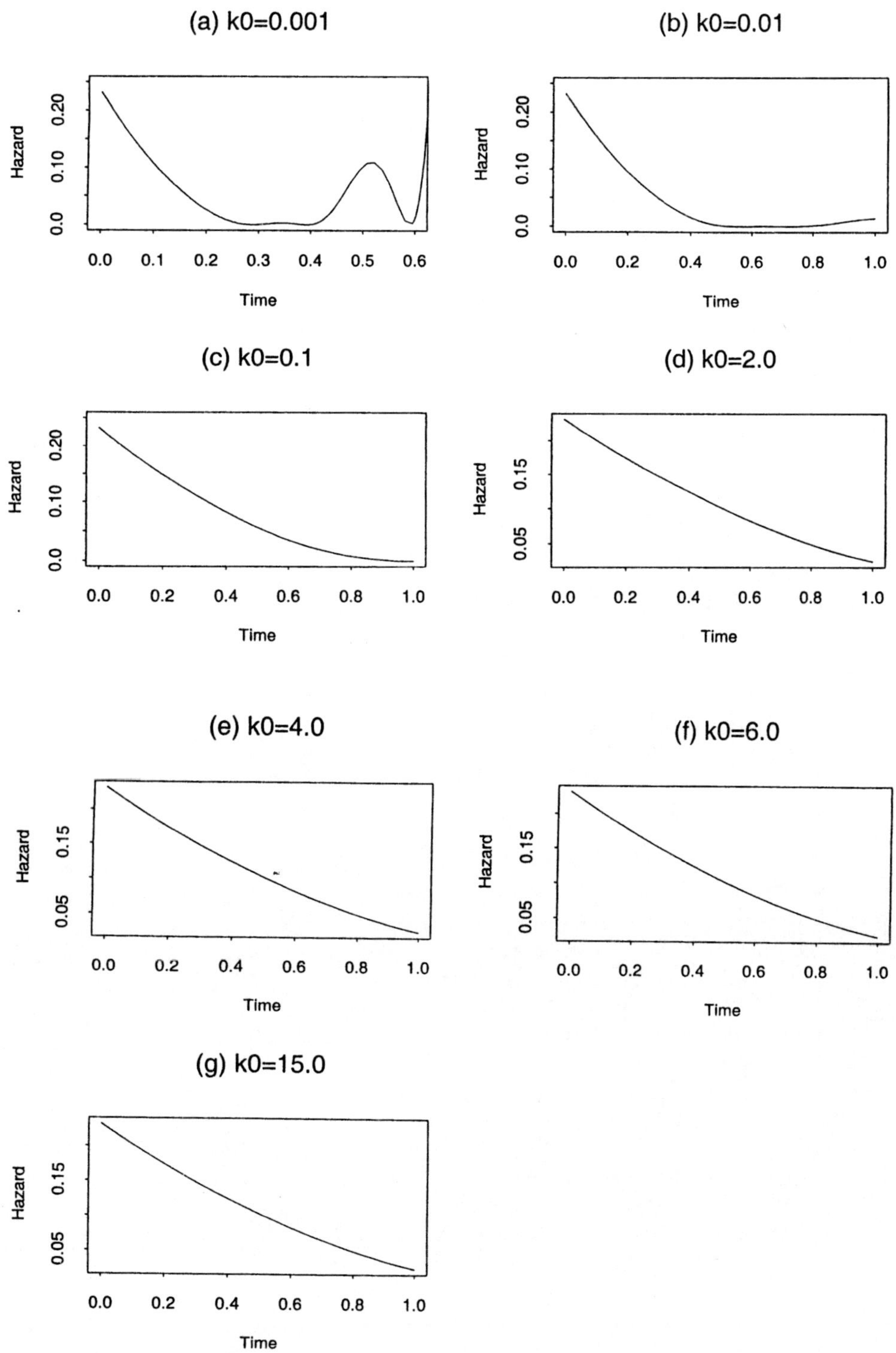
(a) k0=0.001
Hazard
0.0
0.10
0.20
0.0
0.1
0.2
0.3
0.4
0.5
0.6
Time
(b) k0=0.01
Hazard
0.0
0.10
0.20
0.0
0.2
0.4
0.6
0.8
1.0
Time
(c) k0=0.1
Hazard
0.0
0.10
0.20
0.0
0.2
0.4
0.6
0.8
1.0
Time
(d) k0=2.0
Hazard
0.05
0.15
0.0
0.2
0.4
0.6
0.8
1.0
Time
(e) k0=4.0
Hazard
0.05
0.15
0.0
0.2
0.4
0.6
0.8
1.0
Time
(f) k0=6.0
Hazard
0.05
0.15
0.0
0.2
0.4
0.6
0.8
1.0
Time
(g) k0=15.0
Hazard
0.05
0.15
0.0
0.2
0.4
0.6
0.8
1.0
Time

Figure 2. Smoothed 95% Bootstrap Confidence Band for the Hazard, k0=6.0

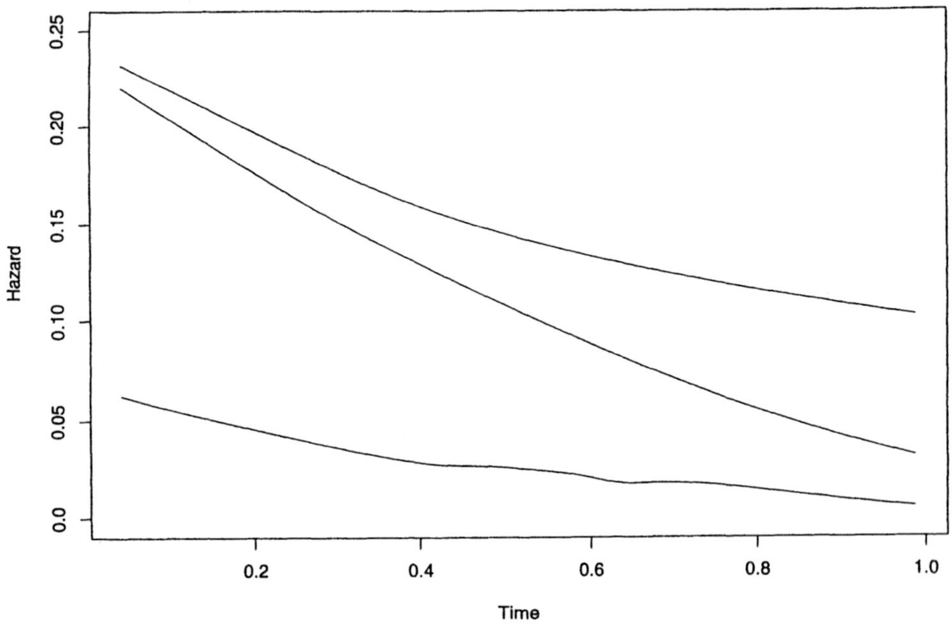

Figure 3. GO Estimate (dot vs. MPL Estimate, k0=6.0 (solid)

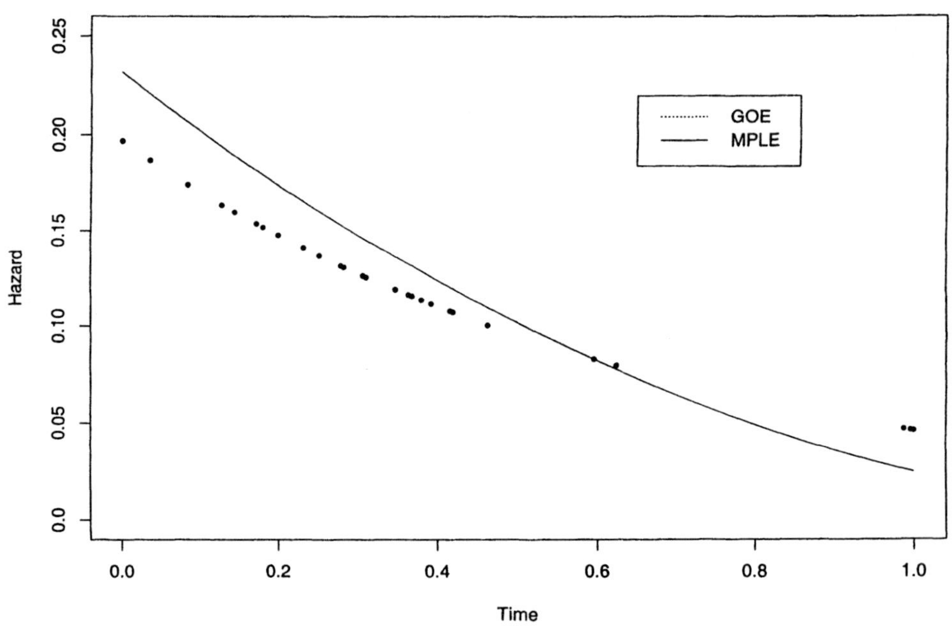

SOME RESULTS ON LOWER RECORDS

S.N.U.A Kirmani
Department 0f Mathematics
University of Northern Iowa
Cedar Falls, IA 50614-0506

M. Ahsanullah
Department of Management Sciences
Rider University
Lawrenceville, NJ 08648-3099

1. INTRODUCTION

Let $X_1, X_2, \ldots$ be a sequence of independent and identically distributed random variables with distribution function $F(t) = P(X_n \leq t)$. An observation X_k is called a lower record if $X_k < X_j$ for every $j < k$. To be more precise, let

$$T_1 = 1$$
$$T_{n+1} = \min\{k : X_k < X_{T_n}, n \geq 1\}$$
$$L_n = X_{T_n}, n \geq 1$$

L_n is called the nth lower record from the sequence $\{X_n, n \geq 1\}$. Analogous definitions hold for upper records. Thus, X_k is an upper record if $X_k > X_i$ for every $i < k$. We refer to Ahsanullah (1995) and Arnold et, al. (1998) for a detailed study of the probabilistic theory of (upper and lower) records.

The objective of this paper is to study some aspects of the probability distributions of L_n, $L_n - L_{n+1}$ and L_{n+1} / L_n. In particular, we are interested in certain properties of these random variables from the point of view of orderings such as hazard rate ordering, stochastic ordering, and dispersive ordering. While the relevant facts about these orderings are given at appropriate places in this paper, we refer to Shaked and Shantikumar (1994) for thorough discussions of various orderings. Section 2 of this paper is concerned with the situation when the distribution function F is logconcave while section 3 deals with implications of dispersive ordering between two random variables. In both of these sections, interest is centered on L_n and $L_n - L_{n+1}$. Section 4 is concerned with behavior of $E(L_{n+1}/ L_n)$. Finally in section 5, we give a characterization of the uniform distribution on (0,1) through the relation $E\{(L_n - L_{n+1})^2 | L_n = x\} = x^2/3, x \in (0,1)$.

2. BEHAVIOR OF LOWER RECORDS AND THEIR GAPS FOR DECREASING HAZARD RATE DISTRIBUTIONS

Let $\{X_n, n\geq 1\}$ be a sequence of independent random variables each having absolutely continuous (with respect to Lebesgue measure) distribution function F. Let L_n denote the nth lower record for this sequence. We shall write F_n to denote the distribution function of L_n. When F has density f, L_n has probability density function (pdf)

$$f_n(t)=\frac{\{-\ln F(t)\}^{n-1}}{(n-1)!}f(t)$$

and

$$F_n(t)=P(L_n\leq t)=\int_{-\infty}^{t}\frac{\{-\ln F(x)\}^{n-1}}{(n-1)!}f(x)\,dx$$

$$=\int_{-\ln F(t)}^{\infty}\frac{u^{n-1}}{(n-1)!}e^{-u}du$$

$$=\overline{\Gamma}_n(-\ln F(t))$$

where $\overline{\Gamma}_n(t)=1-\Gamma_n(t)$ and $\Gamma_n(t)=\int_0^t\frac{u^{n-1}}{(n-1)!}e^{-u}\,du$.

Throughout this paper, a bar on a distribution function indicates the corresponding survival function.

Thus $\overline{\Gamma}_n(t)=P(Z_n>t)=1-\Gamma_n(t)$, where Z has the gamma function with mean = n = variance. This section is concerned with the behavior of L_n and L_n- L_{n+1} when the distribution function F has the decreasing reversed hazard rate (DRHR) property. We will say that F has the DRHR property (loosely, F is DRHR) if F is logconcave. Similarly, F is said to have the increasing reversed hazard rate (IRHR) property (loosely, F is IRHR) if F is logconvex. The motivation for these names comes from the fact that when F has derivative f,

$$\tilde{\lambda}(t)=\frac{d}{dt}\ln F(t)=\frac{f(t)}{F(t)}$$

is called the reversed hazard rate (RHR) function. This function, introduced in Keilson and Sumita(1982) is the dual of the better known hazard rate (or failure rate) function

$$\lambda(t)=-\frac{d}{dt}\ln\overline{F}(t)=\frac{f(t)}{\overline{F}(t)}$$

The RHR function is discussed in detail by Shaked and Shantikumar(1994) and by Block et.al. (1998). It can be easily seen that if F has decreasing hazard rate (more frequently referred to as the decreasing failure rate) then F has decreasing reversed hazard rate. Further, some increasing hazard rate distributions are also DRHR. For example, increasing hazard rate Weibull, gamma and lognormal distributions are also DRHR. A well known theorem of Prekopa, see Prekopa(1980), implies that if f is logconcave (in other words, f is Po'lya Type 2 (PF_2)) then F is lonconcave. Thus, the class of so called PF_2 distributions is a sub class of DRHR distributions. A useful source of information on logconcavity is Klinger and Mangasarian(1968).

We now prove our first result.

Theorem 2.1. If F is DRHR then F_n is DRHR.

Proof. As noted earlier,

$$F_n(t) = \overline{\Gamma}_n(-\ln F(t))$$

Now, $\overline{\Gamma}_n(t)$ is positive, decreasing and logconcave. Hence $1/\overline{\Gamma}_n(t)$ is decreasing and logconvex and therefore, $\frac{1}{F_n(t)} = \frac{1}{\overline{\Gamma}_n(-\ln F(t))}$ is logconvex so that $F_n(t)$is logconcave. It follows from the above theorem that if F_n has decreasing hazard (failure) rate then F_n has DRHR. This deserves comparison with the fact that not all upper record values can possess decreasing hazard rates. See Kochar (1990) in this connection.

The next result provides a strong comparison of $L_n - L_{n+1}$ with $L_{n+1} - L_{n+2}$ when F has the DRHR property. We first recall that a random variable X is said to be smaller than a random variable Y, in the hazard rate ordering, written as $X \leq_{hr} Y$, if and only if,

$$\frac{P(X > t)}{P(Y > t)} \text{ decreases in } t, -\infty < t < \infty.$$

When X and Y have absolutely continuous distributions with respective hazard rates $\lambda_X(t)$ and $\lambda_Y(t)$, the definitions can be restated as follows: $X \leq_{hr} Y$ if $\lambda_X(t) \geq \lambda_Y(t)$ for all $t \in (-\infty, \infty)$. We refer to Shaked and Shantikumar (1994) for detailed discussion of the hazard rate ordering. It is useful to note here that if $X \leq_{hr} Y$, then X is stochastically smaller than Y in hazard rate ordering, written $X \leq_{st} Y$, in the sense that $P(X>t) \leq P(Y>t)$ for all t.

Theorem 2.2. If F is DRHR then $L_{n+1} - L_{n+2} \leq_{hr} L_n - L_{n+1}$.

Proof. Let $S_n(t) = P(L_n - L_{n+1} > t)$. $t \geq 0$ and note the well known fact that

$P(L_n - L_{n+1} > t \mid L_n = x) = F(x - t)/F(x)$, $0 \leq t < x$. Then for all $t \geq 0$,

$$\begin{aligned} S_n(t) &= \int_0^\infty P(L_n - L_{n+1} > t \mid L_n = x)\, dP(L_n \leq x) \\ &= \int_0^\infty \frac{F(x-t)}{F(x)} dP(L_n \leq x) \\ &= E\{\frac{F(L_n - t)}{F(L_n)}\} \end{aligned}$$

We want to show that if F is DRHR then $S_{n+1}(t) / S_n(t)$is decreasing in t. Let $0 < t_1 < t_2$

and define $\alpha_i(x) = F(x - t_i)/F(x)$, $i = 1,2$, so that

$$\frac{S_{n+1}(t_i)}{S_n(t)} = \frac{E\{\alpha_i(L_{n+1})\}}{E\{\alpha_i(L_n)\}}, i = 1,2.$$

When F is DRHR, $\alpha_2(x)/\alpha_1(x)$ is decreasing in x. Further, L_{n+1} is smaller than L_n in likelihood ratio ordering, written $L_{n+1} \leq_{lr} L_n$, in the sense that

$$\frac{f_n(t)}{f_{n+1}(t)} = \frac{\{-\ln F(t)\}^{n-1} f(t)/(n-1)!}{\{-\ln F(t)\}^n f(t)/(n)!} = \frac{-n}{\ln F(t)}$$ is non decreasing in t. (See Shaked and Shantikumar (1994) for a detailed discussion of the likelihood ratio ordering.). It follows by Goldstein (1973) that if F is DRHR then

$$\frac{E\{\alpha_2(L_n)\}}{E\{\alpha_1(L_n)\}} \geq \frac{E\{\alpha_2(L_{n+1})\}}{E\{\alpha_1(L_{n+1})\}},$$ so that

$$\frac{S_{n+1}(t_2)}{S_n(t_2)} = \frac{E\{\alpha_2(L_{n+1})\}}{E\{\alpha_2(L_n)\}} \leq \frac{E\{\alpha_1(L_{n+1})\}}{E\{\alpha_1(L_n)\}} = \frac{S_{n+1}(t_2)}{S_n(t_1)}.$$

Thus $S_{n+1}(t)/S_n(t)$ is decreasing in t.

3. BEHAVIOR OF LOWER RECORDS AND THEIR GAPS UNDER DISPERSIVE ORDERING

Let X and Y be two random variables with distribution functions F and G, respectively. Let F^{-1} and G^{-1} be the right continuous inverses of F and G, respectively, and assume that

$$F^{-1}(\beta) - F^{-1}(\alpha) \leq G^{-1}(\beta) - G^{-1}(\alpha),$$

where $0<\alpha\leq\beta<1$. Then X is said to be smaller than Y in the dispersive order and we write $X \leq_{disp} Y$. It is easily seen that $X \leq_{disp} Y$ if and only if $G^{-1}(F(t)) - t$ increases in t. Clearly, the order $\leq_{disp}$ is a comparison of the variabilities of X and Y. A detailed discussion of this ordering and its various implications are given in Shaked and Shantikumar (1994). In particular, when X and Y have finite variances, the ordering $X \leq_{disp} Y$ implies that $Var(X) \leq Var(Y)$.

Suppose now that $\{X_n, n\geq 1\}$ is a sequence of independent random variables each having the continuous distribution function F and let $L_n^{(X)}$ be the nth lower record of this sequence. Similarly let $L_n^{(Y)}$ denote the nth lower record of a sequence $\{Y_n, n\geq 1\}$ of independent random variables each having the continuous distribution function G. The following theorems provide interesting implications of the dispersive ordering $X \leq_{disp} Y$.

Theorem 3.1. If $X \leq_{disp} Y$ then $L_n^{(X)} \leq_{disp} L_n^{(Y)}$ for all $n\geq 1$.

Proof. As noted earlier,

$$\overline{F}_n(t) = P(L_n^{(X)} > t) = \Gamma_n(\widetilde{\Lambda}_F(t))$$

where

$$\widetilde{\Lambda}_F(t) = -\ln F(t) \text{ and } \Gamma_n(t) = \int_0^t \frac{x^{n-1}}{(n-1)!} e^{-x}\, dx.$$

Similarly,

$$\overline{G}_n(t) = P(L_n^{(Y)} > t) = \Gamma_n(\widetilde{\Lambda}_G(t))$$

where $\widetilde{\Lambda}_G(t) = -\ln G(t)$.

Hence,

$$\overline{G}_n^{-1}\overline{F}_n = (\Gamma_n \widetilde{\Lambda}_G)^{-1}(\Gamma_n \widetilde{\Lambda}_F)$$

$$= \widetilde{\Lambda}_G^{-1}\widetilde{\Lambda}_F$$
$$= G^{-1}(\exp(-\widetilde{\Lambda}_F)$$
$$= G^{-1}F$$

so that

$$G^{-1}F = \overline{G}_n^{-1}\overline{F}_n = \overline{G}_n^{-1}(1-\overline{F}_n) = \overline{G}_n^{-1}F_n.$$

Now,

$$X \leq_{disp} Y \Leftrightarrow G^{-1}F(t) - t \text{ increases in } t$$
$$\Leftrightarrow G_n^{-1}F_n(t) - t \text{ increases in } t$$
$$\Leftrightarrow L_n^{(X)} \leq_{disp} L_n^{(Y)}.$$

We net prove the following Theorem.

Theorem 3.2. If $X \leq_{disp} Y$, then for all $n \geq 1$

$$L_n^{(X)} - L_{n+1}^{(X)} \leq_{st} L_n^{(Y)} - L_{n+1}^{(Y)}$$

in the sense that $P(L_n^{(X)} - L_{n+1}^{(X)} > t) \leq P(L_n^{(Y)} - L_{n+1}^{(Y)} > t)$ for all $t \geq 0$.

Proof. Clearly, $L_n^{(X)} \geq L_{n+1}^{(X)}$ almost surely (a.s.). Hence,

$$X \leq_{disp} Y$$
$$\Rightarrow G^{-1}(F(t)) - t \text{ increases in } t$$
$$\Rightarrow G^{-1}F(L_n^{(X)}) - L_n^{(X)}$$
$$\geq G^{-1}F(L_{n+1}^{(X)}) - L_{n+1}^{(X)}$$
$$\Rightarrow G^{-1}F(L_n^{(X)}) - L_n^{(X)}$$
$$\Rightarrow G^{-1}F(L_{n+1}^{(X)}) - L_{n+1}^{(X)} \quad \text{a.s.}$$
$$\Rightarrow G^{-1}F(L_n^{(X)}) - G^{-1}F(L_{n+1}^{(X)})$$
$$\geq L_n^{(X)} - L_{n+1}^{(X)} \quad \text{a.s.}$$
$$\Rightarrow G_n^{-1}F_n(L_n^{(X)}) - G_{n+1}^{-1}F_{n+1}(L_{n+1}^{(X)})$$

$$\geq L_n^{(X)} - L_{n+1}^{(X)} \quad \text{a.s.}$$

because $G^{-1}F = G_n^{-1}F_n$ for all $n \geq 1$.

Now the joint distribution of $G_n^{-1}F_n(L_n^{(X)})$ and $G_{n+1}^{-1}F_{n+1}(L_{n+1}^{(X)})$ is the same as the joint distribution of $L_n^{(Y)}$ and $L_{n+1}^{(Y)}$. Hence,

$$X \le_{disp} Y$$
$$\Rightarrow L_n^{(Y)} - L_{n+1}^{(Y)} \ge L_n^{(X)} - L_{n+1}^{(X)} \quad \text{a.s.}$$
$$\Rightarrow L_n^{(Y)} - L_{n+1}^{(Y)} \ge_{st} L_n^{(X)} - L_{n+1}^{(X)}$$

4. BEHAVIOR OF EXPECTED VALUES OF RATIOS OF SUCCESIVE LOWER RECORDS

When dealing with lower records from distributions concentrated on $(0,\infty)$, it appears more natural to be interested in L_{n+1}/L_n rather than $L_{n+1} - L_n$. This section provides two results concerning $E(L_{n+1}/L_n)$. As before, L_n denotes the nth lower record from a sequence of independent random variables each having an absolutely continuos (with respect to Lebesgue measure) distribution function F.

Theorem 4.1. Suppose $F(0) = 0$ and $t\tilde{\lambda}(t) = tf(t)/F(t)$ is $\downarrow (\uparrow)$. Then $E\left(\frac{L_{n+1}}{L_n}\right)$ is $\uparrow (\downarrow)$ in n.

Proof. We have

$$E\left(\frac{L_{n+1}}{L_n}\right) = \int_0^\infty P\left(\frac{L_{n+1}}{L_n} > x\right) dx$$
$$= \int_0^1 \int_0^\infty P(L_{n+1} > tx \mid L_n = t)\, dF_n(t)\, dx$$
$$= \int_0^1 \int_0^\infty \left(1 - \frac{F(tx)}{F(t)}\right) dF_n(t)\, dx$$
$$= 1 - \int_0^1 \int_0^\infty \frac{F(tx)}{F(t)}\, dF_n(t)\, dx$$
$$= 1 - \int_0^\infty \int_0^1 \frac{F(tx)}{F(t)}\, dx\, dF_n(t)$$
$$= 1 - \int_0^\infty \psi(t)\, dF_n(t)$$
$$= 1 - E(\psi(L_n)),$$

where

$$\psi(t) = \frac{\int_0^1 F(tx)\, dx}{F(t)}$$
$$= \frac{\int_0^t F(y)\, dy}{tF(t)}$$

Now

$$\psi'(t) = \frac{t F^2(t) - \{F(t) + t f(t)\} \int_0^t F(y)\, dy}{t^2 F^2(t)}$$

$$= \left\{ \frac{tF(t)}{\int_0^t F(y)\, dy} - \frac{F(t) + t f(t)}{F^2(t)} \right\} \frac{\int_0^t F(y)\, dy}{t^2 F(t)}$$

But, $t\tilde{\lambda}(t) \downarrow (\uparrow)$

$$\Rightarrow \frac{F(t) + tf(t)}{F(t)} \downarrow (\uparrow)$$

$$\Rightarrow \frac{tF(t)}{\int_0^t F(y)\, dy}$$

$$= \frac{\int_0^t \left\{ \frac{F(y) + yf(y)}{F(y)} \right\} F(y)\, dy}{\int_0^t F(y)\, dy}$$

$$\geq (\leq) \quad \frac{F(y) + tf(t)}{F(t)}$$

$$\Rightarrow \psi'(t) \geq (\leq) 0$$

Now, $\quad L_n \geq L_{n+1}, \ \psi \uparrow (\downarrow)$

$$\Rightarrow E(\psi(L_n)) \geq (\leq) E(\psi(L_{n+1}))$$

$$\Rightarrow E\left(\frac{L_{n+1}}{L_n}\right) \leq (\geq) E\left(\frac{L_{n+2}}{L_{n+1}}\right).$$

The assumption that F (t)/(tf(t)) is nondecreasing may be restated as requiring that the reciprocal of the reverse hazard rate f(t)/F(t) is starshaped. In other words, we are requiring that t $\tilde{\lambda}(t)$ where $\tilde{\lambda}(t)$ = f(t)/F(t) is nonincreasing. The exponential distribution is an easy example.

To put our next result in perspective, it may help to recall the following result from Bruckner and Ostrow(1962).

Theorem 4.2. (Bruckner and Ostrow (1962)). Let f be a nonnegative continuous function which vanishes at origin. Consider the following six conditions on f:

(i) f is convex

(ii) f is convex on the average (COV) (i.e. $(1/x)\int_0^x f(t)\,dt$ is convex)

(iii) f is star shaped (i.e. $f(\alpha x)\leq \alpha f(x)$ for all α, $0\leq \alpha \leq 1$ and all x)

(iv) f is super additive (i.e. $f(x+y)\geq f(x) + f(y)$, for all x,y)

(v) f is star shaped on average (i.e. . $(1/x)\int_0^x f(t)\,dt$) is star shaped)

(vi) f is superadditive on average (i.e. . $(1/x)\int_0^x f(t)\,dt$) is superadditive).

Then the following chain of implications is valid but none of the reverse implications holds: (i)→ (ii)→(iii)→(iv)→(v) →(vi).

We now prove

Theorem 4.3. If F(0) =0 and F is starshaped (antistarshaped) on the average (i.e. if $\phi(t)=(1/t)\int_0^t F(x)\,dx$ satisfies $\phi(\alpha t)\leq(\geq)\alpha\phi(t)$ for all t and all $0\leq\alpha\leq 1$) then

$$E\left(\frac{L_{n+1}}{L_n}\right)\geq(\leq)\frac{1}{2}.$$

Proof. As seen in Theorem 4.1,

$$E\left(\frac{L_{n+1}}{L_n}\right)=1-E\{\psi(L_n)\}$$

where

$$\psi(t)=\frac{\int_0^t F(y)\,dy}{t\,F(t)}.$$

When F is starshaped (antistarshaped) on the average, the function

$$\frac{(1/t)\int_0^t F(x)\,dx}{t}=\frac{\int_0^t F(x)\,dx}{t^2}$$

is nondcreasing (nonincreasing) so that

$$\frac{t^2F(t)-2t\int_0^t F(x)\,dx}{t^t}\geq(\leq)0$$

or,

$$\psi(t)=\frac{\int_0^t F(x)\,dx}{t\,F(t)}\geq(\geq)\frac{1}{2}.$$

Consequently,

$$E\left(\frac{L_{n+1}}{L_n}\right) = 1 - E\{\psi(L_n)\} \geq (\leq) \frac{1}{2}.$$

5. A CHARACTERIZATION OF THE UNIFORM DISTRIBUTION THROUGH LOWER RECORDS

If L_n is the nth lower record from a sequence of independent random variables each having the uniform distribution $F(t) = t$, $0 \leq t \leq 1$, then it is easy to show that $E\{(L_n - L_{n+1})^2 | L_n = x\} = x^2/3$, for all $x \in (0,1)$. We will prove that the converse is also true.

Theorem 3.1. Suppose that F is absolutely continuous with support (0,1) and pdf f such that f' exists. If $E\{(L_n - L_{n+1})^2 | L_n = x\} = x^2/3$, for all $x \in (0,1)$, then $F(t) = t$, $0 \leq t \leq 1$

Proof The condition $E\{(L_n - L_{n+1})^2 | L_n = x\} = x^2/3$ implies that

$$\int_0^x (x-v)^2 \frac{f(v)}{F(x)} dv = \frac{x^3}{3}$$

i.e.

$$\int_0^x (x-v)^2 f(v) dv = \frac{x^3}{3} F(x)$$

.

Differentiating both sides of the above equation twice with respect to x , we get on simplification

$$4F(x) = 4xf(x) + xf'(x)$$

This is Euler's differential equation and the solution of the equation is $F(x) = cx$ or $F(x) = cx^{-4}$. Since $F(0) = 0$ and $F(1) = 1$, we must have

$$F(x) = x, \; 0 \leq x \leq 1.$$

REFERENCES

1. Ahsanullah, M. (1995). Record Statistics. Nova Science Publishers, Inc. Commack, NY.
2. Arnold, B.C., Balakrishnan, N. and Nagaraja, H.N. (1998). Records. Wiley, New York, NY.
3. Block,H.W., Savits,T.H. and Singh, H. (1998). The reversed hazard rate function. Probability in the Engineering and Informational Sciences, 12,69-90.
4. Brucnker, A.M. and Ostrow,R.(1962), Some function classes related to the class of convex functions. Pacific J. Math.12, 1203-1215
5. Goldstein, M.(1973). An application of ordered expectations.The American Statistician, 27,30.
6. Keilson,J. and Sumita,U. (1982). Uniform stochastic ordering and related inequalities. Canadian Journal of Statistics, 10,181-198.
7. Klinger, A. and Mangasarian, O.L. (1968). Logarithmic convexity and geometric programming. Journal of Mathematical Analysis and Applications, 24, 388-408.
8. Kochar ,S.C.(1990). Some partial ordering results on record values. Communications in Statistics- Theory and Methods, 299-306.
9. Prekopa, A. (1980). Logarithmic concave measures and related topics. In Dempster, M.A.H. (ed.): Stochastic Programming; 63-82. Academic Press, Boston, MA.
10. Shaked,M. and Shantikumar, J.G. (1994), Stochastic Orders and their Applications. Academic Press, San Diego, CA

SIMPLE FORMULAS FOR MEANS, VARIANCES AND COVARIANCES OF ORDER STATISTICS FROM TUKEY'S LAMBDA DISTRIBUTION AND THEIR COMPUTATIONS

MOHAMMAD Z. RAQAB
Department of Statistics
UAE University
United Arab Emirates

ABSTRACT

In this paper, exact expressions for single and product moments of order statistics from Tukey's symmetric distribution are derived. These expressions are written in terms of incomplete beta and hypergeometric functions. Recurrence relations for single moments of order statistics are also obtained. These relations can be used to obtain the higher order moments from those of the lower order. The means, variances and covariances of order statistics from Tukey's lambda distribution are tabulated for $\lambda = 1.0(0.5)4.0$ and for various choices of sample sizes.

Key Words : Order statistics; single moments; product moments; Tukey's symmetric family; hypergeometric function.

AMS Subject Classification: 62G30; 62E05

1. INTRODUCTION

Tukey's (1947, 1960) introduced and discussed the very useful family of distributions defined by the transformation

$$X = F^{-1}(U) = \frac{U^{\lambda} - (1-U)^{\lambda}}{\lambda}, \quad 0 \le u \le 1, \tag{1.1}$$

where U is uniformly distributed on $[0, 1]$. The probability density function (pdf) of X is defined implicitly as

$$f(X(u)) = [u^{\lambda-1} + (1-u)^{\lambda-1}]^{-1}$$

When $\lambda = 1$, the above transformation in (1.1) tends to $X = 2U - 1$ and then X is distributed according to the uniform distribution on $[-1, 1]$. The limiting form of the transformation (1.1) as $\lambda \longrightarrow 0$ is $X = Log(U) - Log(1-U)$, and consequently X is logistic distribution. This limiting form can be seen by applying L'Hopital's rule to (1.1). Tukey (1960) and Van Dyke (1961) have found that this family of distributions can be used to approximate the percentage points of the normal and Student's t distributions. This family has been used also by Filliben (1969) in a approximating symmetric distributions with a wide range of tail weights for studying location estimation problems of symmetric distributions. Joiner and Rosenblatt (1971) gave some properties of this family and investigated the dependence of the mean and coefficient of variation of the sample range. Ramberg and Schmeiser (1974) generalized (1.1) to a four parameter distribution, which is called generalized lambda distribution (GLD), where two parameters are the location and scale parameters and the other parameters are shape parameters. This generalized family includes the family in (1.1) and permits skewed curves to be considered. Recent developments and its uses in fitting data have been discussed in Dudewicz and Karian (1985), Karian, Dudewicz and McDonald (1996), Karian and Dudewicz (1999). In these articles, the authors have shown that the generalized lambda distribution (GLD) play an important role in fitting distributions to a wide variety of data sets.

It is important to mention here that the location and scale parameters can be added to the transformation (1.1). Further, one can estimate the location and scale parameters involved in the generalized lambda model. In this case, we can use the means, variances and covariances of the transformation $X = U^{\alpha} - (1-U)^{\beta}$, where α and β are shape parameters to obtain the coefficients of the best linear unbiased estimators of the location and scale parameters. General expressions of the moments and covariances of the GLD are used in determining the coefficients of the linear estimates of location and scale parameters and linear predictors of the future order statistics from the GLD (Raqab,1996,1999).

Let $X_1, X_2, ..., X_n$ be a random sample of size n from the LD with the transformation as in (1.1). Let $X_{1:n} \leq X_{2:n} \leq ... \leq X_{n:n}$ be the corresponding order statistics. For more details about the importance of order statistics in statistical modelling and its application, see Gnedenko, et al. (1969) and Barlow and Prochan (1981).

2. RECURRENCE RELATIONS OF SINGLE MOMENTS

The useful property of the lambda family is that the percentage points are directly given from the definition (1.1). Consequently, the kth ($k \geq 1$) single moment of the rth order statistic $X_{r:n}$ $(1 \leq r \leq n)$ can be written as

$$\mu_{r:n}^{(k)} = \frac{1}{B(r, n-r+1)} \int_0^1 [\frac{u^{\lambda} - (1-u)^{\lambda}}{\lambda}]^k \, u^{r-1}(1-u)^{n-r} \, du. \qquad 0 \leq u \leq 1,$$

where $B(a, b) = \Gamma(a)\Gamma(b)/\Gamma(a+b)$ is a beta function. When x is a positive integer, $\Gamma(x) = (x-1)!$. Now, we will give two recurrence relations of moments of order statistics from the lambda family.

Lemma 1: For $1 \leq r \leq n$, $k = 1, 2, ...,$ and λ *is* a positive integer,

$$\mu_{r+1:n}^{(k)} = \mu_{r:n}^{(k)} + \frac{kB(r+\lambda, n-r+1)}{rB(r, n-r+1)} \mu_{r+\lambda:n+\lambda}^{(k-1)} + \frac{kB(r+1, n-r+\lambda)}{rB(r, n-r+1)} \mu_{r-1:n+\lambda}^{(k-1)}. \qquad (2.1)$$

Proof. For $k = 1, 2, ..., and\ 1 \le r \le n$, the kth single moment of the rth order statistic is given by

$$\mu_{r+1:n}^{(k)} = \frac{1}{B(r+1, n-r)} \int_0^1 [F^{-1}(u)]^k \, u^r (1-u)^{n-r-1} du. \tag{2.2}$$

Integrating the right hand side of (2.2) by parts treating $(1-u)^{n-r-1}$ for integration and the rest of the integrand for differentiation, we get

$$\begin{aligned} \mu_{r+1:n}^{(k)} &= \frac{1}{B(r, n-r+1)} \int_0^1 [F^{-1}(u)]^k \, u^{r-1} (1-u)^{n-r} du \\ &+ \frac{k}{rB(r, n-r+1)} \int_0^1 [F^{-1}(u)]^k \, u^{r+\lambda-1} (1-u)^{n-r} du \\ &+ \frac{k}{rB(r, n-r+1)} \int_0^1 [F^{-1}(u)]^{k-1} \, u^r (1-u)^{n-r+\lambda-1} du. \end{aligned}$$

Therefore, the kth moment of $(r+1)^{th}$ order statistic can be decomposed into three terms as

$$\mu_{r+1:n}^{(k)} = \mu_{r:n}^{(k)} + \frac{kB(r+\lambda, n-r+1)}{rB(r, n-r+1)} \mu_{r+\lambda:n+\lambda}^{(k-1)} + \frac{kB(r+1, n-r+\lambda)}{rB(r, n-r+1)} \mu_{r+1:n+\lambda}^{(k-1)}.$$

Applying the same technique, we also establish the following recurrence relation.

Lemma 2: For $1 \le r \le n$, $k = 1, 2, ...$, and λ is a positive integer

$$\mu_{\lambda:n+\lambda}^{(k-1)} = \frac{1}{kB(\lambda, n+1)} [\mu_{1:n}^{(k)} - k(n+\lambda)^{-1} \, \mu_{1:n+\lambda}^{(k-1)} - \frac{(-1)^k}{\lambda^k}]. \tag{2.3}$$

Remark 1: The recurrence relation in (2.2) and (2.3) can be used recursively to establish the single moments of order statistics when λ is an integer. For example, for $\lambda = 1$, the recurrence relations (2.1) and (2.3) reduce, respectively, to

$$\mu_{r+1:n}^{(k)} = \mu_{r:n}^{(k)} + \frac{2k}{n+1} \mu_{r+1:n+1}^{(k-1)},$$

and

$$\mu_{1:n+1}^{(k-1)} = \frac{n+1}{2k} [\mu_{1:n}^{(k)} - (-1)^k].$$

3. MOMENTS OF ORDER STATISTICS

From (2.1), the kth single moment of the rth order statistic can be written in terms of beta function (Joiner and Rosenblatt (1971)) as the following:

$$\mu_{r:n}^{(k)} = \frac{1}{\lambda^k B(r, n-r+1)} \sum_{j=o}^{k} (-1)^j \binom{k}{j} B((k-j)\lambda + r, \lambda j + n - r + 1). \tag{3.1}$$

The product moment of any two order statistics $X_{r:n}$ and $X_{s:n}$ ($1 \leq r < s \leq n$) is given by

$$\begin{aligned} \mu_{r,s:n} &= C_{r,s:n}\lambda^{-2} \int_0^1 \int_0^v [u^\lambda - (1-u)^\lambda][v^\lambda - (1-v)^\lambda]\, u^{r-1}(v-u)^{s-r-1} \\ & \qquad (1-v)^{n-s}\, du\, dv, \quad 0 < u < v < \infty, \end{aligned} \tag{3.2}$$

where

$$C_{r,s:n} = \frac{n!}{[(r-1)!(s-r-1)!(n-s)!]}.$$

Expanding $(v-u)^{s-r-1}$ binomially in powers of u and v, we write (3.2) as

$$\mu_{r,s:n} = C_{r,s:n}\lambda^{-2} \sum_{i=0}^{s-r-1} (-1)^i \binom{s-r-1}{i} \int_0^1 [v^\lambda - (1-v)^\lambda]\, v^{s-r-i-1}(1-v)^{n-s} I(v)\, dv,$$

where

$$I(v) = \frac{v^{r+i+\lambda}}{r+i+\lambda} - B_v(r+i, \lambda+1),$$

and

$$B_v(a,b) = \int_0^v t^{a-1}(1-t)^{b-1}\, dt, \ (0 \leq v \leq 1),$$

is incomplete beta function. Substituting the expression of $I(v)$ in (3.2) and simplifying the resulting expression, we immediately write $\mu_{r,s:n}$ in terms of incomplete beta function

$$\begin{aligned} \mu_{r,s:n} &= C_{r,s:n}\lambda^{-2} \sum_{i=0}^{s-r-1} (-1)^i \binom{s-r-1}{i} [\frac{B(s+2\lambda, n-s+1)}{r+\lambda+i} - \frac{B(s+\lambda, n-s+\lambda+1)}{r+\lambda+i} \\ & \quad - \int_0^1 v^{s+\lambda-r-i-1}(1-v)^{n-s} B_v(r+i, \lambda+1)\, dv \\ & \quad + \int_0^1 v^{s-r-i-1}(1-v)^{n-s+\lambda} B_v(r+i, \lambda+1)\, dv]. \end{aligned} \tag{3.3}$$

Now, we write the expression in (3.3) in terms of hypergeometric function ${}_3F_2$, where the generalized hypergeometric function is defined by

$${}_pF_q(\alpha_1, ..., \alpha_p; \beta_1, ..., \beta_q; u) = \sum_{i=0}^{\infty} \frac{(\alpha_1)_i ... (\alpha_p)_i}{(\beta_1)_i ... (\beta_q)_i} \frac{u^i}{i!},$$

where $(\alpha)_i = \alpha(\alpha+1)...(\alpha+i-1)$, $i = 1, 2, ...$, and p and q are nonnegative integers. The hypergeometric functions ${}_2F_1$ and ${}_3F_2$ are special functions of the generalized hypergeometric functions. Using the facts that

$$B_x(a,b) = a^{-1}x^a{}_2F_1\ (a,1-b;a+1;x),$$

(see, Mathai,1993, p.119) and the identity (Raqab,1999)

$$\begin{aligned}\int_0^1 u^{a-1}(1-u)^{b-1}{}_2F_1(c,d;\rho;u)\,du &= B(a,b)\ {}_3F_2(a,c,d;\rho,a+b;1),\\ \text{for } a,b &> 0 \text{ and } d+b-a-c>0,\end{aligned}$$

Therefore, we can write the product moment of any two order statistics $X_{r:n}$ and X_{sn} as the following:

$$\begin{aligned}\mu_{r,s:n} &= C_{r,s:n}\lambda^{-2}\sum_{i=0}^{s-r-1}(-1)^i\binom{s-r-1}{i}\Big[\frac{B(s+2\lambda,n-s+1)}{r+\lambda+i}-\frac{B(s+\lambda,n-s+\lambda+1)}{r+\lambda+i}\\ &\quad -(r+i)^{-1}B(s+\lambda,n-s+1){}_3F_2(s+\lambda,r+i,-\lambda;r+i+1,n+\lambda+1;1)\\ &\quad -(r+i)^{-1}B(s,n-s+\lambda+1){}_3F_2(s,r+i,-\lambda;r+i+1,n+\lambda+1;1)\Big]. \qquad (3.4)\end{aligned}$$

The covariance between the two order statistics $X_{r:n}$ and $X_{s:n}$ ($1 \le r < s \le n$) is given by

$$\sigma_{r,s:n} = Cov(X_{r:n}, X_{s:n}) = \mu_{r,s:n} - \mu_{r:n}\mu_{s:n},$$

where $\mu_{r:n}$ and $\mu_{r,s:n}$ are given respectively in (3.1) and (3.4).

Making use of the exact and explicit expressions for the single and product moments of order statistics, we can immediately obtain the means, variances and covariances of order statistics for any arbitrary sample size n.

4. COMPUTATIONS OF MEANS, VARIANCES AND COVARIANCES

The exact and explicit expression for the single moments of order statistics given in (3.1) allows us to evaluate the means of all order statistics for any sample size n. By using the facts $\mu_{r:n} = -\ \mu_{n-r+1:n}$ (due to the symmetry of the lambda distribution), the amount of computations to be performed for the evaluation of single moments of order statistics can be reduced. When n is odd, $\mu_{(n+1)/2:n} = 0$, but if n is even, $\mu_{n/2:n} = -\ \mu_{n/2+1:n}$. The single moments $\mu_{r:n}$ ($1 \le r \le n$) were computed for sample sizes up to fifteen and $r = [(n+1)/2]+1,...,n$, where $[x]$ is the greatest integer function of x. These computations were checked for their accuracy by the well known identities (see Arnold et al. (1992)), $\sum_{r=1}^{n}\mu_{r:n} = n\mu_{1:1}$ and $\sum_{r=1}^{n}\sigma_{r:n}^2 = n\sigma_{1:1}^2$. Table 1 presents the means of $X_{r:n}$ ($1 \le r \le n = 1(1)10, 15$), and $\lambda = 1.0(0.5)4.0$ to five decimal places.

For the computation of variances and covariances, the product moments $\mu_{r,s:n}$ ($1 \le r < s \le n$) were computed first. These values were used to evaluate the covariances $\sigma_{r,s:n}$ ($1 \le r < s \le n$). For $r = s$,the variances $\sigma_{r,r:n} = \sigma_{r:n}^2$ $(1 \le r \le n)$ were evaluated based on the first and second single moments of order statistics given in (3.1). Due to the symmetry of LD, $\sigma_{r,s:n} = \sigma_{n-s+1,n-r+1:n}$ ($1 \le r \le s \le n$). The variances and covariances were computed for sample sizes up to ten and $r = 1, 2, ..., [(n+1)/2]$ and $j = i, ..., n-r+1$. Finally, these values of the covariances were checked for their accuracy by the well known identity $\sum_{r=1}^{n} \sum_{s=1}^{n} \sigma_{r,s:n} = n\sigma_{1,1:1}$ (see also Arnold et al. (1992)). Table 2 provides the variances and covariances of order statistics to five decimal places for $n = 1(1)10$ and $\lambda = 1.0(0.5)4.0$. All these computations were done by using Mathematica 3.0 to twenty significant digits to get the required accuracy.

Table 1: Means of Lambda Order Statistics for $\lambda = 1(0.5)4.0$ and n up to 15

n	r	$\lambda = 1$	$\lambda = 1.5$	$\lambda = 2$	$\lambda = 2.5$	$\lambda = 3$	$\lambda = 3.5$	$\lambda = 4$
2	2	0.3333	0.22857	0.16667	0.12698	0.10000	0.08081	0.06667
3	3	0.50000	0.34286	0.25000	0.19048	0.15000	0.12121	0.10000
4	3	0.20000	0.13853	0.10000	0.07459	0.05714	0.04476	0.03571
4	4	0.60000	0.41097	0.30000	0.22910	0.18095	0.14670	0.12143
5	4	0.33333	0.23088	0.16667	0.12432	0.09524	0.07459	0.05952
5	5	0.66667	0.45599	0.33333	0.25530	0.20238	0.16472	0.13690
6	4	0.14286	0.09946	0.07143	0.05265	0.03968	0.03048	0.02381
6	5	0.42857	0.29659	0.21429	0.16015	0.12302	0.09665	0.07738
6	6	0.71429	0.48787	0.35714	0.27433	0.21825	0.17834	0.14881
7	5	0.25000	0.17405	0.12500	0.09214	0.06944	0.05335	0.04167
7	6	0.50000	0.34561	0.25000	0.18736	0.14444	0.11397	0.09167
7	7	0.75000	0.51158	0.37500	0.28882	0.23056	0.18907	0.15833
8	5	0.11111	0.07759	0.05556	0.04065	0.03030	0.02297	0.01768
8	6	0.33333	0.23192	0.16667	0.12304	0.09293	0.07157	0.05606
8	7	0.55556	0.38351	0.27778	0.20880	0.16162	0.12810	0.10354
8	8	0.77778	0.52987	0.38889	0.30026	0.24040	0.19778	0.16616
9	6	0.20000	0.13967	0.10000	0.07316	0.05455	0.04135	0.03182
9	7	0.40000	0.27805	0.20000	0.14798	0.11212	0.08668	0.06818
9	8	0.60000	0.41364	0.30000	0.22617	0.17576	0.13993	0.11364
9	9	0.80000	0.54440	0.40000	0.30952	0.24848	0.20501	0.17273
10	6	0.09091	0.06362	0.04546	0.03308	0.02448	0.01838	0.01399
10	7	0.27273	0.19037	0.13636	0.09988	0.07459	0.05667	0.04371
10	8	0.45455	0.31562	0.22727	0.16860	0.12821	0.09954	0.07867
10	9	0.63636	0.43814	0.31818	0.24056	0.18765	0.15003	0.12238
10	10	0.81818	0.55621	0.40909	0.31718	0.25524	0.21112	0.17832
15	9	0.12500	0.08770	0.06250	0.04518	0.03309	0.02453	0.01838
15	10	0.25000	0.17508	0.12500	0.09080	0.06699	0.05013	0.03799
15	11	0.37500	0.26179	0.18750	0.13731	0.10253	0.07788	0.06005
15	12	0.50000	0.34744	0.25000	0.18517	0.14052	0.10884	0.08578
15	13	0.62500	0.43157	0.31250	0.23485	0.18178	0.14408	0.11642
15	14	0.75000	0.51358	0.37500	0.28685	0.22712	0.18462	0.15319
15	15	0.87500	0.59253	0.43750	0.34170	0.27737	0.23150	0.19730

* Missing values can be found by the symmetry relation $\mu_{r:n} = -\mu_{n-r+1:n}$.

Table 2: Variances and Covariances of Lambda Order Statistics $\lambda = 1(0.5)4.0$

n	r	s	$\lambda = 1$	$\lambda = 1.5$	$\lambda = 2$	$\lambda = 2.5$	$\lambda = 3$	$\lambda = 3.5$	$\lambda = 4$
1	1	1	0.33333	0.15677	0.08333	0.04842	0.03016	0.01986	0.01369
2	1	1	0.22222	0.10453	0.05556	0.03230	0.02016	0.01333	0.00925
2	1	2	0.11111	0.05224	0.02778	0.01613	0.01000	0.00653	0.00444
3	1	1	0.15000	0.06962	0.03750	0.02243	0.01454	0.01003	0.00726
3	1	2	0.10000	0.04718	0.02500	0.01441	0.00886	0.00574	0.00387
3	1	3	0.05000	0.02319	0.01250	0.00746	0.00478	0.00322	0.00226
3	2	2	0.20000	0.09597	0.05000	0.02784	0.01640	0.01014	0.00655
4	1	1	0.10667	0.04869	0.02667	0.01652	0.01117	0.00806	0.00609
4	1	2	0.08000	0.03743	0.02000	0.01174	0.00738	0.00491	0.00340
4	1	3	0.05333	0.02495	0.01333	0.00781	0.00489	0.00321	0.00219
4	1	4	0.02667	0.01216	0.00667	0.00411	0.00274	0.00193	0.00141
4	2	2	0.16000	0.07678	0.04000	0.02228	0.01314	0.00814	0.00527
4	2	3	0.10667	0.05118	0.02667	0.01484	0.00872	0.00535	0.00342
5	1	1	0.07937	0.03560	0.01984	0.01271	0.00895	0.00670	0.00525
5	1	2	0.06349	0.02934	0.01587	0.00956	0.00622	0.00428	0.00308
5	1	3	0.04762	0.02220	0.01190	0.00701	0.00441	0.00291	0.00199
5	1	4	0.03175	0.01466	0.00794	0.00477	0.00308	0.00209	0.00147
5	1	5	0.01587	0.00711	0.00397	0.00253	0.00176	0.00129	0.00098
5	2	2	0.12698	0.06047	0.03175	0.01800	0.01089	0.00696	0.00466
5	2	3	0.09524	0.04575	0.02381	0.01321	0.00773	0.00473	0.00301
5	2	4	0.06349	0.03022	0.01587	0.00899	0.00539	0.00339	0.00221
5	3	3	0.14286	0.06925	0.03571	0.01942	0.01106	0.00657	0.00406
6	1	1	0.06122	0.02701	0.01531	0.01012	0.00737	0.00570	0.00459
6	1	2	0.05102	0.02326	0.01276	0.00790	0.05318	0.00379	0.00282
6	1	3	0.04082	0.01886	0.01020	0.00611	0.00392	0.00264	0.00184
6	1	4	0.03061	0.01415	0.00765	0.00458	0.00292	0.00196	0.00135
6	1	5	0.02041	0.00930	0.00510	0.00315	0.00210	0.00147	0.00107
6	1	6	0.01020	0.00450	0.00255	0.00168	0.00121	0.00091	0.00072
6	2	2	0.10204	0.04807	0.02551	0.01483	0.00928	0.00616	0.00429
6	2	3	0.08163	0.03901	0.02041	0.01147	0.00683	0.00427	0.00278
6	2	4	0.06122	0.02925	0.01531	0.00859	0.00510	0.00316	0.00203
6	2	5	0.04082	0.01922	0.01020	0.00592	0.00366	0.00238	0.00160
6	3	3	0.12245	0.05936	0.03061	0.01665	0.00949	0.00564	0.00349
6	3	4	0.09184	0.04451	0.02296	0.01248	0.00709	0.00420	0.00257
7	1	1	0.04861	0.02111	0.01215	0.00826	0.00619	0.00492	0.00405
7	1	2	0.04167	0.01873	0.01042	0.00663	0.00461	0.00340	0.00260
7	1	3	0.03472	0.01589	0.00868	0.00530	0.00349	0.00241	0.00173
7	1	4	0.02778	0.01278	0.00694	0.00418	0.00269	0.00180	0.00125
7	1	5	0.02083	0.00953	0.00521	0.00318	0.00208	0.00142	0.00100
7	1	6	0.01389	0.00624	0.00347	0.00220	0.00152	0.00110	0.00082
7	1	7	0.00694	0.00301	0.00174	0.00118	0.00087	0.00068	0.00055
7	2	2	0.08333	0.03880	0.02083	0.01244	0.00805	0.00555	0.00401
7	2	3	0.06944	0.03290	0.01736	0.00995	0.00609	0.00393	0.00264
7	2	4	0.05556	0.02647	0.01389	0.00784	0.00469	0.00293	0.00190
7	2	5	0.04167	0.01974	0.01042	0.00596	0.00362	0.00230	0.00152
7	2	6	0.02778	0.01292	0.00694	0.00413	0.00264	0.00178	0.00124
7	3	3	0.10417	0.05025	0.02604	0.01434	0.00832	0.00507	0.00322
7	3	4	0.08333	0.04042	0.02083	0.01131	0.00642	0.00379	0.00232
7	3	5	0.06250	0.03014	0.01563	0.00859	0.00496	0.00298	0.00185

n	r	s	$\lambda = 1$	$\lambda = 1.5$	$\lambda = 2$	$\lambda = 2.5$	$\lambda = 3$	$\lambda = 3.5$	$\lambda = 4$
7	4	4	0.11111	0.05419	0.02778	0.01487	0.00829	0.00479	0.00287
8	1	1	0.03951	0.01691	0.00988	0.00688	0.00529	0.00430	0.00360
8	1	2	0.03457	0.01533	0.00864	0.00564	0.00404	0.00306	0.00241
8	1	3	0.02963	0.01341	0.00741	0.00462	0.00312	0.00221	0.00163
8	1	4	0.02469	0.01128	0.00617	0.00377	0.00245	0.00167	0.00117
8	1	5	0.01975	0.00902	0.00494	0.00301	0.00196	0.00133	0.00092
8	1	6	0.01481	0.00670	0.00370	0.00231	0.00155	0.00109	0.00079
8	1	7	0.00988	0.00438	0.00247	0.00161	0.00114	0.00085	0.00065
8	1	8	0.00494	0.00211	0.00123	0.00086	0.00065	0.00052	0.00043
8	2	2	0.06914	0.03180	0.01728	0.01059	0.00708	0.00505	0.00376
8	2	3	0.05926	0.02781	0.01481	0.00868	0.00547	0.00364	0.00254
8	2	4	0.04938	0.02338	0.01235	0.00707	0.00430	0.00274	0.00181
8	2	5	0.03951	0.01870	0.00988	0.00565	0.00342	0.00217	0.00142
8	2	6	0.02963	0.01390	0.00741	0.00433	0.00271	0.00177	0.00120
8	2	7	0.01975	0.00908	0.00494	0.00302	0.00199	0.00139	0.00100
8	3	3	0.08889	0.04256	0.02222	0.01246	0.00742	0.00466	0.00306
8	3	4	0.07407	0.03579	0.01852	0.01014	0.00583	0.00350	0.00219
8	3	5	0.05926	0.02863	0.01481	0.00811	0.00465	0.00277	0.00171
8	3	6	0.04444	0.02127	0.01111	0.00622	0.00367	0.00227	0.00145
8	4	4	0.09877	0.04817	0.02469	0.01322	0.00737	0.00426	0.00255
8	4	5	0.07901	0.03853	0.01975	0.01057	0.00588	0.00339	0.00201
9	1	1	0.03273	0.01382	0.00818	0.00583	0.00458	0.00378	0.00322
9	1	2	0.02909	0.01274	0.00727	0.00486	0.00357	0.00277	0.00223
9	1	3	0.02545	0.01139	0.00636	0.00406	0.00281	0.00205	0.00154
9	1	4	0.02182	0.00988	0.00545	0.00338	0.00224	0.00156	0.00111
9	1	5	0.01818	0.00826	0.00455	0.00279	0.00182	0.00124	0.00086
9	1	6	0.01455	0.00658	0.00364	0.00225	0.00149	0.00102	0.00073
9	1	7	0.01091	0.00488	0.00273	0.00174	0.00119	0.000859	0.00064
9	1	8	0.00727	0.00318	0.00182	0.00121	0.00088	0.000674	0.00053
9	1	9	0.00364	0.00153	0.00091	0.00065	0.00050	0.000409	0.00034
9	2	2	0.05818	0.02644	0.01455	0.00914	0.00629	0.00462	0.00353
9	2	3	0.05091	0.02364	0.01273	0.00763	0.00495	0.00340	0.00244
9	2	4	0.04364	0.02050	0.01091	0.00635	0.00395	0.00258	0.00175
9	2	5	0.03636	0.01715	0.00909	0.00524	0.00320	0.00204	0.00135
9	2	6	0.02909	0.01367	0.00727	0.00423	0.00262	0.00169	0.00113
9	2	7	0.02182	0.01013	0.00545	0.00326	0.00210	0.00142	0.00099
9	2	8	0.01455	0.00660	0.00364	0.00228	0.00155	0.00111	0.00083
9	3	3	0.07636	0.03625	0.01909	0.01093	0.00670	0.00435	0.00296
9	3	4	0.06545	0.03144	0.01636	0.00910	0.00535	0.00329	0.00211
9	3	5	0.05456	0.02630	0.01364	0.00750	0.00433	0.00261	0.00162
9	3	6	0.04364	0.02096	0.01091	0.00606	0.00354	0.00216	0.00136
9	3	7	0.03273	0.01553	0.00818	0.00467	0.00284	0.00180	0.00119
9	4	4	0.08727	0.04242	0.02182	0.01179	0.00666	0.00392	0.00240
9	4	5	0.07273	0.03548	0.01818	0.00973	0.00541	0.00311	0.00185
9	4	6	0.05818	0.02828	0.01455	0.00785	0.00442	0.00258	0.00155
9	5	5	0.09091	0.04452	0.02273	0.01204	0.00660	0.00374	0.00219
10	1	1	0.02755	0.01149	0.00689	0.00500	0.00400	0.00336	0.00290
10	1	2	0.02479	0.01073	0.00620	0.00423	0.00318	0.00252	0.00206
10	1	3	0.02204	0.00976	0.00551	0.00359	0.00254	0.00190	0.00146

n	r	s	$\lambda=1$	$\lambda=1.5$	$\lambda=2$	$\lambda=2.5$	$\lambda=3$	$\lambda=3.5$	$\lambda=4$
10	1	4	0.01928	0.00865	0.00482	0.00304	0.00206	0.00146	0.00106
10	1	5	0.01653	0.00746	0.00413	0.00256	0.00169	0.00116	0.00082
10	1	6	0.01377	0.00622	0.00344	0.00213	0.00140	0.00096	0.00067
10	1	7	0.01102	0.00494	0.00275	0.00173	0.00117	0.00082	0.00059
10	1	8	0.00826	0.00366	0.00207	0.00134	0.00094	0.00070	0.00053
10	1	9	0.00551	0.00238	0.00138	0.00094	0.00067	0.00055	0.00044
10	1	10	0.00275	0.00115	0.00069	0.00050	0.00040	0.00033	0.00028
10	2	2	0.04959	0.02227	0.01240	0.00797	0.00564	0.00424	0.00332
10	2	3	0.04408	0.02026	0.01102	0.00675	0.00450	0.00319	0.00235
10	2	4	0.03857	0.01797	0.00964	0.00572	0.00364	0.00244	0.00170
10	2	5	0.03306	0.01550	0.00826	0.00482	0.00299	0.00194	0.00130
10	2	6	0.02755	0.01291	0.00689	0.00401	0.00248	0.00160	0.00107
10	2	7	0.02204	0.01026	0.00551	0.00326	0.00206	0.00137	0.00094
10	2	8	0.01653	0.00759	0.00413	0.00253	0.00167	0.00116	0.00084
10	2	9	0.01102	0.00494	0.00275	0.00177	0.00123	0.00091	0.00069
10	3	3	0.06612	0.03110	0.01653	0.00967	0.00610	0.00408	0.00286
10	3	4	0.05785	0.02758	0.01446	0.00818	0.00493	0.00313	0.00207
10	3	5	0.04959	0.02380	0.01240	0.00690	0.00404	0.00248	0.00157
10	3	6	0.04132	0.01983	0.01033	0.00574	0.00336	0.00205	0.00129
10	3	7	0.03306	0.01576	0.00826	0.00467	0.00279	0.00174	0.00113
10	3	8	0.02479	0.01166	0.00620	0.00362	0.00226	0.00148	0.00101
10	4	4	0.07714	0.03729	0.01928	0.01056	0.00609	0.00368	0.00231
10	4	5	0.06612	0.03217	0.01653	0.00890	0.00500	0.00292	0.00176
10	4	6	0.05510	0.02681	0.01377	0.00742	0.00416	0.00241	0.00144
10	4	7	0.04408	0.02130	0.01102	0.00603	0.00345	0.00206	0.00127
10	5	5	0.08264	0.04047	0.02066	0.01095	0.00601	0.00340	0.00199
10	5	6	0.06887	0.03373	0.01722	0.00912	0.00500	0.00282	0.00164

*Missing values can be found by the symmetry relation $\sigma_{r,s:n} = \sigma_{n-s+1,n-r+1:n}$

REFERENCES

1. Arnold, B. C., Balakrishnan, N. and Nagaraja, H. N. (1992). *A First Course in Order Statistics*, John Wiley, New York.

2. Barlow, R.E. and Prochan, F. (1981). *Statistical Theory of Reliability and Life Testing,* To Begin With, Silver Spring, MD.

3. Dudewicz, E. J. and Karian, Z. A. (1985). Modern Design and Analysis of Discrete Event Computer Simulations, *IEEE Computer Society,* New York.

4. Filliben, J. J. (1969). *Simple and Robust Linear Estimation of the Location Parameters of A Symmetric Distribution*, P.h.D Thesis, Princeton University.

5. Gnedenko, B. V., Belyayev, Yu. K. and Solovyev, A. D. (1969). *Mathematatial Methods of Reliability Theory*, Academic Press, New York.

6. Hastings, C., Mosteller, F., Tukey, J. W. and Winsor, C. P. (1947). Low Moments of Small Samples: A Comparative Study of Order Statistics, *Annals of Mathematical Statistics* 18, 413-26.

7. Joiner, B. L. and Rosenblatt, J. R. (1971). Some Properties of the Range in Samples from Tukey's Symmetric Lambda Distributions, *JASA* 66, 394-399.

8. Karian, Z. A. and Dudewicz, E. J. (1999). *Modern Statistical, Systems, and GPSS Simulation*, 2nd Edition, CRC Prss, Boca Raton Florida.

9. Karian, Z. A., Dudewicz, E. J. and McDonald, P. (1996). The Extended Generalized Lambda Distribution System for Fitting Distributions to Data: History, Completion of Theory, Tables, Applications, The Final Word on Moment Fits, *Commun. Statist.:Simulation and Computations* 25(3), 611-642.

10. Mathai, A. M. (1993). *A Handbook of Generalized Special Functions for Statistical and Physical Sciences*, Oxford Univ. Press, New York.

11. Filliben, J. J. (1969). *Simple and Robust Linear Estimation of the Location Parameters of A Symmetric Distribution*, Ph.D Thesis, Princeton Univ.

12. Ramberg, J. S. and Schmeiser, B. W. (1974). An Approximate Method for Generating Asymmetric Random Variables, *Commun. ACM* 17, 78-82.

13. Ramberg, J. S., Dudewicz, E. J., Tadikamalla, P. R. and Mykytka, E. F. (1979). A Probability Distribution and Its Uses in Fitting Data, *Technometrics* 21, 201-214.

14. Raqab, M. Z. (1996). Linear Predictors of Future Order Statistics, *Commun. Statist.-Theory & Meth.* 25(7), 1553-1568.

15. Raqab, M. Z. (1999). Best Linear Estimates of the Location and Scale Parameters from the Generalized Lambda Distribution, Submitted to Publication.

16. Tukey, J. W. (1960). The Practical Relationship between the Common Transformations of Percentages or Fractions and Amounts, *Technical Report* 36, Statistical Research Group, Princeton.

17. Van Dyke, J. (1961). Numerical Investigation of the Random Variable $y = C(u^\lambda - (1-u)^\lambda)$. Unpublished Working Paper, National Bureau of Standards Statistical Engineering Laboratory.

IMMIGRATION-BRANCHING DIFFUSIONS AND THEIR EXTINCTION

I. RAHIMOV

KFUPM, Box 1339, Dhahran, 31261, Saudi Arabia ,
and Institute of Mathematics, Hodjaev St., 29, 700143,
Tashkent, Uzbekistan.
E-mail rahimov@kfupm.edu.sa

ABSTRACT

The Markov branching diffusion in which offspring move at random in a bounded space X and allowing time-dependent immigration of particles from an external source is considered. It have not been known an explicit form of the limit distribution for such processes. In the paper, using a combination of analitic methods and direct probabilistic arguments, we obtain explicit limit distributions for the vector of population sizes in disjoint subsets of the space X. Asymptotic behavior of the non-extinction probability and the expected size of the population are also studied when first moment functional of the immigration process decreases regularly.

Key words: branchin diffusion,Markov process, immigration, time-dependent immigration, extinction, explicit distribution, partial process, moment functional
AMS 1991 Subject classification: Primary 60J80,
Secondary 60G70.

1 INTRODUCTION

We consider a population evolving as a branching process where the offspring move at random on a bounded space X. Typical models that have been studied by many authors are branching random walks and branching diffusions. In branching random walks the offspring jump when they are born, and retain their position until they die or reproduce. In branching diffusions, which we are going to deal with, the offspring move throughout their lives according to diffusions, independently of each other and of the family tree, each offspring starting from the location of its birth. Each particle at the end of its life independently of others generates a population whose members may be located at any point of X. The lifetime of a particle is defined as the hitting time of either of two barriers ∂ and Δ. If the particle hits ∂, it is instantaneously replaced by the empty population, that is disappers. If the particle hits Δ, then it is replaced by a population of new particles distributed in X according to a probability law.

In addition a random number of new particles produced by an external source may immigrate into the population and initial positions of these new particles may also be arbitrary points of X. The state of the process at any given time is characterized by the total number n of particles at present and their positions $x_1, x_2, ..., x_n \in X$. The development of the process in the state space $\hat{X}$ of all finite populations of particles located in X is Markovian. In other words, the motion of a particle up to the time of reproduction is a Markov process on X.

In the literature many papers have been published on branching or immigration branching diffusions. Mathematical foundations of a general theory of stochastic population processes were given by Moyal (1962). Existence and probabilistic constructions of branching diffusions are due to Ikeda, Nagasawa and Watanabe (1965, 1966) (see also Conner(1961, 1967), Savits (1969), Hering(1973, 1978), Asmussen and Hering(1977), Kageyama and Ogura (1980)). Convergence problems for systems of critical branching Markov chains are considered by Cox (1994). It should be noted the book by Asmussen and Hering(1983) as a convenient source on branching diffusions. In that book using analitic methods limit theorems for the continuous and discrete time immigration-branching diffusions with non-stationary immigration were proved (see [2], p.231-239). However the limit distributions obtained there have no explicit form. In this paper we demonstrate that, if we use a technique developed to study Galton-Watson processes with decreasing immigration (Rahimov (1995), Ch III), it is possible to obtain explicit limit distributions for such processes when the first moment functional of the immigration decreases. The method based on a combination of analitic and probabilistic arguments. Namely the proofs of the limit theorems will be carried out in the following scheme. First using analitic methods we prove limit theorems for some "partial processes" counting only descendants of particles immigrated in the beginning of the process or immigrated "recently". Then from these theorems by direct probabilistic arguments obtain limit distributions for the basic process.

For simpilicity we consider here the case of immigration at discrete time points. However it will be seen further that all statements of the paper can be obtained by the same arguments in the case of Poisson immigration process as well.

Let (X, U) be a measurable space, $X^{(n)}$ be the symmetrization of the direct product of n copies of X (see Moyal(1961)). We put $X^{(0)} = \theta$, where θ is some extra point,

$$\hat{X} = \bigcup_{n=0}^{\infty} X^{(n)}$$

and denote $\hat{U}$ the σ-algebra on $\hat{X}$ induced by U. An element $x \in X$ we understand as position, energy or type of a particle diffusing in the bounded space. Then, it is clear that $\hat{X}$ is the space of all possible populations and θ denotes an empty population.

We denote $\boldsymbol{\beta}$ as the Banach algebra of all bounded, complex-valued U-measurable functions ξ on X with supremum norm

$$\| \xi \| = \sup_{x \in X} | \xi(x) |$$

and put $\boldsymbol{\beta}_+ = \{\xi \in \boldsymbol{\beta}: \xi > 0\}$. Let ∇ be the open unit ball in $\boldsymbol{\beta}$ and $\bar{\nabla}$ its closure.

For $\hat{x} \in \hat{X}$ and $\xi \in \bar{\nabla}$ we define

$$\hat{x}[\xi] = \begin{cases} 0 & \text{if } \hat{x} = \theta \\ \sum_{i=1}^{n} \xi(x_i) & \text{if } \hat{x} = < x_1, ..., x_n > \end{cases} \tag{1}$$

For example, if $\xi(x) = 1_A(x), A \subset X$, then $\hat{x}_t[1_A]$ is the number of particles in A at time t. Introduce the union operation for populations $\hat{x}, \hat{y} \in \hat{X}$ as following:

$$\hat{x} + \hat{y} = \begin{cases} \theta & \text{if } \hat{x} = \hat{y} = \theta \\ < x_1, ..., x_n, y_1, ..., y_n > & \text{otherwise,} \end{cases} \tag{2}$$

that is, if $\hat{x} = < x_1, ..., x_n >, \hat{y} = < y_1, ..., y_n >$. Then the Markov branching diffusion $\{\hat{x}_t, P^{\hat{x}}\}$ is defined by the relation

$$\hat{x}_{t+s} = \sum_{i=1}^{\hat{x}_t[1]} \hat{x}_{t+s}^{t,i}, t, s > 0 \tag{3}$$

with $\hat{x}_{t+s}^{t,i}, i = 1, 2, ..., \hat{x}_t[1]$ are conditionally independent, given $\mathcal{F}_t := \sigma(\hat{x}_u, u \leq t)$ and

$$P^{\hat{x}}\{\hat{x}_{t+s}^{t,j} \in \hat{A} | \mathcal{F}_t\} = P^{<x_j>}\{\hat{x}_s \in \hat{A}\}$$

a.s.$P^{\hat{x}}$, $\hat{x}_t = < x_1, ..., x_{\hat{x}_t[1]} >$. Here and later on $P^{\hat{x}}$ and $E^{\hat{x}}$ denote the probability and expectation when the initial population is $\hat{x}$ and $\mathcal{F}_t$ is the σ-algebra generated by the branching diffusion up to time t.

Let $t \in N_0 = \{0, 1, ...\}$ and $\hat{y}_t = < y_{t1}, ..., y_{tl} >$ be the population of immigrating at time t particles, where $y_{tj} \in X$ is the position (or energy) of j-th particle from $\hat{y}_t$. If we denote $\hat{x}_t^{<y_{ki}>}$ the branching diffusion initiated by particle y_{ki}, then the discrete time immigration-branching diffusion can be given by the following relation:

$$\hat{Z}_t = \sum_{k=0}^{t} \sum_{i=1}^{\hat{y}_k[1]} \hat{x}_{t-k}^{<y_{ki}>}. \tag{4}$$

The content of the paper is as following. In part 2 we state basic assumptions and some known results which will be used further. Part 3 is devoted to study of asymptotic behavior of the "non-extinction" probability. Results obtained there will well be used in part 5, but they are of independent interest as well. Asymptotic behavior of the expected size of population is studied in part 4. In part 5 the class of all possible explicit limit distributions is obtained.

2 BASIC ASSUMPTIONS

Let for $\hat{x} \in \hat{X}, \hat{A} \in \hat{U}, s, t \in N_0$

$$P_t(\hat{x}, \hat{A}) = P\{\hat{x}_{t+s} \in \hat{A} | \hat{x}_s = \hat{x}\}$$

be the branching transition function of $\hat{x}_t$. For $\eta \in \bar{\nabla}$ and $\hat{x} \in \hat{X}$ we define

$$\tilde{\eta}(\hat{x}) = \begin{cases} 1, & \hat{x} = \theta \\ \prod_{i=1}^{n} \eta(x_i), & \hat{x} =< x_1, ..., x_n > \end{cases} \tag{5}$$

Then the generating functional

$$F_t(\hat{x}, \eta) = \int_{\hat{X}} \tilde{\eta}(\hat{y}) P_t(\hat{x}, d\hat{y})$$

of $P_t(\hat{x}, .)$ is well-defined on $\bar{\nabla}$.

It is known (Asmussen, Hering(1983), p.139) that the mapping $F_t : \bar{\nabla} \mapsto \bar{\nabla}$ defined by $F_t[\eta](x) := F_t(< x >, \eta), \eta \in \bar{\nabla}, x \in X$ satisfies the semigroup relation $F_{t+s}[\eta] = F_s[F_t[\eta]], t, s \geq 0, \eta \in \bar{\nabla}$ and it is called a generating semigroup.

Assume that there exist a linear-bounded functional of ξ on $\boldsymbol{\beta}$ defined by the relation

$$M_t(\hat{x}, \xi) = \int_{\hat{X}} \hat{y}[\xi] P_t(\hat{x}, d\hat{y}).$$

It is also known, Asmussen, Hering (1983), that $M_{t+s} = M_t M_s$ for any $s, t \in N_0$, that is $\{M_t\}$ is a semigroup. It is called the moment semigroup of the process.

Now we introduce so called (M) and (R) assumptions which are important for our further considerations.

Assumption (M). The moment semigroup $\{M_t\}$ can be represented as

$$M_t = \rho^t P + \Delta_t, t \geq 0, \tag{6}$$

where $\rho \in (0,\infty)$, $P\xi = \Phi^\star[\xi]\varphi, \xi \in \boldsymbol{\beta}$, with $\Phi^\star : \boldsymbol{\beta} \mapsto \boldsymbol{C}$ is linear-bounded, non-negative on $\boldsymbol{\beta}_+, \varphi \in \boldsymbol{\beta}_+$; further $\Delta_t : \boldsymbol{\beta} \mapsto \boldsymbol{\beta}$ such that for all $t > 0$

$$P\Delta_t\xi = \Delta_t P\xi = 0, \xi \in \boldsymbol{\beta},$$

$$-\gamma_t P\xi \leq \Delta_t\xi \leq \gamma_t P\xi, \xi \in \boldsymbol{\beta}_+$$

with $\gamma_t : (0,\infty) \mapsto \boldsymbol{R}_+$ satisfying $\rho^{-t}\gamma_t \downarrow 0, t \uparrow \infty$.

When X is a finite set and M_t is primitive, satisfaction of (M) follows from the well-known Perron's theorem (see for details Sevastyanov (1971), Ch IV). In the general case it can also be satisfied by a wide class of branching diffusions. Examples of such processes can be seen in Hering (1978).

Suppose $\{M_t\}$ exists as a semigroup of bounded operators. Then

$$1 - F_t[\eta] = M_t[1-\eta] - R_t(\eta)[1-\eta] \tag{7}$$

where the mapping $R_t(.)[\eta] : \bar{\nabla} \otimes \boldsymbol{\beta} \mapsto \boldsymbol{\beta}$ is non-increasing in the first variable and linear-bounded in the second, such that

$$0 = R_t(1)\xi \leq R_t(\eta)\xi \leq M_t\xi, (\eta,\xi) \in \bar{\nabla}_+ \otimes \boldsymbol{\beta}.$$

Assumption (R). For every $t > 0$, there exist a mapping $g_t : \bar{\nabla}_+ \mapsto \boldsymbol{\beta}$ such that

$$R_t(\xi)[1-\xi] = g_t[\xi]\rho^t\Phi^\star[1-\xi]\varphi, i \in \bar{\nabla}_+,$$

$$\lim_{\|1-\xi\| \to \infty} \| g_t[\xi] \| = 0.$$

Under the assumption (M) the (R) is automatically satisfied if X is finite. In the general case it may follow from (M) under some restrictions on φ and $\Phi^\star$. Some sufficient conditions for fulfillment of (R) are given, for example, in the book by Asmussen and Hering(1983, p.160).

We also assume that the second moment functional of $\hat{x}_t$

$$M_2^{(2)}(< x >, \xi, \eta) = E^{<x>}\hat{x}_t^{(2)}[\xi,\eta]$$

is finite. Here

$$\hat{x}^{(2)}[\xi,\eta] = \begin{cases} 0, & \hat{x}[1] \leq 1, \\ \sum\sum_{i\neq j}\xi(x_i)\eta(x_j), & \hat{x} = < x_1, ..., x_n >, n \geq 2. \end{cases}$$

We denote

$$\mu = \frac{1}{2t}\Phi^\star[M_t^{(2)}[\varphi]], M_t^{(2)}[\xi](x) = M_t^{(2)}(< x >, \xi, \xi)$$

It is known that

$$q(x) = \lim_{t\to\infty} F_t[0](x)$$

exists for any $x \in X$, and given (M) with $\rho = 1$ the quantity μ is constant as a function of t.

3 APPROXIMATION OF NON-EXTINCTION PROBABILITY

Now we proceed to consider the immigration-branching diffusion. Let F_k^I be the generating functional of $P\{\hat{y}_k \in \hat{A}\}, \hat{A} \in \hat{U}$. If F_k^I has a bounded first moment functional

$$M_k^I[\xi] = \int_{\hat{X}} \hat{y}[\xi] P\{\hat{y}_k \in d\hat{y}\},$$

then it follows that (7) holds for F_k^I. Assume that $\alpha(t) = M_t^I[\varphi]$ is a regularly varying function as $t \to \infty$ and

$$r(\varepsilon) = \sup_{s>0} \frac{R_s^I((1-\varepsilon)1)[\varphi]}{\alpha(s)} \to 0, \varepsilon \to 0. \tag{8}$$

Condition (8) holds, for example, if the second factorial moment

$$M_k^{I(2)}[\xi,\eta] = \int_{\hat{X}} \hat{y}^{(2)}[\xi,\eta] P\{\hat{y}_k \in d\hat{y}\}$$

of the immigration process is finite as a symmetric bilinear-bounded functional on $\boldsymbol{\beta}^{(2)} = \boldsymbol{\beta} \otimes \boldsymbol{\beta}$ and

$$\sup_k \frac{M_k^{I(2)}[1,1]}{M_k^I[\varphi]} < \infty.$$

In the case of stationary immigration process, it is clear that, the state $\theta \in \hat{X}$ is a reflecting screen for the process. If the immigration process has an decreasing intensity, then the state θ may reflect or may also absorb the process. Hence it is important to know the behavior of $P\{\hat{Z}_t \neq \theta\}$ as $t \to \infty$. The first result gives us conditions under which the above probability tends to 1, to zero or to a positive limit less than 1.

Theorem **1** *Let* (M) , (R) *and (8) be satisfied,* $\rho = 1, 0 < \mu < \infty, q = 1, \alpha(t) \to 0, t \to \infty$.

1°. *If* $\alpha(t)\ln t \to 0$, *then*

$$\lim_{t\to\infty} P\{\hat{Z}_t \neq \theta\} = 0.$$

2°. *If* $\alpha(t)\ln t \to C \in (0,\infty)$, *then*

$$\lim_{t\to\infty} P\{\hat{Z}_t \neq \theta\} = 1 - \exp\{-\frac{C}{\mu}\}.$$

3°. *If* $\alpha(t)\ln t \to \infty$, *then*

$$\lim_{t\to\infty} P\{\hat{Z}_t \neq \theta\} = 1.$$

Proof. First we prove part 2°. If we denote $H_t[\xi]$ the generating functional of the process $\hat{Z}_t$, then it follows from (4) that

$$H_t[\xi] = \prod_{k=0}^{t} F_k^I[F_{t-k}[\xi]]. \tag{9}$$

We consider the sum

$$A = \sum_{k=0}^{t} M_k^I[1 - F_{t-k}[0]] = A_1 + A_2 + A_3, \tag{10}$$

where $A_i, i = 1,2,3$, are sums with respect to $0 \le k < t/\ln t, t/\ln t \le k \le t - \ln t$ and $t - \ln t < k \le t$.

Further we use the following results from Asmussen, Hering (1983,pp. 184, 202).

Lemma **1** *If (M) and (R) are satisfied and* $q = 1$, *then for every* $t > 0$ *there exist a mapping* $h_t : \bar{\nabla}_+ \to \boldsymbol{\beta}$ *such that*

$$1 - F_t[\xi] = (1 + h_t[\xi])\Phi^\star[1 - F_t[\xi]]\varphi, \xi \in \bar{\nabla}_+, \tag{11}$$

where $\lim_{t\to\infty} \| h_t[\xi] \| = 0$ *uniformly in* $\xi \in \bar{\nabla}_+$.

The next result is a generalization of so called basic lemma in the theory of Galton-Watson processes (see Jagers(1983), p.25).

Lemma **2** *If (M) and (R) are satisfied,* $0 < \mu < \infty, \rho = 1, q = 1$ *then for any* $\delta > 0$

$$\lim_{n\to\infty} \frac{1}{n\delta} \left\{ \Phi^\star[1 - F_{n\delta}[\xi]]^{-1} - \Phi^\star[1 - \xi]^{-1} \right\} = \mu \tag{12}$$

uniformly in $\xi \in \bar{\nabla}_+ \bigcap \{\Phi^\star[1 - \xi] > 0\}$.

It follows from (11) and (12) that

$$(n+1)(1 - F_n[0]) = (1 + \varepsilon_n)\frac{\varphi}{\mu}, \tag{13}$$

where $\lim_{n\to\infty} \| \varepsilon_n \| = 0$.

Since M_t^I is a linear-bounded functional, using (13), we have

$$A_1 = \sum_{k\in E_1} \frac{1}{t-k+1} M_k^I[(1+\varepsilon_{t-k})\frac{\varphi}{\mu}],$$

where $E_1 = \{k : 0 \leq k < t/\ln t\}$. Using here the simple estimates

$$1- \| \varepsilon \| \leq 1+\varepsilon \leq 1+ \| \varepsilon \| \tag{14}$$

we obtain that A_1 is non-greater than

$$\mu^{-1}(1+\sup_{k\in E_1} \| \varepsilon_{t-k} \|) \sum_{k\in E_1} \alpha(k)(t-k+1)^{-1}$$

which tends to zero as $t \to \infty$.

Now we considerA_2. Again using (13) and (14), we obtain

$$A_2 \leq \mu^{-1}(1+\sup_{k\in E_2} \| \varepsilon_{t-k} \|) \sum_{k\in E_2} \alpha(k)(t-k+1)^{-1},$$

where $E_2 = \{k : t/\ln t < k \leq t - \ln t\}$. Since under the our conditions

$$\lim_{t\to\infty} \sum_{k\in E_2} \alpha(k)(t-k+1)^{-1} = C,$$

we have $\limsup_{t\to\infty} A_2 \leq C/\mu$. If we use the left side of (14), we obtain that $\liminf A_2$ is non-less than C/μ. Thus A_2 tends to C/μ as $t \to \infty$.

By similar arguments it can be verified that $A_3 \to 0$ as $t \to \infty$.

If we use (7) written for F_t^I, we have that under the condition (8)

$$\lim_{t\to\infty} \sum_{k=0}^{t} \{1 - F_k^I[F_{t-k}[0]] \ = \lim_{t\to\infty} A. \tag{15}$$

Thus assertion of the part 2° follows from (15) and the simple relation

$$\ln H_t[0] = \sum_{k=0}^{t} \ln F_k^I[F_{t-k}[0]].$$

Now we prove part 1°. For any $\varepsilon > 0$ and fixed μ we can find such a $C > 0$ that $1 - \exp\{-C/\mu\} < \varepsilon$. Let $\hat{Z}_t^\star$ be the immigration-branching diffusion with the same branching transition function and with immigration process such that $\alpha^\star(t)\ln t \to 0$. Then it is clear that for any $\varepsilon > 0$

$$\limsup_{t\to\infty} P\{\hat{Z}_t^\star \neq \theta\} \leq \lim_{t\to\infty} P\{\hat{Z}_t \neq \theta\} \leq \varepsilon.$$

From the last relation we have the assertion of the part 1°.

Part 3° can also be derived from part 2°. To do this one has to choose $C > O$ such that $\exp\{-C/\mu\} < \varepsilon$ and have to compare the "non-extinction" probability of the process $\hat{Z}_t$ and a new process whose immigration process satisfies conditions of the part 3°. The theorem is proved.

We now consider the asymptotic behavior of the non-extinction probability in the case 1° i.e. when $\alpha(t)\ln t \to 0, t \to \infty$. Introduce the following notation:

$$\alpha(t) = M_t^I[\varphi] = \frac{l(t)}{t^\alpha + 1}, \alpha \geq 0, a(t) = \sum_{k=0}^{t} \alpha(k)$$

$$\beta(t) = P^{<x>}\{\hat{x}_t \neq \theta\}, b(t) = \sum_{k=0}^{t} \beta(k).$$

Here $l(t)$ is a slowly varying as $t \to \infty$ function. It follows from (13) that under the our assumptions

$$t\mu\beta(t) \sim \varphi, t \to \infty. \tag{16}$$

Note that $a(t)$, when $\alpha \geq 1$, and $b(t)$ are slowly varying functions. Therefore, it follows from the Karamata representation of slowly varying functions that there exist positive, integer valued functions $L_i(t), i = 1, 2$, such that $L_i(t) \to \infty, L_i(t) = o(t), t \to \infty$ and (see also Lemma 3 below)

$$a(L_1(t)) \sim a(t), b(L_2(t)) \sim b(t), t \to \infty. \tag{17}$$

Introduce the following "partial" processes:

$$\hat{Z}_t^{(j)} = \sum_{k \in I_j^t} \sum_{i=1}^{\hat{y}_k(1)} \hat{x}_{t-k}^{<y_{ki}>},$$

where

$$I_1^t = \{k \in N : 0 \leq k \leq L_1(t)\}, I_3^t = \{k \in N : t - L_2(t) \leq k \leq t\},$$

$$I_2^t = \{k \in N : L_1(t) + 1 \leq k \leq t - 1 - L_2(t)\}.$$

It is clear that

$$\hat{Z}_t = \hat{Z}_t^{(1)} + \hat{Z}_t^{(2)} + \hat{Z}_t^{(3)} \tag{18}$$

Theorem **2** *Let assumptions of Theorem 1 are satisfied. Then*

1°. *If* $\alpha \geq 1$, *then*

$$P\{\hat{Z}_t^{(1)} \neq \theta\} \sim (\mu t)^{-1} a(t),$$

if $\alpha < 1$, *then*

$$P\{\hat{Z}_t^{(1)} \neq \theta\} = o(\frac{a(t)}{t});$$

2°.

$$P\{\hat{Z}_t^{(3)} \neq \theta\} \sim \mu^{-1} \alpha(t) \ln t;$$

3°.

$$P\{\hat{Z}_t \neq \theta\} \sim \frac{1}{\mu t} a(t) + \frac{1}{\mu} \alpha(t) \ln t.$$

Remark. It should be noted that part 3° of the theorem shows that event $\{\hat{Z}_t \neq \theta\}$ may occur because of either $\{\hat{Z}_t^{(1)} \neq \theta\}$ or $\{\hat{Z}_t^{(2)} \neq \theta\}$ and, when $\alpha(t) \sim t^{-1} const$ these two possibilities are asymptotically equiprobable.

Proof of Theorem 2. First we prove part 1°. It follows from (7) and (13) that

$$0 \leq 1 - F_k^I[F_{t-k}[0]] \leq \frac{1}{t-k+1} M_k^I[(1 + \varepsilon_{t-k}) \frac{\varphi}{\mu}] \to 0 \tag{19}$$

as $t \to \infty$ uniformly in $k \in I_1$. Therefore

$$-\sum_{k \in I_1} \ln F_k^I[F_{t-k}[0]] \sim \sum_{k \in I_1} (1 - F_k^I[F_{t-k}[0]]).$$

Using (7), (13) and that M_k^I and R_k^I are linear-bounded functionals we obtain that the last sum as $t \to \infty$ is equivalent to

$$\delta_t = \frac{1 + \parallel \varepsilon_t \parallel}{\mu} \sum_{k \in I_1} \frac{M_k^I[\varphi]}{t-k+1} [1 - r(k,t)],$$

where

$$r(k,t) = \frac{R_k^I[F_{t-k}[0]][\varphi]}{M_k^I[\varphi]}.$$

Taking into account the choice of $L_1(t)$ and condition (8) we have that δ_t is equivalent to $(\mu t)^{-1} a(t)$ when $\alpha \geq 1$. If $\alpha < 1$, then $a(L_1(t)) = o(a(t))$, and thus in this case $\delta_t = o(t^{-1} a(t))$. The assertion of part 1^0 follows from this and the simple relation $1 - e^{-x} \sim x, x \to 0$.

Let us prove part 2°. It is not difficult to see that (19) holds as $t \to \infty$ uniformly in $k \in I_3$ also. Thus we have to consider

$$A = \sum_{k \in I_3} (1 - F_k^I[F_{t-k}[0]]) = A_1 + A_2, \tag{20}$$

where

$$A_1 = \sum_{k\in I_3} M_k^I[1 - F_{t-k}[0]], A_2 = \sum_{k\in I_3} R_k^I(F_{t-k}[0])[1 - F_{t-k}[0]].$$

Using again (13), (14) and that M_k^I is linear-bounded functional we obtain that $A_1 \sim \mu^{-1}\alpha(t)\ln t, t \to \infty$. Now we consider A_2. Since R_k^I is linear-bounded functional, using (13) and (14) we have that it is non-greater than

$$const \cdot \sum_{k\in I_3} (t-k+1)^{-1} R_k^I(F_{t-k}[0])[\varphi].$$

It follows from condition (8) that for any $\varepsilon > 0$ there exist $\delta > 0$ such that $r(\Delta) < \varepsilon$ for all $\Delta < \delta$. Since $F_k[0] \to 1$, there exist $m \in N$ such that $1 - F_{t-k}[0] < \delta$ for $t - k > m$. Now we partition I_3 as $I_3 = I_3^{(1)} \bigcup I_3^{(2)}$, where

$$I_3^{(1)} = \{k : t - L_2(t) \le k \le t - m\}, I_3^{(2)} = \{k : t - m < k \le t\}.$$

Then we have

$$\sum_{k\in I_3^{(1)}} (t-k+1)^{-1} R_k^I(F_{t-k}[0])[\varphi] \le \varepsilon\alpha(t)\ln t.$$

Since $R_t^I(\eta)\xi \le M_t^I\xi$, the second part of the sum is non-greater than

$$\sum_{k\in I_3^{(2)}} (t-k+1)^{-1}\alpha(k) = o(\alpha(t)\ln t)$$

as $t \to \infty$ for any fixed m. Therefore we have $A_2 = o(\alpha(t)\ln t), t \to \infty$. Thus we obtain the assertion of part 2° from (20) and estimates for A_1 and A_2. Now we prove part 3°. First we consider

$$P_2 = P\{\hat{Z}_t^{(2)} \ne \theta\} = 1 - \prod_{k\in I_2} F_k^I[F_{t-k}[0]].$$

Using simple inequality $\prod_i(1-\varepsilon_i) \ge 1 - \sum_i \varepsilon_i, \varepsilon_i \ge 0$ and relation (7) we obtain that P_2 is non-greater than

$$\sum_{k\in I_2} M_k^I[1 - F_{t-k}[0]].$$

Since M_k^I is a linear functional we have from (13) and (14) that

$$P_2 \le \frac{1}{\mu}(1 + \sup_{k\in I_2} \| \varepsilon_{t-k} \| \sum_{k\in I_2} (t-k+1)^{-1} M_k^I[\varphi].$$

It follows from properties $M_k^I[\varphi]$ that there exist $C(\alpha) > 0$ such that

$$\min_{0\le k\le t} \left\{ \frac{t-k+1}{t+1} + \frac{M_t^I[\varphi]}{M_k^I[\varphi]} \right\} \ge C(\alpha) > 0.$$

Therefore

$$\sum_{k\in I_2}\frac{1}{t-k+1}\alpha(k)\leq\frac{1}{C(\alpha)}\left\{\frac{1}{t}\sum_{k\in I_2}\alpha(k)+\alpha(t)\sum_{k\in I_2}\frac{1}{t-k}\right\},$$

where

$$\frac{1}{t}\sum_{k\in I_2}\alpha(k)\leq\frac{const}{t}[a(t)-a(L_1(t))]=o(\frac{1}{t}a(t)),$$

$$\alpha(t)\sum_{k\in I_2}\frac{1}{t-k}\leq const\cdot\alpha(t)[b(t)-b(L_2(t))]=o(\alpha(t)\ln t).$$

Thus we can conclude that

$$P_2=o(\frac{1}{t}a(t)+\alpha(t)\ln t). \tag{21}$$

Further we use the following equality

$$\{\hat{Z}_t\neq\theta\}=\bigcup_{i=1}^{3}\{\hat{Z}_t^{(i)}\neq\theta\} \tag{22}$$

where events on the right hand side are independent. If we denote $P_i, i=1,2,3$ the probability of the i-th event, then

$$P\{\hat{Z}_t\neq\theta\}=\sum_{i=1}^{3}P_i-\sum_{i\neq j}P_iP_j+P_1P_2P_3. \tag{23}$$

The assertion of the part 3° follows from (23), if we use results of parts 1° and 2° and relation (21). Theorem 2 is proved.

4 THE LIMIT THEOREMS

Now we turn our attention to limit theorems for the vector of population sizes in disjoint subsets of the space X. Let $\{A_i, i=1,...,j\}$ be a measurable decomposition of X, that is

$$X=\cup_{i=1}^{j}A_i, A_l\cap A_m=\phi.$$

If $\chi(A)$ is the indicator function of the set $A\subseteq X$, then it is clear that $\hat{Z}_t[\chi(A)]$ is the number of particles in A at time t.

Theorem **3** *Let (M), (R) and (8) be satisfied,* $\rho = 1, 0 < \mu < \infty$ *and* $q = 1$. *If* $\alpha(t) \to 0$ *such that* $\alpha(t)\ln t \to \infty$, *then for every finite measurable decomposition* $\{A_i, i = 1, ..., j\}$ *of* X

$$\lim_{t\to\infty} P\left\{(\frac{\hat{Z}_t[\chi(A_i)]}{t})^{\alpha(t)/\mu} \le x_i, i = 1, ..., j\right\} = P\{\nu_1 \le x_1, ..., \nu_j \le x_j\},$$

where $\nu_1 = ... = \nu_j$ *with probability 1 and* ν_k *has the uniform distribution on* $[0, 1]$.

Further we need the following result, which cam be proved using the well known representation for slowly varying functions.

Lemma **3** *If* $L(x)$ *is a slowly varying as* $x \to \infty$ *function, then there is another slowly varying function* $l(x) \sim L(x), x \to \infty$ *for which it is possible to find function* $\lambda_l(x) \to \infty, \lambda_l(x) = o(x)$ *such that as* $x \to \infty$

$$\sup_{u\in\Delta(x)} \mid \frac{l(u)}{l(x)} - 1 \mid \to 0, \Delta(x) = [\frac{x}{\lambda_l(x)}, x].$$

In some sense the Lemma 3 is an extension of so called uniform convergence theorem for slowly varying functions.

Proof of Theorem 4. It follows from the condition $\alpha(t)\ln t \to \infty$ that $\alpha(t)$ is a slowly varying function as $t \to \infty$. Thus there is a function $\lambda_\alpha(t)$ for which the extended uniform convergence theorem holds (see Lemma 3). If $0 < x_i < 1, i = 1, ..., j$ and $x = \min\{x_1, ..., x_j\}$, then the function $L(t) = tx^{\mu/\alpha(t)} \to \infty, L(t) = o(t)$ as $t \to \infty$. Now we consider the sum

$$\sum_{k=0}^{t} M_k^I[1 - F_{t-k}[\xi_t]], \xi_t = \exp\left\{-\frac{\sum_{i=1}^{j}\lambda_i\chi(A_i)}{tx_i^{\mu/\alpha(t)}}\right\},$$

where $\lambda_i \ge 0, i = 1, ..., j$. We decompose the sum as following

$$A = A_1 + A_2 + A_3 \tag{24}$$

where $A_i, i = 1, 2, 3$ are sums of the terms when k belongs to $K_1 = \{k : 0 \le k < t/\lambda_\alpha(t)\}, K_2 = \{k : t/\lambda_\alpha(t) \le k < t - L(t)\}$ and $K_3 = \{k : t - L(t) \le k \le t\}$ respectively. In this case we obtain from (11) and (12) the following relation

$$1 - F_n[\xi] = \frac{\varphi}{\mu n + \Phi^\star[1 - \xi]^{-1}}(1 + \varepsilon_n^\star[\xi]), \tag{25}$$

where $\| \varepsilon_n^\star \| \to 0$ as $n \to \infty$ uniformly in $\xi \in \bar{\nabla}_+ \cap \{\Phi^\star[1-\xi] > 0\}$. The following estimates are also true for any ξ from the above set:

$$1 - \| \varepsilon_n^\star[\xi] \| \leq 1 + \varepsilon_n^\star[\xi] \leq 1 + \| \varepsilon_n^\star[\xi] \| \tag{26}$$

First we consider A_2. Using (25), right side of (26) and Lemma 3 we get

$$\limsup_{t\to\infty} A_2 \leq \limsup_{t\to\infty} \alpha(t) \sum_{k\in K_2} \left\{\mu k + \Phi^\star[1-\xi_t]^{-1}\right\}^{-1}.$$

It follows from the choice of $L(t)$ and linearity of $\Phi^\star$ that

$$\lim_{t\to\infty} L(t)\Phi^\star[1-\xi_t] = \delta(E_0) \equiv \sum_{i\in E_0} \lambda_i \Phi^\star[\chi(A_i)], \tag{27}$$

where $E_0 = \{i : x_i = x\}$. Thus we have from the above estimate that

$$\limsup_{t\to\infty} A_2 \leq \limsup_{t\to\infty} \frac{\alpha(t)}{\mu} \ln(1 + \mu\delta(E_0)x^{-\mu/\alpha(t)}) = -\ln x.$$

If we use the left side of (26), by the same arguments obtain that the lim inf of A_2 is non-less than $-\ln x$. Consequently $A_2 \to -\ln x$ as $t \to \infty$.

Using (25) and right side of (26) we can see that $\limsup \lambda_\alpha(t) A_1 < \infty$, that is A_1 tends to zero as $t \to \infty$.

Now we consider A_3. Again using (25), (26) and the fact that $\alpha(t)$ slowly varies, we obtain, for some positive constant C, the following estimate

$$A_3 \leq C\alpha(t) \sum_{k\in K_3} \{\mu(t-k) + \Phi^\star[1-\xi_t]^{-1}\}^{-1} < C\delta(E_0)\alpha(t),$$

which shows that A_3 also tends to zero as $t \to \infty$. Thus it follows from (24) that $A \to -\ln x$ as $t \to \infty$.

Using relation (7) written for $F_k^I[\xi]$ we have

$$\sum_{k=0}^{t} \{1 - F_k^I[F_{t-k}[\xi_t]]\} = A - \sum_{k=0}^{t} R_k^I[F_{t-k}[\xi_t]][1 - F_{t-k}[\xi_t]].$$

The relations (9), (24) and the simple approximation $\ln(1-x) \sim -x, x \to 0$, gives us that

$$\lim_{t\to\infty} \ln H_t[\xi_t] = -\lim_{t\to\infty} A = \ln x.$$

Now we use the standard arguments allowing to obtain the limit theorem with large deviations from the behavior of the Laplace transform (see [15], for example). The number $0 < x < 1$ is the Laplace transform of the random vector $\tau = (\tau_1, ..., \tau_j)$

such that $P\{\tau = \mathbf{0}\} = \mathbf{x}, \mathbf{P}\{\cup_{i=1}^{j}\{\tau_i = \infty\}\} = \mathbf{1} - \mathbf{x}$ and $P\{\cap_{i=1}^{j}\{\tau_i < \infty\}, \tau \neq \mathbf{0}\} = 0$. Hence by the continuity theorem for Laplace transforms

$$\lim_{t\to\infty} P\left\{\hat{Z}_t[\chi(A_i)]t^{-1}x_i^{-\mu/\alpha(t)} \le y_i, i = 1, ..., j\right\} = x$$

for any finite positive numbers $y_1, ..., y_j$. The assertion of the theorem follows from the last relation if we put $y_1 = ... = y_j = 1$. The theorem is proved.

The following theorem gives a limit distribution when $\alpha(t) \sim C(\ln t)^{-1}$.

Theorem **4** *Let (M), (R) and (8) be satisfied,* $\rho = 1, 0 < \mu < \infty$ *and* $q = 1$. *If* $\alpha(t) \to 0$ *such that* $\alpha(t)\ln t \to C \in (0, \infty)$ *as* $t \to \infty$, *then for every finite measurable decomposition* $\{A_i, i = 1, ...j\}$ *of X*

$$\lim_{t\to\infty} P\left\{\frac{\hat{Z}_t[\chi(A_i]^{\alpha(t)/\mu} - 1}{e^{C/\mu} - 1} \le x_i, i = 1, ..., j\right\} = P\{\nu_i \le x_i, i = 1, ..., j\},$$

where $\nu_1 = ... = \nu_j$ *with probability 1 and* $P\{\nu_i \le x\} = e^{-C/\mu} + x(1 - e^{-C/\mu}), 0 \le x \le 1$.

Proof. We consider the sum from (24) with

$$\xi_t = \exp\{-\sum_{i=1}^{j} \lambda_i \chi(A_i / C_t(x_i)\},$$

where $C_t(x_i) = (1 + x_i(e^{C/\mu} - 1))^{\mu/\alpha(t)}, 0 < x_i < 1$. It is not difficult to see that under the conditions of Theorem 5 $C_t(x_i) \to \infty, C_t(x_i) = o(t)$ as $t \to \infty$. If we decompose the sum as in (24) with $\lambda_\alpha(t) = \ln t$ and $L(t) = \ln\ln t$, using again (25) and (26) obtain that as $t \to \infty$

$$A_2 \sim \frac{\alpha(t)}{\mu} \ln \frac{\mu t \Phi^\star[1 - \xi_t] + 1}{\mu \Phi^\star[1 - \xi_t] \ln\ln t + 1}. \tag{28}$$

Taking into account the fact that $\lim_{t\to\infty} C_t(x)\Phi^\star[1 - \xi_t] = \delta(E_0), where x = \min\{x_1, ..., x_j\}$ and $\delta(E_0)$ the same as in (27), we have from (28) that A_2 tends to $C\mu^{-1} - \ln(1 + x(e^{C/\mu} - 1))$. The same arguments as in the proof of Theorem 4 shows that A_1 and A_3 tend to zero as $t \to \infty$. Thus

$$\lim_{t\to\infty} H_t[\xi_t] = e^{-C/\mu} + x(1 - e^{-C/\mu}).$$

The assertion of the theorem follows from the last relation by continuity theorem for the Laplace transforms. The theorem is proved.

We proceed a further study in the following scheme. We consider the "partial processes" introduced in Section 3. It is clear from the definition that $Z_t^{(1)}$ and $Z_t^{(3)}$

are branching diffusions generated by the particles immigrated in the beginning of the process and by the recent immigrants, respectively. First, we prove limit theorems for partial processes using analytic technique as in the proof of theorems 4 and 5. After that we deduce the explicit limit theorem for the basic process from those results by the direct probability arguments.

We denote by $H_t^{(i)}[\xi]$ generating functionals of processes $\hat{Z}_t^{(i)}, i = 1, 2, 3$, that is

$$H_t^{(i)}[\xi] = \prod_{k \in I_i} F_k^I[F_{t-k}[\xi]].$$

Theorem **5** *Let $(M), (R)$ and (8) be satisfied, $\rho = 1$, $0 < \mu < \infty$, $q = 1$. If $\alpha(t) \to 0$ such that $\alpha(t) \ln t \to 0$, $\alpha \geq 1$, then for any finite measurable decomposition $\{A_i, i = 1, \ldots, i\}$ of X*

$$\lim_{t\to\infty} P\left\{ \frac{\hat{Z}_t^{(1)}[\chi[A_i]]}{t} \leq x_i,\ i = 1, \ldots, j | \hat{Z}_t^{(1)} \neq \theta \right\} = \\ = P\left\{ \nu_i \Phi^\star[\chi(A_i)] \leq x_i, i = 1, \ldots, j \right\},$$

where $\nu_1 = \cdots = \nu_j$ with probability 1 and $P\{\nu_i \leq x\} = 1 - e^{-x/\mu}, x \geq 0$

Theorem **6** *Let $(M), (R)$ and (8) be satisfied, $\rho = 1$, $0 < \infty, q = 1$. If $\alpha(t) \to 0$ such that $\alpha(t) \ln t \to 0$, then for any finite measurable decomposition $\{A_i, i = 1, \ldots, j\}$ of X*

$$\lim_{t\to\infty} P\left\{ \frac{\ln \hat{Z}_t^{(3)}[\chi(A_i)]}{\ln t} \leq x_i, i = 1, \ldots, j | \hat{Z}_t^{(3)} \neq \theta \right\} = P\left\{ \nu_i \leq x_i, i = 1, \ldots, j \right\}$$

where $(\nu_1, \ldots, \nu_j)$ are the same as in Theorem 4.

Proof of Theorem 6. We consider the sum

$$A = \sum_{k \in I_1} \left(1 - F_k^I[F_{t-k}[\xi_t]]\right), \xi_t = \exp\left\{ -t^{-1} \sum_{i=1}^{j} \lambda_i \chi(A_i) \right\},$$

where $\lambda_i \geq 0, i = 1, \ldots, j$ and I_1 is the set defined in Section 3. Using (7) we can write

$$A = A_1 + A_2 \tag{29}$$

where

$$A_1 = \sum_{k \in I_1} M_k^I[1 - F_{t-k}[\xi_t]], A_2 = \sum_{k \in I_1} R_k^I(F_{t-k}[\xi_t])[1 - F_{t-k}[\xi_t]].$$

Let us consider A_1. Using relations (25), (26) and the definition of the set I_1, we get that A_1 is equivalent as $t \to \infty$ to $a(L_1(t))\left\{\mu t + \Phi^\star[1-\xi_t]^{-1}\right\}^{-1}$ where $a(t)$ is defined in the Section 3. Since $\Phi^\star[\cdot]$ is linear-bounded and nonnegative on β_+ (see Assumption (M)), $t\Phi^\star[1-\xi_t] \to \delta(E)$ as $t \to \infty$, where $E = \{1,\ldots,j\}$ and δ is the same as in (27). Thus from here and from (17) we conclude that

$$A_1 \sim \frac{\delta(E)}{1+\mu\delta(E)}\,a(t)t^{-1},\; t \to \infty. \tag{30}$$

Now, we consider A_2. Since $R_k^I(\cdot)[\cdot]$ is a non-increasing function of the first argument and a linear bounded functional with respect to the second argument, using relations (25), (26) we find that for some positive constant C and sufficiently large t

$$A_2 \le C \sup_k \frac{R_k^I((1-\varepsilon_t)1)[\varphi]}{\alpha(k)} t^{-1}a(t),$$

where $\varepsilon_t > 0$ such that $\varepsilon_t \to 0$ as $t \to \infty$. The last inequality shows that under condition (8), $A_2 = o(t^{-1}a(t))$ as $t \to \infty$.

It is not difficult to see that the Laplace transform of the conditional distribution in Theorem 6 equals

$$1 - \frac{1-H_t^{(1)}[\xi_t]}{P\{\hat{Z}_t^{(1)} \ne \theta\}}.$$

Using the approximation $\ln(1-x) \sim -x$, $x \to 0$ we have that $1-H_t^{(1)}[\xi_t] \sim A$, $t \to \infty$ and, consequently, taking into account (29), (30) and part 1^0 of Theorem 2, we obtain that the limit of the above above Laplace transform is $(1+\mu\delta(E))^{-1}$. The theorem is proved.

Proof of Theorem 7. First we consider the sum

$$C_t(m) = \sum_{k=0}^{m}\left\{\mu k + \Phi^\star[1-\xi_t]^{-1}\right\}^{-1},\; \xi_t = \exp\left\{-\sum_{i=1}^{j}\lambda_i\chi(A_i)t^{-x_i}\right\}.$$

It is not difficult to see that $C_t(t) - C_t(L_2(t)) \le b(t) - b(L_2(t))$ and the last difference has the order $o(\ln t)$ according to relation (17). If we denote $q(x,t) = \{\mu x + \Phi^\star[1-\xi_t]^{-1}\}^{-1}$, then $C_t(t)$ can be written as follows:

$$C_t(t) = \int_0^{t-1} q(x,t)dx + C^{(1)}(t), 0 \le C^{(1)}(t) \le \sum_{k=0}^{t-1}[q(k)t) - q(k+1,t)] \tag{31}$$

and the sum on the right side tends to zero as $t \to \infty$. Since $t^x\Phi^\star[1-\xi_t] \to \delta(E_0)$, where $x = \min\{x_1,\ldots,x_j\}$ and $\delta(E_0)$ the same as in (27), we find that the integral in (31) is equivalent to $\mu^{-1}\ln(1+\mu\delta(E_0)t^{1-x})$ as $t\to\infty$. Hence we have

$$C_t(L_2(t))\sim\frac{1-x}{\mu}\ln t, t\to\infty. \tag{32}$$

Now we consider

$$C^{(2)}(t)=\sum_{k\in I_3}\|\varepsilon^\star_{t-k}[\xi_t]\|q(t-k,t),$$

where $\varepsilon^\star_t[\cdot]$ is the functional from (25). Since $\|\varepsilon^\star_k[\xi]\|\to 0$ as $k\to\infty$ uniformly in $\xi\in\overline{\nabla}_+\cap\{\Phi^\star[1-\xi]>0\}$, there exists a positive integer m such that $\|\varepsilon^\star_{t-k}[\xi_t]\|<\varepsilon$ for $t-k\geq m$.

We partition I_3 as $I_3=I_3^{(1)}\cup I_3^{(2)}$, where $I_3^{(1)}=\{k:t-L_2\leq k\leq t-m\}$ and $I_3^{(2)}=\{t-m<k<t\}$. It follows from the choice of m that the sum of terms with $k\in I_3^{(1)}$ is less than $\varepsilon C_t(L_2(t))$ and the sum of terms with $k\in I_3^{(2)}$ tends to zero as $t\to\infty$ for any fixed m. Thus using (25) and (26), we obtain that as $t\to\infty$

$$A_1\equiv\sum_{k\in I_3}M_k^I[1-F_{t-k}[\xi_t]]\sim\alpha(t)C_t(L_2(t)). \tag{33}$$

The same arguments as in the proof of Theorem 6, if we take into account (32) and (33), give us that under condition (8), $1-H_t^{(3)}[\xi_t]\sim\mu^{-1}(1-x)\alpha(t)\ln t$ as $t\to\infty$. Consequently, using part 2^0 of Theorem 2, we conclude that the Laplace transform of the conditional distribution in Theorem 7 tends to $x=\min\{x_1,\dots,x_j\}$ as $t\to\infty$. How from this fact the assertion of the theorem follows can be shown as in the proof of Theorem 5. The theorem is proved.

We now formulate the limit theorem for the basic branching diffusion. It turns out that the form of limiting distributions and the normalizing functions depends on the behavior at infinity of the function $\Theta(t)=t\alpha(t)\ln t/a(t)$.

Theorem **7** *Let* $(M),(R)$ *and (8) be satisfied,* $\rho=1, 0<\mu<\infty, q=1, \alpha(t)\ln t\to 0, t\to\infty$ *and* $\{A_i, i=1,\dots,j\}$ *be a finite measurable decomposition of* X.

1^0. If $\liminf_{t\to\infty}\Theta(t)>0$, *then for any* $x_i\in[0,1]$, $i=1,\dots,j, x=\min\{x_1,\dots,x_j\}$ *as* $t\to\infty$

$$\begin{aligned} P\left\{\frac{\ln\hat Z_t[\chi[A_i]]}{\ln t}\right. &= \left.\leq x_i, i=1,\dots,j|\hat Z_t\neq\theta\right\} \\ &= \frac{x\Theta(t)}{1+\Theta(t)}(1+o(1)). \end{aligned}$$

2⁰. If $\limsup_{t\to\infty} \Theta(t) < \infty$, *then for any* $x_i \geq 0, i = 1, \ldots, j$ *as* $t \to \infty$

$$P\left\{\frac{\hat{Z}_t[\chi[A_i]]}{t} \leq x_i, i = 1, \ldots, j | \hat{Z}_t \neq \theta\right\} = \frac{\Theta(t) + G(x_1, \ldots, x_j)}{1 + \Theta(t)}(1 + o(t)),$$

where $G(x_1, \ldots, x_j)$ *is the limiting distribution in Theorem 6.*

Examples. 1^0. If $\alpha < 1$, then $\Theta(t) \to \infty$ and we obtain from part 1^0 of the theorem that the limit distribution is the same as in Theorem 7. If $\alpha > 1$, then $\Theta(t) \to 0$ and it follows from part 2^0 of the theorem that the limit distribution is the same as in Theorem 6.

2^0. Let now $\alpha = 1$, $l(t) \equiv C_0 \in (0, \infty)$. Then $\Theta(t) \to 1$ as $t \to \infty$. In this case it follows from parts 1^0 and 2^0 of Theorem 8 that the limit distributions there have atoms of the same mass 2^{-1} at points 1 and zero respectively. The limit Theorems 6 and 7 obtained for the partial processes explain the cause of the appearance of these atoms.

Proof of Theorem 8. Let $B_i = B_i(t)$ be the event that i-th partial process is not equal to θ at time t. We consider the following events

$$C_i = B_i \cap \overline{B}_j \cap \overline{B}_k, C_{ij} = B_i \cap B_j \cap \overline{B}_k, C_{ijk} = B_i \cap B_j \cap B_k,$$

where $i, j, k = 1, 2, 3, i \neq j, i \neq k, j \neq k$.

If we denote $D = \{\hat{Z}_t \neq \theta\}$, then it is not difficult to see that

$$D = \bigcup_{i=1}^{3} C_i \cup \bigcup_{\substack{i,j=1 \\ i \neq j}}^{2} C_{ij} \cup C_{123}. \tag{34}$$

Since the events on the right side of (39) are disjoint, for any event A we have

$$P\{A|D\} = \sum_{i=1}^{3} P\{A|C_i\}\frac{P\{C_i\}}{P(D)} + \sum_{\substack{i,j=1 \\ i \neq j}}^{2} \frac{P\{A \cap C_{ij}\}}{P\{D\}} + \frac{P\{A \cap C_{123}\}}{P\{D\}}. \tag{35}$$

It follows from the definition of the process that the partial processes $\hat{Z}_t^{(i)}, i = 1, 2, 3$, are independent. Therefore, using the assertion of Theorem 1, we get, under the conditions of Theorem 8, that

$$P\{C_i\} \sim P\{B_i\},\ P\{C_{ij}\} \sim P\{B_i\}P\{B_j\},\ P\{C_{123}\} \sim P\{B_1\}P\{B_2\}P\{B_3\}.$$

Let us prove part 1^0. In order to do it we put in (40) that

$$A = \left\{\frac{\ln \hat{Z}_t[\chi(A_i)]}{\ln t} \leq x_i,\ \ i = 1, \ldots, j\right\}, 0 < x_i < 1.$$

Using Theorems 6 and 7 we have that $P\{A|C_3\} \to x = \min\{x_1,\dots,x_j\}$ and $P\{A|C_1\} \to 0$ as $t \to \infty$. On the other hand, as it was shown in the proof of Theorem 2, $P\{C_2\} \sim P\{B_2\} = o(P\{D\})$, $t \to \infty$. Thus the first sum in (35) is equivalent to $xP\{B_3\}/P\{D\}$ as $t \to \infty$.

Taking into account results of Theorem 2 again, we see that $P\{C_{ij}\} = o(P\{D\})$ and $P\{C_{123}\} = o(P\{D\})$ as $t \to \infty$. Hence, it follows from these relations and from (35) that as $t \to \infty$

$$P\{A|D\} = \frac{x\Theta(t)}{1+\Theta(t)} + o(1)$$

which gives the assertion of part 1^0 when liminf $\Theta(t) > 0$. Part 1^0 of the theorem is proved.

Now we prove part 2^0. We now put in (35) $A = \left\{t^{-1}\hat{Z}_t[\chi(A_i)] \le x_i, i = 1,\dots,j\right\}$. In this case we obtain from Theorems 6 and 7 that $P\{A|C_1\}$ tends to $G(x_1,\dots,x_j)$ and $P\{A|C_3\} \to 1$ as $t \to \infty$.

It follows from the condition limsup$\Theta(t) < \infty$ that $\alpha \ge 1$. Consequently we can use part 1^0 of Theorem 2 and thus, the first sum in (35) as $t \to \infty$ is equivalent to $(G(x_1,\dots,x_j) + \Theta(t))(1+\Theta(t))^{-1}$. The theorem is proved.

References

[1] Asmussen S., Hering H. (1977) Some modified branching diffusion models. Math.Biosci., 35, p. 281-299.

[2] Asmussen S., Hering H. (1983) Branching Processes, Birkhauser, Boston, Basel, Stuttgart.

[3] Athreya K., Ney P.(1972) Branching Processes, Springer-Verlag.

[4] Conner H. E. (1961) A limit theorem for a position-dependent branching process, Journal Math. Analysis and Appl., 3, p. 560-591.

[5] Conner H. E. (1967) Asymptotic behavior of averaging processes for a branching process of restricted Brownian partcles, J. Math. Analysis Appl., V. 20, 464-479.

[6] Cox J. T. (1994) On the ergodic theory of critical branching Markov chains, Stochastic Process. Appl. V. 50, No 1, 1-20.

[7] Hering H.(1973) Asymptotic behaviour of immigration-branching processes with general set of types. I: Critical branching part, Adv.Appl. Probab.,5, p. 391-416.

[8] Hering H. (1978) Multigroup branching diffusions. Branching Processes, Edit. A. Joffe and P. Ney, Marcel Dekker, Inc., New York, Basel, p. 177-217

[9] Ikeda N., Nagasawa M., Watanabe S. (1965) On branching Markov processes, Proc. Japan Acad. Sci., V. 41, No 9, 816-821.

[10] Ikeda N., Nagasawa M., Watanabe S. (1966) A construction of branching Markov processes, Proc. Japan Acad. Sci., V. 42, No 4, 380-384.

[11] Ikeda N., Nagasawa M., Watanabe S. (1966) On branching semi-groups I, II, Proc. Japan Acad. Sci., V. 42, No 8, 1016-1026 .

[12] Kageyama S., Ogura Y (1980) On a limit theorem for branching one-dimentional diffusion processes, Publ.RIMS Kyoto Univ. 16, 355-376.

[13] Moyal (1962) The general theory of stochastic population processes. Acta Math. 108, p.1-31.

[14] Jagers P. (1975) Branching Processes with Biological Applications, J. Wiley & Sons XIII, London.

[15] Rahimov I.(1995) Random Sums and Branching Stochastic Processes, Springer-Verlag, Ser. LNS, 96, New York.

[16] Savits T. (1969) The explosion problem for a branching Markov process, Osaka J. Math., 6, 375-395.

[17] Seneta E. (1976) Regularly Varying Functions, Springer-Verlag, Ser. LNM 508, Berlin.

[18] Sevast'yanov B.A.(1971) Branching Processes, Nauka, Moscow.

Inference Concerning About a Process Capability Index

S. E. Ahmed
University of Regina
Regina, Saskatchewan
S4S 0A2 Canada
E-mail: ahmed@math.uregina.ca
and
M. H. Rahbar
The Aga Khan University
Karachi 74800, Pakistan

Abstract

Capability indices are widely used in practice. This article considers two large sample methods of obtaining confidence intervals and asymptotic test statistics for a process capability index. A simulation study is carried out to appraise the performance of the log transformed method as compared to the classical method of drawing inferences. Simulation study may suggest that the log transformed procedures has an edge over the classical method. Based on these results recommendations are made for practical use of these procedures.

Key Words and Phrases: Process capability index, Asymptotic confidence intervals and tests, log transformation, Monte Carlo simulation.

1. INTRODUCTION

A *Process Capability* (C_p) index has been introduced in the statistical quality control literature as an indispensable tool to help in the assessment of process performance.

This C_p index is defined as

$$C_p = \frac{\text{allowable process spread or tolerance}}{\text{actual process spread or variation}}$$

Let LSL = *lower specification limit* and USL = *upper specification limit* and σ be the standard deviation associated with the process, further, let $d = \frac{USL-LSL}{2}$. The above formula may be rewritten in the following compact form.

$$C_p = \frac{d}{3\sigma}, \quad \sigma > 0.$$

This index has two advantages over other indices. First, it is tractable analytically and second it has the interpretation. A value of C_p greater than or equal to one implies that the tolerance $USL - LSL$ is at least 6σ of the process distribution. If we define the μ as the location, i.e., the mid point of $[LSL, USL]$ of the process distribution. Then by Chebyshev's rule at least 89% of the products are within 3σ. On the other hand if the process distribution is normal then by empirical rule approximately 99.73% of the products will be within 3σ (so if $C_p = 1$, this means within the specification limits).

Kane (1986) presented a detailed study of C_p. A Bayes solution to the problem is given in Cheng and Spiring (1989). For a review we refer to Pearn, Kotz and Johnson (1992). However, to our knowledge, almost all of the theory in the literature is developed for samples from normal populations only.

In the present investigation we discuss the point and asymptotic interval estimation for C_p. Large sample test statistics are also proposed and their asymptotic properties are studied. A simulation study is provided to examine the performance of the proposed tests and estimation procedures. We hope to present a start on an important problem.

A plan of this paper is as follows. In the next section, along with the preliminary notations and basic assumptions, we propose asymptotic pivotal quantities for estimation of C_p. In section 3 we use a variance stabilizing transformation (log transformation) to study the large sample behavior of a log transformed estimator for C_p. The results of sections 2 and 3 lead to introduction of four asymptotic test statistics which are given in section 4. Our simulation result for the two type of confidence intervals and the four test statistics are presented in section 5. Finally, a comparison and a discussion of these results is given in the final section.

2. ESTIMATION PROCEDURES

Suppose $Y_1, \cdots, Y_n$ denote independent measurements from a process population having a distribution with mean parameter μ and the variance of Y is σ^2. In addition, we assume that the forth moment μ_4 exists.

Let $\hat{\mu}_n$ and $\hat{\sigma}_n^2$ be the estimator of μ and σ^2 respectively, where

$$\hat{\mu}_n = \frac{1}{n}\sum_{j=1}^{n} Y_j$$

and

$$\hat{\sigma}_n^2 = \frac{1}{n-1}\sum_{j=1}^{n}(Y_j - \hat{\mu}_n)^2.$$

For large n,

$$n^{1/2}(\hat{\mu}_n - \mu) \xrightarrow[D]{} \mathcal{N}(0, \sigma^2) \tag{2.1}$$

and

$$n^{1/2}(\hat{\sigma}_n^2 - \sigma^2) \xrightarrow[D]{} \mathcal{N}(0, \mu_4 - \sigma^4) \tag{2.2}$$

where the notation $\xrightarrow[D]{}$ means *convergence in distribution.*

The parameter of interest to be estimated here is $C_p = \frac{d}{3\sigma}$. A natural estimator of C_p, is obtained by replacing the unknown σ by a consistent estimator $\hat{\sigma}_n$. This yields

$$\hat{C}_p = \frac{d}{3\hat{\sigma}_n}. \tag{2.3}$$

Since the sample size n is taken to be large in the present investigation and the standard deviation is assumed to be positive, it is reasonable to assume that sample standard deviation, $\hat{\sigma}_n$ is also going to be positive.

To develop the asymptotic confidence interval and test statistic for C_p, we first state the following result (Chan, Xiong and Zhang (1990)).

For $n \to \infty$

$$n^{1/2}\{\hat{C}_p - C_p\} \xrightarrow[D]{} \mathcal{N}\{0, \tau^2\}, \tag{2.3}$$

where

$$\tau^2 = \frac{(USL - LSL)^2}{144\sigma^2} \frac{(\mu_4 - \sigma^4)}{\sigma^4}.$$

Since τ^2 is not in a convenient form for inference purposes about C_p, we rewrite τ^2 in the following form.

$$\tau^2 = \frac{\beta - 1}{4} C_p^2,$$

where β is the kurtosis of the underlying distribution for the measurements from the process population.

Thus, the asymptotic distribution for the estimator of parameter C_p is

$$Z_1[\hat{C}_p] = \frac{n^{1/2}\{\hat{C}_p - C_p\}}{\tau} \xrightarrow[D]{} \mathcal{N}(0, 1).$$

Since the quantity $Z_1[\hat{C}_p]$ depends on the unknown quantity C_p, it will not be useful for constructing confidence intervals for C_p. However, it can be used as a test statistic for test of hypothesis regarding C_p.

Another pivotal quantity can be defined by replacing τ by its empirical estimate $\hat{\tau}_n$ in the denominator of $Z_1[\hat{C}_p]$ and is given by

$$Z_2[\hat{C}_p] = \frac{n^{1/2}\{\hat{C}_p - C_p\}}{\hat{\tau}_n} \xrightarrow[D]{} \mathcal{N}(0, 1),$$

where

$$\hat{\tau}_n^2 = \frac{\hat{\beta}_n - 1}{4} \hat{C}_p^2, \quad \hat{\beta}_n = \frac{\hat{\mu}_{n4}}{\hat{\sigma}_n^4} \quad \text{and} \quad \hat{\mu}_{n4} = \frac{1}{n} \sum_{i=1}^{n} (Y_i - \hat{\mu}_n)^4.$$

Further, $\hat{\tau}_n$ is a consistent estimator of τ Thus, $Z_2[\hat{C}_p] \xrightarrow[D]{} \mathcal{N}(0, 1)$. Thus, for large n, it is appropriate to approximate the distribution of $Z_2[\hat{C}_p]$ by the standard normal distribution.

An asymptotic $(1 - \alpha)100\%$ confidence interval for C_p is

$$Pr\left\{\hat{C}_p - z_{\frac{\alpha}{2}} \left(\frac{\hat{\tau}_n^2}{n}\right)^{1/2} \leq C_p \leq \hat{C}_p + z_{\frac{\alpha}{2}} \left(\frac{\hat{\tau}_n^2}{n}\right)^{1/2}\right\} \to 1 - \alpha \quad \text{as} \quad n \to \infty, \tag{2.4}$$

where $z_{\frac{\alpha}{2}}$ is the $\left(1 - \frac{\alpha}{2}\right)$ 100 percentile point of the standard normal distribution.

Recall that the asymptotic distribution of $\hat{C}_p$ depends on two unknown parameters C_p and the kurtosis β. In the following section we present a variance stabilizing transformation which will not depend on C_p, however it will be a function of β which need to be estimated. Thus, the proposed procedures based on the log transformation will have a different asymptotic behavior.

3. VARIANCE STABILIZING TRANSFORMATION

Generally speaking, the results on the large samples distribution of functions of asymptotically normal statistics are of special interest for the variance stabilizing transformations. The asymptotic distribution of the quantity $n^{1/2}\{\hat{C}_p - C_p\}$ follows a normal distribution with mean 0 and variance $\frac{\beta-1}{4}C_p^2$. Since the asymptotic variance depends on unknown parameters C_p, the question is whether we could find a suitable transformation of $\hat{C}_p$ for which the asymptotic variance does not depend on the unknown parameter C_p. To obtain a variance-stabilizing transformation we must define a function $g[\hat{C}_p]$ such that $n^{1/2}\{g[\hat{C}_p] - g[C_p\} \xrightarrow[D]{} N(0, k^2)$, where k is independent of C_p. In the following we show that this may be achieved by taking a *log* transformation.

Theorem 3.1. For large n, $n^{1/2}\{log[\hat{C}_p] - log[C_p\} \xrightarrow[D]{} \mathcal{N}(0, \frac{(\beta-1)}{4})$

The following lemma is used for the proof of the theorem. The proof of the Lemma is given in Dudewicz and Mishra (1988, page 327) hence omitted in this paper.

Lemma 3.1 Let $\hat{\theta}$ be an estimator of the parameter θ. Suppose that $n^{1/2}\{\hat{\theta} - \theta\} \xrightarrow[D]{} \mathcal{N}(0, \sigma^2)$ and let g be a continuos function such that $g'(\theta)$ exists and $g'(\theta) \neq 0$. Then it follows that $n^{1/2}\{g(\hat{\theta}) - g(\theta)\}/\sigma g'(\theta) \xrightarrow[D]{} \mathcal{N}(0,1)$.

Proof of the Theorem: Letting $g(x) = log(x)$, we have $g'(x) = x^{-1}$ and $g'(C_p) = \frac{1}{C_p} \neq 0$. Thus, by the above lemma the result of the theorem follows.

The results of Theorem 3.1, allow us to construct interval estimation procedures and test statistics when making inferences for C_p. Note that,

$$Pr\left\{log[\hat{C}_p] - z_{\frac{\alpha}{2}}\left(\frac{\beta-1}{4n}\right)^{1/2} \leq log[C_p] \leq log[\hat{C}_p] + z_{\frac{\alpha}{2}}\left(\frac{\beta-1}{4n}\right)^{1/2}\right\}$$

converges to $1-\alpha$ $n \to \infty$. This in turn implies that

$$Pr\left\{e^{log[\hat{C}_p] - z_{\frac{\alpha}{2}}\left(\frac{\beta-1}{4n}\right)^{1/2}} \leq C_p \leq e^{log[\hat{C}_p] + z_{\frac{\alpha}{2}}\left(\frac{\beta-1}{4n}\right)^{1/2}}\right\}$$

converges to $1-\alpha$.

4. ASYMPTOTIC TESTS AND POWER

We propose the following test statistics to test for $H_o : C_p = C_p^o$ against $H_a : C_p \neq C_p^o$ or $C_p < C_p^o$ or $C_p > C_p^o$. These statistics are given by

$$Z_1^*[\hat{C}_p] = \frac{n^{1/2}\{\hat{C}_p - C_p^o\}}{kC_p^o}, \quad k^2 = \frac{\beta - 1}{4}, \quad \beta \text{ is known under } H_o,$$

$$Z_2^*[\hat{C}_p] = \frac{n^{1/2}\{\hat{C}_p - C_p^o\}}{\hat{k}\hat{C}_p}, \quad \hat{k} = \frac{\hat{\beta}_n - 1}{4}$$

$$Z_3^*[\hat{C}_p] = \frac{n^{1/2}\{log\hat{C}_p - logC_p^o\}}{k}, \quad \beta \text{ is known under } H_o,$$

$$Z_4^*[\hat{C}_p] = \frac{n^{1/2}\{log\hat{C}_p - logC_p^o\}}{\hat{k}}.$$

For large n and under the null hypothesis $Z_i^*[\hat{C}_p]$, $(i = 1,2,3,4)$ could be approximated by a standard normal distribution, which provide the asymptotic critical values.

It is important to note that for a fixed alternative that is different from the null hypothesis the power of all four tests statistics proposed earlier will converge to one as $n \to \infty$. Thus, to study the asymptotic power properties of $Z_i^*[C_p]$, we confine ourselves to a sequence of local alternative $\{K_n\}$. When C_p is the parameter of interest, such a sequence may be specified by

$$K_n : C_p = C_p^o + \frac{\xi}{n^{\frac{1}{2}}}. \tag{4.1}$$

where ξ is a fixed real number. Evidently, C_p approaches C_p^o at a rate to $n^{-1/2}$. Stochastic convergence of $\hat{C}_p$ to C_p ensures that $\hat{C}_p \xrightarrow[p]{} C_p$ under local alternatives as well.

Theorem 4.1: Under the local alternatives in (4.1) and as $n \to \infty$

$$n^{1/2}\{\hat{C}_p - C_p\} \xrightarrow[D]{} \mathcal{N}\{\xi,\ k^2C_p^2\},$$

$$n^{1/2}\{log\hat{C}_p - logC_p\} \xrightarrow[D]{} \mathcal{N}(\xi, k^2).$$

Hence, under the local alternatives we have the following corollary.

Corollary 4.1: Under the local alternatives in (4.1) and as $n \to \infty$

$$Z_i^*[\hat{C}_p] \xrightarrow[D]{} \mathcal{N}\left\{\frac{k^2\xi}{C_p}, 1\right\}, \quad i = 1, 2$$

$$Z_i^*[\hat{C}_p] \xrightarrow[D]{} \mathcal{N}\left\{\frac{\xi}{k^2}, 1\right\}, \quad i = 3, 4.$$

Therefore, each $Z_i^{*2}[\hat{C}_p]$ has asymptotically a noncentral chi-square distribution with 1 degree of freedom and noncentrality parameter λ_i, where $\lambda_1 = \lambda_2 = \frac{\xi^2}{k^2 C_p^2}$, and $\lambda_3 = \lambda_4 = \frac{\xi^2}{k^2}$.

Hence, the power calculations of the proposed test statistic can be done by using noncentral chi-square distribution.

We realize that $\hat{C}_p$ is a nonlinear function of asymptotic normal estimators which may not be well approximated by a normal unless the sample sizes are very large. A simulation study is carried out to observe the effect of the sample sizes on the asymptotic results.

5. SIMULATION RESULTS

5.1. Interval Estimation

In this sub-section we present the result of our simulation study which was conducted to assess the performance of the two proposed interval estimators for the capability index given in this paper. We have considered a normal distribution with mean 15 and standard deviation one ($N(15, 1)$). And for an exponential distribution with mean one ($EXP(1)$). Some other values of the parameters are also considered for the both populations. The similar analysis is obtained, hence not appended here to save the space. Finally, in Both cases the true value of the parameter of interest is 4/3, i.e., $C_p = 4/3$

In addition, two scenarios were considered for the computation of kurtosis β and τ, asymptotic variance of $\hat{C}_p$, one using the exact value of kurtosis from the known underlying distribution and the other by estimating the sample kurtosis from the given data. The simulation is conducted at various sample sizes 25, 36, 49, 64 and 100. In each Case, one hundred samples were randomly generated using the statistical package SAS on the mainframe computer MAX at the university of Regina, Canada. The coverage probability and the mean width of the generated confidence intervals are presented in Table 1. As expected the coverage probability and the mean width of the confidence intervals were closer to their actual values in the case of normal distribution as compared to the exponential distribution. Importantly, it was noted that when the kurtosis is estimated the coverage probability may be significantly different from the nominal value. A possible explanation for this is a high level of variation among various estimators of the kurtosis. For example, in the case of the exponential distribution the expected value for kurtosis is 9. However, there were few samples yielding an estimate of less than one for kurtosis. In such cases the estimate for the asymptotic variance is negative. Therefore, to remedy this we used 1.01 as an estimate of the kurtosis.

5.2. Testing Hypothesis

For testing $H_o : C_p = C_p^o$ against $H_a : C_p \neq C_p^o$, another simulation study was performed to examine the performance of $Z_1^{*2}[\hat{C}_p]$, $Z_2^{*2}[\hat{C}_p]$, $Z_3^{*2}[\hat{C}_p]$ and $Z_4^{*2}[\hat{C}_p]$.

To asses the significance level and compare the power of these tests, 1000 repeated samples were taken from the two aforementioned distributions.

In order to assess the effect of the sample size on the results, simulation was conducted at various sample sizes 25, 36, 49, 64 and 100. For each sample the four test statistics $Z_i^{*2}[\hat{C}_p]$ $i =, 1, 2, 3, 4$ were computed.

The simple hypothesis $H_o : C_p = 4/3$ is tested against the global alternative $H_a : C_p \neq 4/3$ at five percent level of significance. The relative frequencies of the nominal α levels, that is, the proportion of samples for which H_o is rejected when $C_p = 4/3$ are recorded. These proportions are called empirical α levels. The empirical powers of the test, that is, the proportion of the tests for which the null hypothesis is rejected when $C_p \neq 4/3$, are also computed. Some simulation results are given in Tables 2-3 which present the empirical powers and significance levels corresponding to the nominal level $\alpha = 0.05$ for selected value of δ, where δ is the normalized difference between hypothesized and true value of the parameter.

It is important to note that in some cases the observed level of significance reported in these tables is noticeably different from the nominal level 0.05. At this level of significance, for all four test statistics used for testing $C_p = 4/3$, the variance in the proportion of samples which lead to rejection of the null hypothesis is $(0.05)(0.95)/1000$. However, in some cases specially when the sample size is relatively small the observed values in the table do not fall within three standard deviations of their expected values. These findings will be discussed in the following section.

Finally, we would like to comment that present simulation study may not provide completely convincing evidence to draw the conclusion. Obviously an estimated kurtosis will result in less valid confidence intervals than a known kurtosis. Further, the use of two sampling situations may not to be considered complete. However, present investigation may lead to more general problems. First author is considering to extend this problem in to multi-sample situation.

6. CONCLUDING REMARKS

Large sample confidence intervals and test statistics are very popular in statistical analysis. For example, most of the statistics textbooks suggest that at least 30 observations are required for estimation of the mean of a population. The results of this paper indicate that even though the same large sample theory is used for estimation of C_p, the behavior of the estimator introduced in this paper and its variance is heavily dependent on the Kurtosis of the underlying distribution. For example, in the case of normal distribution the Kurtosis is 3 but in the case of exponential the expected value of kurtosis is 9. That is why the performance of the estimation procedures and the test statistics is significantly different between the two distributions used in our simulation study especially for small sample sizes. These results strongly suggest that these procedures need to be used cautiously. We used only two sampling situations, normal and exponential, which by no means is complete. However, our limited simulation study gives some evidence to our findings.

It is interesting to note that the test statistics based on log transformation seems to perform better as compared to the other method. However, for the interval estimation both procedures provide comparable results.

In parallel lines it is feasible to establish the similar inferential results for some other useful process capability indices.

REFERENCES

1. Chan, L. K., Xiong, Z, and Zhang, D. (1190). On the asymptotic distribution of some process capability indices. *Comm. Stat.-Theor. Meth.* **19**, 11-18.

2. Kane, V.E. (1986). Process capability indices. *Journal of Quality Technology* **18**, 41-52.

3. Cheung, S. W. and Spiring, F. A. (1991). Assessing process capability: A Bayesian approach. *IEE Transactions* **21**, 97-98.

4. Pearn, W.L., Kotz, S. and Johnson, N.L. (1992). Distributional and inferential properties of process capability indices. *Journal of Quality Technology* **24**, 216-231.

Table 1: The average coverage probability and mean width of the generated confidence intervals, with nominal coverage probability is 0.95.

		Without Transformation		With Log Transformation	
Distribution	Sample Size n	Average Coverage Probability	Mean Width	Average Coverage Probability	Mean Width
Normal when true value of τ is used	25	0.95	0.741	0.96	0.750
	36	0.93	0.621	0.93	0.626
	49	0.96	0.538	0.97	0.541
	64	0.96	0.470	0.95	0.472
	100	0.96	0.377	0.96	0.378
Normal when τ is estimated	25	0.94	0.663	093	0.670
	36	0.94	0.549	0.93	0.553
	49	0.95	0.482	0.94	0.484
	64	0.94	0.448	0.93	0.450
	100	0.95	0.757	0.97	0.766
Exponential when true value of τ is used	25	0.97	1.675	0.97	1.762
	26	0.94	1.324	0.95	1.372
	49	0.95	1.108	0.96	1.138
	100	0.95	0757	0.97	0.766
Exponential when τ is estimated	25	0.93	1.247	0.73	1.172
	36	0.96	1.075	0.82	1.156
	49	0.94	0.878	0.85	0.830
	64	0.91	0.826	0.88	0.821
	100	0.93	0.678	0.91	0.676

Table 2: The empirical powers and significance levels corresponding to the nominal level $\alpha = 0.05$ for exponential distribution.

Test	n	$\delta = 0$	$\delta = 1.5$
Z_1^*	25	.086	0.994
Z_2^*	25	.227	0.946
Z_3^*	25	.046	0.690
Z_4^*	25	.234	0.925
Z_1^*	36	.067	1.000
Z_2^*	36	.172	0.989
Z_3^*	36	.040	0.847
Z_4^*	36	.186	0.978
Z_1^*	49	.068	0.998
Z_2^*	49	.166	0.992
Z_3^*	49	.043	0.948
Z_4^*	49	.169	0.988
Z_1^*	64	.059	1.000
Z_2^*	64	.140	0.999
Z_3^*	64	.036	0.990
Z_4^*	64	.147	0.999
Z_1^*	100	.079	1.000
Z_2^*	100	.114	1.000
Z_3^*	100	.044	1.000
Z_4^*	100	.113	1.000

Table 3: The empirical powers and significance levels corresponding to the nominal level $\alpha = 0.05$ for a standard normal distribution.

Test	n	$\delta = 0$	$\delta = 0.5$
Z_1^*	25	.079	0.868
Z_2^*	25	.122	0.732
Z_3^*	25	.077	0.581
Z_4^*	25	.119	0.676
Z_1^*	36	.061	0.944
Z_2^*	36	.087	0.826
Z_3^*	36	.059	0.736
Z_4^*	36	.092	0.774
Z_1^*	49	.074	0.971
Z_2^*	49	.092	0.891
Z_3^*	49	.065	0.847
Z_4^*	49	.096	0.868
Z_1^*	64	.066	0.991
Z_2^*	64	.074	0.951
Z_3^*	64	.059	0.927
Z_4^*	64	.090	0.934
Z_1^*	100	.066	1.000
Z_2^*	100	.063	0.993
Z_3^*	100	.053	0.992
Z_4^*	100	.070	0.991

ESTIMATION OF MIXING PROPORTIONS FOR OL-MIXTURES VIA EM ALGORITHM

Prashant Sansgiry and Olcay Akman
Department of Mathematics and Statistics
Coastal Carolina University

SUMMARY

This article deals with the EM algorithm to estimate the parameters in a population where the samples follow a length-biased version of the assumed distribution. The proposed modifications are applied to the mixture inverse Gaussian lifetime model which arises as a discrete mixture of original and length-biased inverse Gaussian densities.

Key Words and Phrases: Length-biased, inverse Gaussian, maximum likelihood, lifetime models.

INTRODUCTION

The Expectation-Maximization (EM) algorithm is a general purpose computing tool for maximum likelihood estimation for wide variety of situations including mixture models, incomplete-data problems, and outliers. Dempster et al. (1977) established the foundation of the algorithm, and remains to be one of the most important articles in the area although since then a spate of articles appeared in the literature. Over the years a number of modifications and extensions of the algorithm have been developed to overcome some of the initial limitations, and a whole battery of EM-based algorithms are still being developed. McLachan and

Krishnan (1997) is one of the latest publications in the area which reviews the entire formation of the algorithm with its extensions. In this article we propose to modify the EM algorithm to tackle the estimation properties of length-biased distributions. A length-biased distribution arises as a selection model where the sampling distribution follows a weighted version of the original population due to a stochastic mechanism which distorts the probability of an item being selected. Statistical analysis based on length-biased samples has been studied in detail by many authors including Rao (1977), Zelen (1974), Zelen and Feinleib (1977), Day (1983), Gupta and Akman (1995, 1998), Akman and Sansgiry (1999).

MAXIMUM LIKELIHOOD ESTIMATION VIA EM ALGORITHM

Let Y be a random vector corresponding to the observed data y with pdf $g(y;\Theta)$ where $\Theta = (\theta_1, \ldots, \theta_d)'$ is a vector of unknown parameters with parameter space Ω. In the context of the EM algorithm the observed data vector y is viewed as incomplete and is considered as an observable function of the complete data. The notion of *incomplete data* includes the conventional sense of missing data, but it also applies to cases where the complete data represent what would be available from some hypothetical experiment. Within this framework we let x denote the complete-data vector, and we let z denote the vector containing the additional data, commonly referred to as the unobservable or missing data. if we let $g_c(x;\Theta)$ be the pdf of the random vector X then the complete data log likelihood function is given by

$$\log L_c(\Theta) = \log g_c(x;\Theta). \tag{1}$$

The EM algorithm approaches the problem of solving the incomplete data likelihood equation $\partial L(\Theta)/\partial\Theta = 0$ indirectly by proceeding iteratively in terms of the complete-data log likelihood function, $\log_c L(\Theta)$. As it is unobservable, it is replaced by its conditional expectation given y, using the current fit for Θ. More specifically, let $\Theta^{(0)}$ be some initial value for Θ. Then on the first iteration , the E-step requires the calculation of

$$Q(\Theta;\Theta^{(0)}) = E_{\Theta^{(0)}}\{\log_c(\Theta)|y). \tag{2}$$

On the other hand the M-step requires the maximization of $Q(\Theta;\Theta^{(0)})$ with respect to Θ over Ω. In other words, we choose $\Theta^{(1)}$ such that

$$Q(\Theta^{(1)};\Theta^{(0)}) \geq Q(\Theta;\Theta^{(0)}) \tag{3}$$

for all $\Theta \in \Omega$. The E- and M-steps are carried out iteratively until the difference

$$L(\Theta^{(k+1)}) - L(\Theta^{(k)}) \tag{4}$$

changes by an arbitrarily small amount in the case of convergence of the sequence of likelihood values $\{L(\Theta^{(k)})\}$. For the convergence properties of the algorithm as well as the details of E- and M-steps, the reader is encouraged to see McLachlan and Krishnan (1997), and Dempster et al. (1977).

LENGTH-BIASEDNESS

A distribution function G_F defined on R^+ is called *length-biased* distribution corresponding to a df F (also defined on R^+), if

$$G_F(y) = \mu_F^{-1}\left\{\int_0^y x \cdot dF(x)\right\}, \quad \forall y \in R^+, \tag{5}$$

where $\mu_F = \int_0^\infty x \cdot dF(x)$ Note that the Radon-Nikodym derivative of G_F wrt F is given by

$$\frac{dG(x)}{dF(x)} = \frac{x}{\mu_F} \tag{6}$$

Hence, the length-biased pdf can be written as

$$g(x) = \frac{x \cdot f(x)}{\mu_F}, \tag{7}$$

where μ_F is assumed to be finite and nonzero. Thus when observations are selected with probability proportional to their "length", the resulting distribution is called *length-biased* The concept of length-biased sampling has been mainly developed by Rao (1977), Zelen and Feinleib (1969). Such models are also called *selection models*, and random samples obtained under sampling restrictions are called *selection samples.* The reader is encouraged to see Frazer (1966), and Frazer (1952) for detailed studies involving such samples. The length-biased distribution occurs naturally for some sampling plans in reliability, biometry, wildlife studies, and survival analysis, among others. It is interesting to note that even if a random sample can be obtained, the experimenter may not choose to use it since a carefully chosen length-biased sample may turn out to be more informative (e.g. Bayarri and DeGroot(1989)).

The possible bias due to the nature of the selection process can be remedied by considering a random variable, X_p, whose probability density function is a discrete mixture of the assumed (original) pdf and its length-biased version, that is;

$$f_p(x|\theta) = (1-p)\cdot f(x|\theta) + p\cdot g(x|\theta), \quad 0 \le p \le 1. \tag{8}$$

Using (7), we observe

$$f_p(x|\theta) = f(x|\theta)(1-p+p\frac{x}{E(X)}). \tag{9}$$

We call a pdf defined in (9) an *original length-biased* (OL) mixture pdf.

EM-ALGORITHM FOR ORIGINAL LENGTH-BIASED MIXTURES

EM Algorithm can also be used in likelihood estimation for mixture models, where

$$f_\phi(\mathbf{X}) = \sum_{i=1}^{m} \pi_i \cdot f_i(\mathbf{X};\theta)$$

where $\sum \pi_i = 1$. The likelihood equation, $\partial L(\phi)/\partial\phi = 0$ can be so manipulated that the likelihood estimate $\hat{\phi}$ satisfies

$$\hat{\pi}_i = \sum_{j=1}^{n} a_{ij}/n, \quad (i = 1,\dots,m)$$

and

$$\sum_{i=1}^{m}\sum_{j=1}^{n} a_{ij}\partial \log f_i(x_j;\hat{\theta})/\partial\hat{\theta} = 0.$$

Corresponding to the formulation of finite mixture models, we let the vector of indicator variables $z_j = (z_{1j}\dots,z_{mj})'$ be defined by

$$z_{ij} = \begin{cases} 1 & x_j \in G_i \\ 0 & x_j \notin G_i \end{cases}$$

where $z_1, \ldots, z_n \overset{iid}{\sim} M_m(1, \pi)$.

(Under necessary assumptions) x_j given z_j has log density

$$\sum_{i=1}^{m} z_{ij} \log f_i(x_j; \theta) \quad (j = 1 \ldots, n).$$

Hence the log-likelihood for the complete data X, and Z is given by

$$L(\phi) = \sum_i \sum_j z_{ij} \{\log \pi_i + \log f_i(x_j; \theta)\}.$$

The EM Algorithm is applied to the mixture model by treating Z as missing data.

In particular

$$f(x) = \sum_i \pi_i f_i(x|\theta_i) = \sum_i \pi_i f_i(x)$$

. The log-likelihood is

$$\begin{aligned} L &= \sum_j \log f(x_j) \\ &= \sum_i \sum_j a_{ij} (\log(\pi_i f_i(x_j)) - \log(a_{ij})) \end{aligned}$$

where $a_{ij} = \frac{\pi_i f_i(x_j)}{f(x_j)}$. Using Lagrange multipliers, we get

$$\begin{aligned} a_{ij}^* &= \frac{\pi_i f_i(x_j)}{f(x_j)} \\ \pi_i^* &= \frac{1}{n} \sum_{j=1}^{n} a_{ij} \\ \phi_i^* &= \arg\max_{x \in \mathcal{D}} \sum_j a_{ij} \log f(x_j). \end{aligned}$$

In particular, for the OL-Mixtures of the form

$$f(\mathbf{X}; \psi) = \sum_{i=1}^{2} p_i \cdot f_i(x) \tag{10}$$

the loglikelihood is

$$\mathcal{L} = \sum_{j=1}^{n} \log f(x_j|\psi) \tag{11}$$

Substituting (10) in (11), we get

$$\mathcal{L} = \sum_{j=1}^{n} \log(\sum_{i=1}^{2} p_i \cdot f_i(x)) \tag{12}$$

For each j, we maximize

$$\sum_{i=1}^{2} a_{ij}\{\log(p_i f_i(x_j)) - \log(a_{ij})\}$$

subject to $\sum_{i=1}^{2} a_{ij} = 1$. That is, maximize (using Lagrange multiplier)

$$\sum_{i=1}^{2} a_{ij}\{\log(p_i f_i(x_j)) - \log(a_{ij})\} + \lambda(\sum_{i=1}^{2} a_{ij} - 1);$$

this results in

$$a_{ij} = \frac{p_i \cdot f_i(x_j)}{\sum_{i=1}^{2} p_i \cdot f_i(x_j)}$$

Similarly maximizing w.r.t p_i gives

$$p_i = \frac{\sum_{j=1}^{n} a_{ij}}{n}.$$

Hence we have maximized

$$\begin{aligned} f_\Theta &= \sum_j (a_{1j} \log f_\theta(x_j) + a_{2j} \log f_\theta(x_j)) \\ &= \sum_j (a_{1j} \log f_\theta(x_j) + a_{2j} \log(\frac{x_j}{E(X)} f_\theta(x_j)) \\ &= \sum_j \log f_\theta(x_j) - \sum_j a_{2j} \log(E(X)) \\ &+ \text{(term with no } \theta) \end{aligned}$$

We can summarize the maximization process for OL mixtures by the following algorithm.

Initial guess for a_{ij}

1. Calculate p_i and θ

2. Update a_{ij}

3. Go to step 1 until $|a_{ij}^{(k+1)} - a_{ij}^{(k)}| < \epsilon$

AN EXAMPLE

Let X be an inverse Gaussian random variable with pdf

$$f(x|\mu,\lambda) = (\frac{\lambda}{2\pi x^3})^{1/2} \exp^{-\frac{\lambda}{2\mu^2}\frac{(x-\mu)^2}{x}}, \quad x > 0,\ \lambda,\mu > 0. \qquad (13)$$

In the OL-mixture setting, we use the mixture model (MIG) given by

$$f(x|\mu,\lambda,p) = (\frac{\lambda}{2\pi x^3})^{1/2} \exp^{-\frac{\lambda}{2\mu^2}\frac{(x-\mu)^2}{x}} (1 - p + p\frac{x}{\mu}), \quad 0 \le p \le 1. \qquad (14)$$

One can show that an MLE of μ is the only positive solution of the quadratic

$$\mu^2(\sum_j a_{2,j})(\sum_j 1/x_j) + \mu(n^2 - 2n\sum_j a_{2,j}) + (\sum_j x_j)(\sum_j a_{2,j}) - n\sum_j x_j = 0.$$

Hence the MLE of λ is obtained from

$$\lambda = \frac{\mu^2 \sum_j a_{2,j}}{\sum_j x_j - n\mu}.$$

The Figures 1 and 2 present an estimate of the distribution of $\hat{p}$. Table 1 contains simulation results for the mean and standard error for the three parameters of an inverse Gaussian mixture and compares with the results of comparable methods.

CONCLUDING REMARKS

The EM algorithm with the proposed modification presented here is a viable alternative to usual computing methods, such as search method, Akman (1994), simulated annealing (Goffe et. al 1994), in obtaining MLE of parameters in OL-Mixtures of lifetime distributions. In particular the mixing proportion, p, which provides the length-biasedness of the sampling distribution can be estimated with this algorithm as fast as other comparable methods.

Table 1: Estimates of p for Mixture Inverse Gaussian

$\hat{p}$ (Mean)					
Methods	**p = 0.0**	**p = 0.25**	**p = 0.5**	**p = 0.75**	**p = 1.0**
Modified EM	2.218E-01	3.084E-01	5.057E-01	6.970E-01	7.778E-01
Sim.Annealing	2.224E-01	3.088E-01	5.061E-01	6.974E-01	7.783E-01
Search	2.217E-01	3.083E-01	5.057E-01	6.970E-01	7.778E-01

$\hat{p}$ (Median)					
Methods	**p = 0.0**	**p = 0.25**	**p = 0.5**	**p = 0.75**	**p = 1.0**
Modified EM	0.15603	0.29834	0.49321	0.71924	0.84351
Sim. Annealing	0.15605	0.29835	0.49327	0.71924	0.84354
Search	0.15700	0.29850	0.49300	0.71950	0.84400

Fig 1: Distribution of $\hat{p}$ under different values of p.

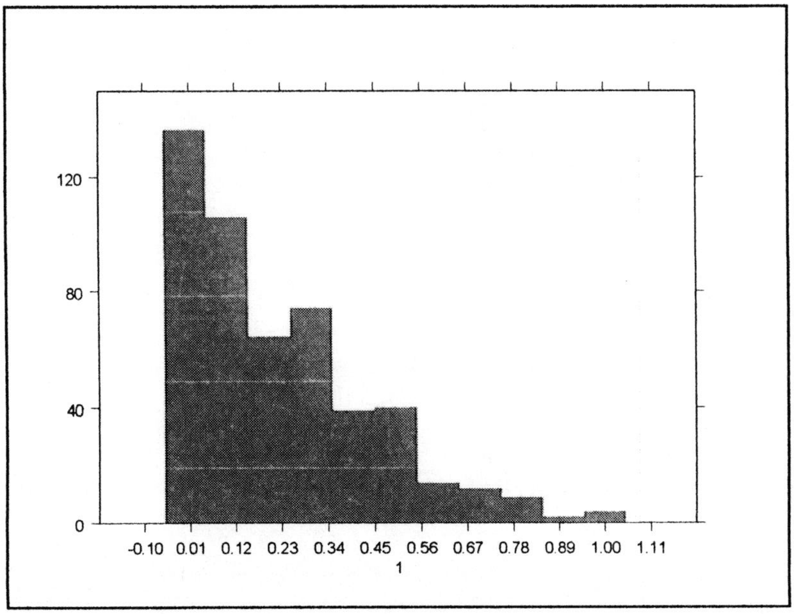

P=0. (Original)

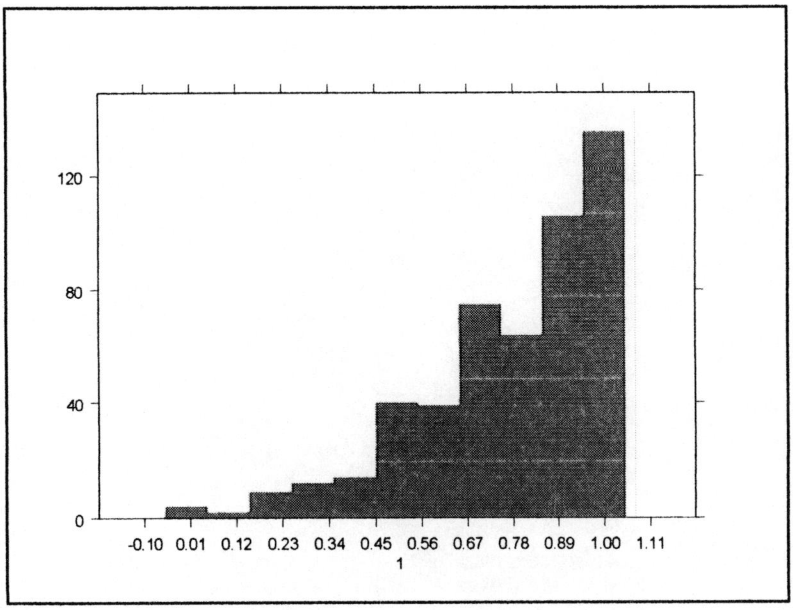

P=1. (Length -biased)

Fig 2: Distribution of $\hat{p}$ under different values of p.

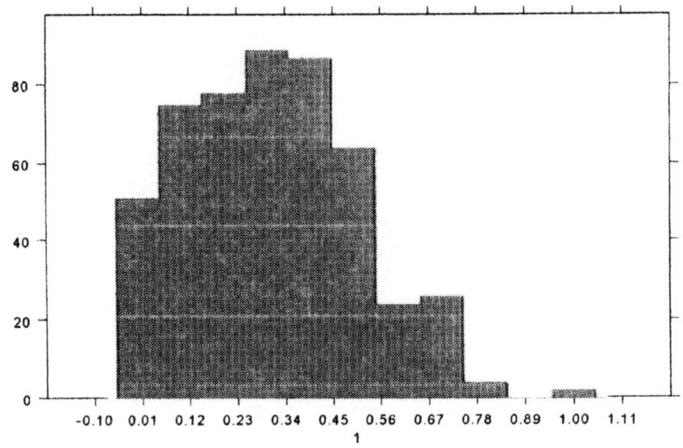

P=0.25

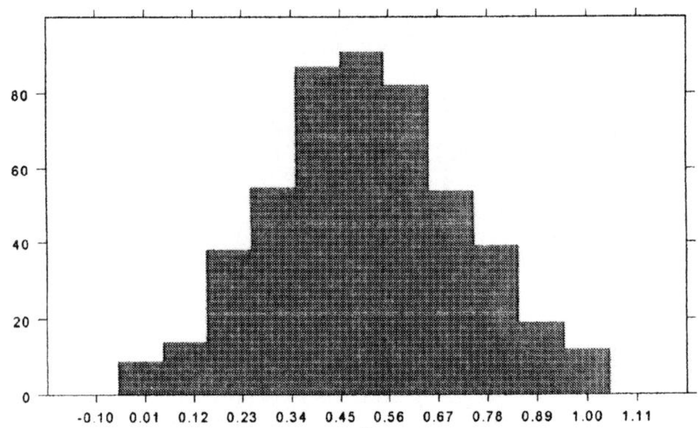

P=0.50

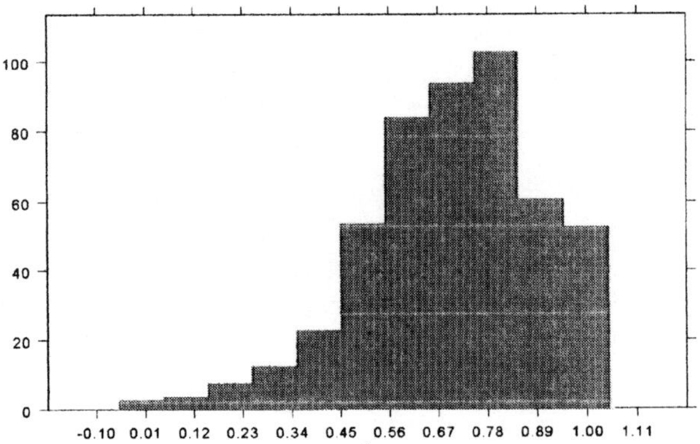

P=0.75

REFERENCES

Akman, O. (1994). "On the Reliability Studies of a Weighted Inverse Gaussian Model" Ph.D. Thesis, University of Maine.

Akman, O., Sansgiry, P. (1999). "On the Estimation of Reliability Based on Mixture Inverse Gaussian Distributions" *Applied Statistical Science* **4**, 221-228.

Bayarri, M.J. and DeGroot, M.H. (1989). "Information in Selection Models", *Probability and Bayesian Statistics*, (R. Viertl, ed.) New York: Plenum Press.

Day, N. E. and Walter, S. D. (1983). "Estimation Of The Duration Of A Pre-Clinical Disease State Using Screening Data " *American Journal Of Epidemiology*, **18 (6)** , 865-886.

Dempster, A. P., Laird, N. M. and Rubin, D. B. (1977). "Maximum Likelihood from Incomplete Data via the EM Algorithm (with discussion)" *J. R. Statist. Soc. B*, **39**, 1-38.

Frazer, D.A.S.(1966). "Sufficiency for Selection Models", *Sankhyā*, **29**, 329-334.

Frazer, D.A.S. (1952). "Sufficient Statistics and Selection Depending on the Parameters", *Annals. of Math. Stat.*, **23**,417-425.

Goffe,W.L. Ferrier, G.D. and Rogers, J. (1994). "Global Optimization of Statistical Functions with Simulated Annealing" , *Journal of Econometrics*, **60,(1/2)**, Jan./Feb. , 65-100.

Gupta, R.C., Akman, O. (1995). "On the Reliability Studies of a Weighted Inverse Gaussian Model", *J. Statistical Planning and Inference*, **48**, 68-93.

Gupta, R.C. and Akman, O. (1998). "Statistical Inference based on the Length-Biased Data for the Inverse Gaussian Distribution",*Statistics*, **31** , 325-337.

Rao, C.R. (1977). "A Natural Example of a Weighted Distribution", *Ameri. Stat.*, **31**, 24-26.

Zelen, M.,(1974). "Length-Biased Sampling and Biomedical Problems", *Biometric Society Meeting* , Dallas, Texas.

Zelen, M., Feinleib, M. (1969). "On the Theory of Screening for Chronic Diseases", *Biometrika*, **56**, 601-614.

On Evaluation of the Mean Service Cycle Time in Tandem Queueing Systems

N.K. Krivulin* V.B. Nevzorov†

Abstract

The problem of exact evaluation of the mean service cycle time in tandem systems of single-server queues with both infinite and finite buffers is considered. It is assumed that the interarrival and service times of customers form sequences of independent and identically distributed random variables with known mean values. We start with tandem queues with infinite buffers, and show that under the above assumptions, the mean cycle time exists. Furthermore, if the random variables which represent interarrival and service times have finite variance, the mean cycle time can be calculated as the maximum out from the mean values of these variables. Finally, obtained results are extended to evaluation of the mean cycle time in particular tandem systems with finite buffers and blocking.

Keywords: tandem queueing systems, mean cycle time, recursive equations, independent random variables, bounds on the mean value

1 Introduction

We consider tandem systems of single-server queues with both infinite and finite buffers. The interarrival and service times of customers are assumed to form sequences of independent and identically distributed random variables. Given the mean values of interarrival and service times, we are interested in evaluating the mean cycle time of the system. In what follows, the mean cycle time is used in reference to the mean value of the time interval between two successive departures of customers from the system as the number of customers tends to infinity. The inverse of the mean cycle time, which implies the mean number of customers leaving the system per unit time, is referred to as system throughput.

Among other characteristics including, in particular, the mean waiting time of customers, both the mean cycle time and the throughput present system

---

*School of Management, St.Petersburg State University, Dekabristov Lane 16, 199155 St.Petersburg, Russia, e-mail: Nikolai.Krivulin@pobox.spbu.ru. The work was partially supported by the Russian Foundation of Basic Research, Grant #00-01-00760

†Faculty of Mathematics and Mechanics, St.Petersburg State University, Bibliotechnaya Sq. 2, Petrodvorets, 198904 St.Petersburg, Russia, e-mail: Valery.Nevzorov@pobox.spbu.ru. The work was partially supported by the Russian Foundation of Basic Research, Grant #99-01-00732

performance measures commonly used in the analysis of queues. Note that the problem of evaluating the mean waiting time is known as rather difficult; it allows for the exact solution in an explicit form only for particular queueing systems under certain restrictions on customer arrival and service processes. One can find an overview of related results in [1] (see, also, [2] for more recent results and references).

In many cases, the mean cycle time can be calculated exactly under fairly general assumptions. As an illustration, one can consider results obtained in [3, 4] in the context of investigation of stability of queueing systems. It has been shown in [4] that for a general single-server system with infinite buffer capacity, regardless of whether it is stable or not, the mean cycle time can be calculated as the maximum out from the mean values of customer interarrival and service times. Moreover, if a tandem system of queues with infinite buffers is stable, the intensities of both customer arrival and departure processes coincide, and therefore, the mean cycle time is equal to the mean interarrival time of customers.

For more complicated queueing systems including tandem queues with finite buffers and blocking, and fork-join networks there are some techniques which allow one to derive bounds on the mean cycle time. Specifically, an efficient approach which relies on results of the theory of large deviations as well as on the Perron-Frobenius spectral theory has been proposed in [5, 6]. As another example, one can consider simple bounds in [7, 8], obtained by using an approach essentially based on pure algebraic manipulations together with application of bounds on extreme values, derived in [9, 10].

In this paper, we first give quite general conditions for the mean cycle time in tandem queueing systems with infinite buffers to exist, and show how it can be calculated through the mean values of the interarrival and service times of customers. The obtained results are then extended to evaluation of the mean cycle time in particular tandem systems with finite buffers, which operate under the manufacturing and communication blocking rules.

As the starting point, we take obvious recursive equations which describe tandem system dynamics, and then examine their related explicit solution. Our approach is based on simple algebraic manipulations with the solution, combined with some classical results providing bounds on the mean value of maxima of independent and identically distributed random variables. The approach does not involve taking account of stability of the entire system, and therefore, offers a unified way of examining both stable and unstable systems.

The rest of the paper is organized as follows. In Section 2, we introduce some notations, and consider recursive equations describing the dynamics of tandem queueing systems with infinite buffers. Section 3 presents preliminary results including an existence theorem and some inequalities. Our main result which provides general existence conditions and a simple expression for calculating the mean cycle time is included in Section 4. The obtained results are then applied to the examination of the mean cycle time in particular tandem systems with finite buffers in Section 5. Finally, Section 6 offers some concluding remarks and discussion.

2 Tandem Queues with Infinite Buffers

We consider a series of M single-server queues with infinite buffers and customers of a single class. Each customer that arrives into the system is initially placed in the buffer at the 1st server and then has to pass through all the queues one after the other. Upon the completion of his service at server i, the customer is instantaneously transferred to queue $i+1$, $i = 1, \ldots, M-1$, and occupies the $(i+1)^{\text{st}}$ server provided that it is free. If the customer finds this server busy, he is placed in its buffer and has to wait until the service of all his predecessors is completed.

Denote the time between the arrivals of nth customer and his predecessor by τ_{0n}, and the service time of the nth customer at server i by τ_{in}, $i = 1, \ldots, M$, $n = 1, 2, \ldots$. Furthermore, let $D_0(n)$ be the nth arrival epoch to the system, and $D_i(n)$ be the nth departure epoch from the ith server. We assume that for each i, $i = 0, 1, \ldots, M$, the sequence $\{\tau_{in} | \, n = 1, 2, \ldots\}$ consists of nonnegative random variables (r.v.'s).

With the condition that the tandem queueing system starts operating at time zero, and it is free of customers at the initial time, we put $D_i(0) = 0$ for all $i = 0, \ldots, M$. The recursive equations representing the system dynamics can readily be written as

$$\begin{aligned} D_0(n) &= D_0(n-1) + \tau_{0n}, \\ D_m(n) &= \max(D_{m-1}(n), D_m(n-1)) + \tau_{mn}, \quad m = 1, \ldots, M, \end{aligned}$$

for all $n = 1, 2, \ldots$.

The above recursions can be resolved to get

$$D_m(n) = \max_{1 \le k_1 \le \cdots \le k_m \le n} \left\{ \sum_{j=1}^{k_1} \tau_{0j} + \sum_{j=k_1}^{k_2} \tau_{1j} + \cdots + \sum_{j=k_m}^{n} \tau_{mj} \right\} \tag{1}$$

for all $m = 1, \ldots, M$.

We consider the evolution of the system as a sequence of service cycles: the 1st cycle starts at the initial time, and it is terminated as soon as the Mth server completes its 1st service, the 2nd cycle is terminated as soon as this server completes its 2nd service, and so on. Clearly, the completion time of the nth cycle can be represented as $D_M(n)$.

In many applications, one is interested in evaluating the mean service cycle time of the tandem system, which can also be treated as the mean interdeparture time of customers from the system. It is defined as

$$\gamma = \lim_{n \to \infty} \frac{1}{n} D_M(n) \tag{2}$$

provided that the above limit exists. The system throughput π presents another performance measure of interest, which is calculated as the inverse of the mean cycle time; that is, $\pi = 1/\gamma$.

3 Preliminary Results

In order to examine the existence of the mean cycle time for the tandem queueing system, we will apply the next classical theorem which has been proved in [11].

Theorem 1 *Let* $\{\zeta_{ln} \mid l, n = 0, 1, \ldots; l < n\}$ *be a family of r.v.'s which satisfy the following properties:*

Subadditivity: $\zeta_{ln} \leq \zeta_{lk} + \zeta_{kn}$ *for all* $l < k < n$;

Stationarity: the joint distributions are the same for both families $\{\zeta_{ln} \mid l < n\}$ *and* $\{\zeta_{l+1,n+1} \mid l < n\}$;

Boundedness: for all $n = 1, 2, \ldots$, *there exists* $\mathbb{E}[\zeta_{0n}] \geq -cn$ *for some positive constant* c.

Then there exists a constant γ, *such that it holds*

1. $\lim_{n\to\infty} \zeta_{0n}/n = \gamma$ *with probability one (w.p.1),*
2. $\lim_{n\to\infty} \mathbb{E}[\zeta_{0n}]/n = \gamma$.

Let us now consider some useful inequalities which will be exploited in evaluation of the mean cycle time in the next section. In what follows, we assume $\xi_1, \ldots, \xi_n$ to be independent r.v.'s.

We start with a classical result from [12], providing an upper bound on the mean value of the maximum of cumulative sums

$$\zeta_k = \xi_1 + \cdots + \xi_k$$

of independent r.v.'s with zero means.

Lemma 2 *If* $\mathbb{E}[\xi_k] = 0$, *and* $\mathbb{E}|\xi_k|^p < \infty$ *for some* $p > 1$, $k = 1, \ldots, n$, *then it holds*

$$\mathbb{E}\left[\max_{1\leq k\leq n} |\zeta_k|\right]^p \leq 2\left(\frac{p}{p-1}\right)^p \mathbb{E}|\zeta_n|^p.$$

The next inequality has been derived in [13]. Note that it actually remains valid under somewhat weaker conditions than that of independence between the r.v.'s $\xi_1, \ldots, \xi_n$.

Lemma 3 *If* $\mathbb{E}[\xi_k] = 0$, *and* $\mathbb{E}|\xi_k|^p < \infty$ *for some* p, $1 \leq p \leq 2$, $k = 1, \ldots, n$, *then it holds*

$$\mathbb{E}|\zeta_n|^p \leq \left(2 - \frac{1}{n}\right) \sum_{k=1}^n \mathbb{E}|\xi_k|^p.$$

With Lemmas 2 and 3, one can prove the following statement.

Lemma 4 *If* $\mathbb{E}[\xi_k] = 0$, *and* $\mathbb{E}[\xi_k^2] < \infty$, $k = 1, \ldots, n$, *then it holds*

$$\mathbb{E}\left[\max_{1\leq k\leq n} \zeta_k\right] \leq 2\sqrt{\frac{2(2n-1)}{n}} \left(\sum_{k=1}^n \mathbb{E}[\xi_k^2]\right)^{1/2}.$$

Proof. First note that

$$\mathbb{E}\left[\max_{1\le k\le n}\zeta_k\right]\le\mathbb{E}\left[\max_{1\le k\le n}|\zeta_k|\right]\le\left(\mathbb{E}\left[\max_{1\le k\le n}|\zeta_k|\right]^2\right)^{1/2}.$$

By applying Lemma 2 with $p=2$, and then Lemma 3, we get

$$\mathbb{E}\left[\max_{1\le k\le n}|\zeta_k|\right]^2\le 8\mathbb{E}[\zeta_n^2]\le 8\left(2-\frac{1}{n}\right)\sum_{k=1}^{n}\mathbb{E}[\xi_k^2].$$

Finally, extracting square root leads us to the desired result. □

Now suppose that $\xi_1,\dots,\xi_n$ present independent and identically distributed (i.i.d.) r.v.'s. With this condition, in particular, the inequality in Lemma 4 takes the form

$$\mathbb{E}\left[\max_{1\le k\le n}\zeta_k\right]\le 2\sqrt{2(2n-1)\mathbb{E}[\xi_1^2]}.$$

The next result obtained in [9, 10] offers an upper bound for the expected value of maximum of i.i.d. r.v.'s.

Lemma 5 *If* $\mathbb{E}[\xi_1]<\infty$ *and* $\mathbb{D}[\xi_1]<\infty$, *then it holds*

$$\mathbb{E}\left[\max_{1\le k\le n}\xi_k\right]\le\mathbb{E}[\xi_1]+\frac{n-1}{\sqrt{2n-1}}\sqrt{\mathbb{D}[\xi_1]}.$$

Assuming $\xi_1,\dots,\xi_n$ to be i.i.d. r.v.'s, let us introduce the notation

$$\zeta_{lk}=\xi_l+\xi_{l+1}+\dots+\xi_k$$

with $1\le l\le k\le n$, and consider the following statement.

Lemma 6 *If* $\mathbb{E}[\xi_1]=a\le 0$, *and* $\mathbb{D}[\xi_1]<\infty$, *then it holds*

$$\mathbb{E}\left[\max_{1\le l\le k\le n}\zeta_{lk}\right]\le\mathbb{E}[\xi_1]+\left(4\sqrt{2(2n-1)}+\frac{n-1}{\sqrt{2n-1}}\right)\sqrt{\mathbb{D}[\xi_1]}.$$

Proof. Simple algebraic manipulations give

$$\begin{aligned}\max_{1\le l\le k\le n}\zeta_{lk}&=\max_{1\le l\le k\le n}\left\{\sum_{i=1}^{k}\xi_i+\sum_{i=1}^{l-1}(-\xi_i)\right\}\\&\le\max_{1\le l\le k\le n}\left\{\sum_{i=1}^{k}(\xi_i-a)+\sum_{i=1}^{l}(-\xi_i+a)+\xi_l\right\}\\&\le\max_{1\le k\le n}\sum_{i=1}^{k}(\xi_i-a)+\max_{1\le k\le n}\sum_{i=1}^{k}(-\xi_i+a)+\max_{1\le k\le n}\xi_k.\end{aligned}$$

Proceeding to expectation, with $\mathbb{E}(\xi_1 - a)^2 = \mathbb{D}[\xi_1]$, we have from Lemmas 4 and 5

$$\begin{aligned}\mathbb{E}\left[\max_{1\le l\le k\le n} \zeta_{lk}\right] &\le 4\sqrt{2(2n-1)\mathbb{D}[\xi_1]} + \mathbb{E}[\xi_1] + \frac{n-1}{\sqrt{2n-1}}\sqrt{\mathbb{D}[\xi_1]} \\ &= \mathbb{E}[\xi_1] + \left(4\sqrt{2(2n-1)} + \frac{n-1}{\sqrt{2n-1}}\right)\sqrt{\mathbb{D}[\xi_1]}.\end{aligned}$$

□

4 Exact Evaluation of the Mean Cycle Time

We are now in a position to prove our main result which can be formulated as follows.

Theorem 7 *Suppose that $\{\tau_{in} | \ n = 1, 2, \ldots\}$, $i = 0, 1, \ldots, M$, are mutually independent sequences of i.i.d. r.v.'s with $0 \le \mathbb{E}[\tau_{i1}] < \infty$.*

Then the limit at (2) exists w.p.1, and if $\mathbb{D}[\tau_{i1}] < \infty$, it is given by

$$\gamma = \max_{0\le i\le M} \mathbb{E}[\tau_{i1}]. \tag{3}$$

Proof. First, we have to verify the existence of the limit at (2). In order to apply Theorem 1, let us denote

$$\zeta_{ln} = \max_{l<k_1\le\cdots\le k_M\le n} \left\{ \sum_{j=l+1}^{k_1} \tau_{0j} + \sum_{j=k_1}^{k_2} \tau_{2j} + \cdots + \sum_{j=k_M}^{n} \tau_{Mj} \right\} \tag{4}$$

for each l, n, $0 \le l < n$, and note that we can now write

$$D_M(n) = \zeta_{0n}.$$

With simple algebraic manipulations, it is not difficult to verify that the family $\{\zeta_{ln} | \ l < n\}$ defined by (4) is subadditive. Since $\tau_{i1}, \tau_{i2}, \ldots$, are i.i.d. r.v.'s for each $i = 0, 1, \ldots, M$, the family also possesses the stationarity property. Finally, boundedness follows from the condition $0 \le \mathbb{E}[\tau_{i1}] < \infty$ which immediately results in $\mathbb{E}[\zeta_{0n}] = \mathbb{E}[D_M(n)] \ge 0$.

Therefore, one can apply Theorem 1 so as to conclude that the limit at (2) exists w.p.1, and it can be calculated as

$$\gamma = \lim_{n\to\infty} \frac{1}{n}\mathbb{E}[D_M(n)].$$

Suppose that the maximum at (3) is achieved at some $i = m$. Consider the completion time $D_M(n)$ and represent it in the form

$$D_M(n) = \max_{1\le k_1\le\cdots\le k_M\le n} \left\{ \sum_{j=1}^{k_1} \tau_{0j} + \sum_{j=k_1}^{k_2} \tau_{1j} + \cdots + \sum_{j=k_M}^{n} \tau_{Mj} \right\} = \sum_{j=1}^{n} \tau_{mj} + \mu,$$

where

$$\mu = \max_{1\le k_1\le\cdots\le k_M\le n}\Bigg\{\sum_{j=1}^{k_1}(\tau_{0j}-\tau_{mj}) + \sum_{j=k_1}^{k_2}(\tau_{1j}-\tau_{mj})+\cdots+\sum_{j=k_M}^{n}(\tau_{Mj}-\tau_{mj}) + \tau_{mk_1}+\cdots+\tau_{mk_M}\Bigg\}. \tag{5}$$

Now we can write

$$\frac{1}{n}\mathbb{E}[D_M(n)] = \mathbb{E}[\tau_{m1}] + \frac{1}{n}\mathbb{E}[\mu].$$

Let us examine the expected value $\mathbb{E}[\mu]$. With $k_1 = \cdots = k_m = 1$, and $k_{m+1} = \cdots = k_M = n$, we have from (5)

$$\mu \ge \tau_{01}+\tau_{11}+\cdots+\tau_{m-1,1}+\tau_{m+1,n}+\cdots+\tau_{Mn}\ge 0,$$

and so $\mathbb{E}[\mu]\ge 0$.

On the other hand, simple algebra gives us an obvious upper bound for μ in the form

$$\begin{aligned}\mu \le\ & \max_{1\le k_1\le n}\sum_{j=1}^{k_1}(\tau_{0j}-\tau_{mj}) \\ & + \max_{1\le k_1\le k_2\le n}\sum_{j=k_1}^{k_2}(\tau_{1j}-\tau_{mj})+\cdots+\max_{1\le k_M\le n}\sum_{j=k_M}^{n}(\tau_{Mj}-\tau_{mj}) \\ & + M\max(\tau_{m1},\ldots,\tau_{mn}).\end{aligned}$$

With the condition that $\mathbb{E}(\tau_{i1}-\tau_{m1})\le 0$ for all $i = 0,1,\ldots,M$, one can apply Lemma 6 to the first $M+1$ terms on the right-hand side, and then Lemma 5 to the last one so as to get

$$\begin{aligned}\mathbb{E}[\mu] \le\ & \sum_{i=0}^{M}\left(\mathbb{E}(\tau_{i1}-\tau_{m1}) + \left(4\sqrt{2(2n-1)}+\frac{n-1}{\sqrt{2n-1}}\right)\sqrt{\mathbb{D}(\tau_{i1}-\tau_{m1})}\right) \\ & + M\left(\mathbb{E}[\tau_{m1}]+\frac{n-1}{\sqrt{2n-1}}\sqrt{\mathbb{D}[\tau_{m1}]}\right) \\ =\ & \sum_{\substack{i=0\\ i\ne m}}^{M}\mathbb{E}[\tau_{i1}] + \left(4\sqrt{2(2n-1)}+\frac{n-1}{\sqrt{2n-1}}\right)\sum_{\substack{i=0\\ i\ne m}}^{M}\sqrt{\mathbb{D}(\tau_{i1}-\tau_{m1})} \\ & + M\frac{n-1}{\sqrt{2n-1}}\sqrt{\mathbb{D}[\tau_{m1}]},\end{aligned}$$

and therefore,

$$\mathbb{E}[\mu] \leq \sum_{\substack{i=0 \\ i \neq m}}^{M} \mathbb{E}[\tau_{i1}] + O(\sqrt{n}).$$

Finally, we have the double inequality

$$\mathbb{E}[\tau_{m1}] \leq \frac{1}{n}\mathbb{E}[D_M(n)] \leq \mathbb{E}[\tau_{m1}] + \frac{1}{n}\sum_{\substack{i=0 \\ i \neq m}}^{M} \mathbb{E}[\tau_{i1}] + \frac{O(\sqrt{n})}{n},$$

and with $n \to \infty$, immediately arrive at (5). □

Corollary 8 *Under the same conditions as in Theorem 7, if at least one of the expectations* $\mathbb{E}[\tau_{i1}]$, $i = 0, 1, \ldots, M$, *is positive, then it holds*

$$\pi = \left(\max_{0 \leq i \leq M} \mathbb{E}[\tau_{i1}] \right)^{-1}.$$

5 Tandem Queues with Finite Buffers

In this section, we show how the above approach can be applied to the analysis of tandem systems which include queues with finite buffers. Because of limited buffer capacity, servers in the systems may be blocked according to one of the blocking rules [14]. Below we present examples of systems with manufacturing blocking and communication blocking which are most commonly encountered in practice.

Let us consider a system which consists of two queues in tandem. Suppose that the buffer at the first server is infinite, while that at the second server is finite. The customers arriving to the system have to pass through the queues consecutively, and then leave the system.

First we suppose that the system operates under the manufacturing blocking rule. With this type of blocking, if upon completion of a service, the first server sees the buffer of the second one is full, it cannot be freed and has to remain busy until the second server completes its current service to provide a free space in its buffer.

For simplicity, let us assume the finite buffer to have capacity 0. With the notations introduced above, one can represent the dynamics of the system by the equations

$$\begin{aligned} D_0(n) &= D_0(n-1) + \tau_{0n}, \\ D_1(n) &= \max(\max(D_0(n), D_1(n-1)) + \tau_{1n}, D_2(n-1)), \\ D_2(n) &= \max(D_1(n), D_2(n-1)) + \tau_{2n}, \end{aligned} \tag{6}$$

for all $n = 1, 2, \ldots$.

Note that from the second equation, we have $D_1(n) \geq D_2(n-1)$, and therefore, the third equation can be reduced to

$$D_2(n) = D_1(n) + \tau_{2n}.$$

Clearly, under appropriate conditions, both $\mathbb{E}[D_1(n)]/n$ and $\mathbb{E}[D_2(n)]/n$ have a common limit γ as n tends to ∞, which can be considered as the mean cycle time of the system.

By resolving the recursive equations, we get

$$D_1(n) = \max_{1 \leq k \leq n} \left\{ \sum_{j=1}^{k} \tau_{0j} + \tau_{1k} + \sum_{j=k}^{n-1} \max(\tau_{1,j+1}, \tau_{2j}) \right\}.$$

As it is easy to verify, $D_1(n)$ satisfies the double inequality

$$L(n) - \max(\tau_{1,n+1}, \tau_{2n}) \leq D_1(n) \leq U(n), \tag{7}$$

where

$$\begin{aligned} L(n) &= \max_{1 \leq k \leq n} \left\{ \sum_{j=1}^{k} \tau_{0j} + \sum_{j=k}^{n} \max(\tau_{1,j+1}, \tau_{2j}) \right\}, \\ U(n) &= \max_{1 \leq k \leq n} \left\{ \sum_{j=1}^{k} \tau_{0j} + \sum_{j=k}^{n} \max(\tau_{1j}, \tau_{2,j-1}) \right\}. \end{aligned}$$

Taking into account that both $L(n)$ and $U(n)$ actually have the form of (1), one can see that, under the same conditions as in Theorem 7, it holds

$$\lim_{n\to\infty} \frac{1}{n}\mathbb{E}[L(n)] = \lim_{n\to\infty} \frac{1}{n}\mathbb{E}[U(n)] = \max(\mathbb{E}[\tau_{01}], \mathbb{E}\max(\tau_{11}, \tau_{21})).$$

Finally, proceeding to mean value in both sides of (7), divided by n, we conclude that the mean cycle time is given by

$$\gamma = \max(\mathbb{E}[\tau_{01}], \mathbb{E}\max(\tau_{11}, \tau_{21})).$$

Let us now assume the system to follow the communication blocking rule. This type of blocking requires the first server not to initiate service of a customer if the buffer of the second server is completed. With the finite buffer having capacity 0, the system dynamics is described by the same recursions as above, except for equation (6) which now takes the form

$$D_1(n) = \max(D_0(n), D_1(n-1), D_2(n-1)) + \tau_{1n}.$$

Resolving the recursive equations leads us to the expression

$$D_2(n) = \max_{1 \leq k \leq n} \left\{ \sum_{j=1}^{k} \tau_{0j} + \sum_{j=k}^{n} (\tau_{1j} + \tau_{2j}) \right\}.$$

Under the same conditions as in Theorem 7, we get the mean cycle time

$$\gamma = \max(\mathbb{E}[\tau_{01}], \mathbb{E}[\tau_{11}] + \mathbb{E}[\tau_{21}]).$$

6 Concluding Remarks

First note that with similar arguments as used in the proof of Theorem 7 and related lemmas, one can verify that the statement of the theorem remains valid with the condition $\mathbb{E}[\tau_{i1}^{\alpha}] < \infty$ for some $\alpha > 1$, instead of $\mathbb{D}[\tau_{i1}] < \infty$.

Furthermore, the proof of the theorem does not actually require that for each n, r.v.'s τ_{in} with $i = 0, 1, \ldots, M$, be independent. This allows one to apply obtained results to tandem queueing systems with dependence for each customer between his interarrival and service times, including tandem queues with identical service times at each server.

Theorem 1 actually offers more general existence conditions for the mean cycle time, which imply stationarity of the sequence $\{(\tau_{0n}, \tau_{1n}, \ldots \tau_{Mn}) | \ n = 1, 2, \ldots\}$ in place of independence conditions in Theorem 7.

Let us consider recursive equations describing the dynamics of the tandem queues above, and note that the symbol τ_{in} can be thought of as the nth service time at server i rather than the service time of the nth customer at the server. Since in this case, the order in which customers are selected from a queue for service is of no concern, the equation also describes the dynamics of systems with any queueing disciplines not permitting the preempting of service, and therefore, Theorem 7 can trivially be extended to such systems.

Finally note that the proof of the theorem provides us with bounds on $\mathbb{E}[D_M(n)]$, as well as an upper bound of order $n^{-1/2}$ for the convergence rate of $\mathbb{E}[D_M(n)]/n$ to the mean cycle time as n tends to ∞.

References

[1] Kleinrock, L. *Queueing Systems, Volume II: Computer Applications.* John Wiley, New York, 1976.

[2] Disney, R.L. and König, D. Queueing Networks: A Survey of Their Random Processes. *SIAM Review* 27 (1985), no. 3, 335–403.

[3] Sacks, J. Ergodicity of Queues in Series. *Ann. Math. Statist.* 31 (1960), 579–588.

[4] Loynes, R.M. The Stability of a Queue with Non-independent Inter-arrival and Service Times. *Proc. Camb. Phil. Soc.* 58 (1962), no. 3, 497–520.

[5] Baccelli, F. and Konstantopoulos, P. Estimates of Cycle Times in Stochastic Petri Nets. *Applied Stochastic Analysis*, Lecture Notes in Control and Information Sciences, 177. Springer, New York, 1991, 1–20.

[6] Glasserman, P. and Yao, D.D. Stochastic Vector Difference Equations With Stationary Coefficients. *J. Appl. Probab.* 32 (1995), 851–866.

[7] Krivulin, N.K. Bounds on Mean Cycle Time in Acyclic Fork-Join Queueing Networks, in *Proc. Intern. Workshop on Discrete Event Systems, Cagliari, Italy, August 26-28, 1998.* IEE, London, 1998, 469–474.

[8] Krivulin, N.K. Monotonicity Properties and Simple Bounds on the Mean Cycle Time in Acyclic Fork-Join Queueing Networks, in *Recent Advances in Information Science and Technology*, N. Mastorakis, Ed., World Scientific, 1998, 147-152.

[9] Gumbel, E.J. The Maxima of the Mean Largest Value and of the Range *Ann. Math. Statist.* 25 (1954), 76–84.

[10] Hartly, H.O. and David, H.A. Universal Bounds for Mean Range and Extreme Observations. *Ann. Math. Statist.* 25 (1954), 85–99.

[11] Kingman, J.F.C. Subadditive Ergodic Theory. *Ann. Probab.* 1 (1973), 883–909.

[12] Marcinkiewicz, J. and Zigmund, A. Sur les Fonctions Indépendantes. *Fund. Math.* 29 (1937), 60–90.

[13] von Bahr, B. and Esseen, C.-G. Inequalities for the rth Absolute Moment of a Sum of Random Variables, $1 \leq r \leq 2$. *Ann. Math. Statist.* 36 (1965), 299–303.

[14] Krivulin, N.K. A Max-algebra Approach to Modeling and Simulation of Tandem Queueing Systems. *Math. Comput. Modelling* 22 (1995), no. 3, 25–37.

AN ALGORITHM MODELLING STATIONARY LOAD CONDITION OF A NODE IN A TRANSPUTER NETWORK

G.MISHKOY B.RYBAKIN D.MISHKOY

Institute of Mathematics and Informatics of the Academy of Science of Moldova,
Free International University of Moldova
Academy 5, Kishinev, Moldova
e-mail: gmishkoy@hotmail.com
e-mail: rybakin@math.md

SUMMARY

The growing interest in creation and utilization of multiprocessor systems gives rise to the necessity to investigate load balance of nodes in such a systems. This work concerns algorithms implementing a mathematical model of stationary operation mode of multiprocessor system.

Key Words and Phrases: Parallel computing, loading of a node, priority queues, busy period,non-zero switching, Laplace-Stieltjes transform.

1. INTRODUCTION

When organizing parallel computations in transputer networks much attention is paid to load control algorithms. Methods and tools of queueing theory are successfully used for this purpose as effective means to model and study these algorithms. Stochastic nature of the processes, of their handling, and of interaction of information flows, which are, as a rule, inhomogeneous, can result in overload of some parts or nodes of the network, whereas another parts of the system may be idle.

There could be proposed various algorithms to control information flows in the transputer system. They could be divided into the three following classes:

- Algorithms controlling the nodes in their stationary load state;
- Dynamic load control algorithms;
- Extra-system algorithms minimizing cost function for waiting a communication channel and/or processor to be freed.

This article discusses only algorithms of the first class.

The main idea of all such algorithms is based on a formalization and analysis of processes having place in transputer systems by means of queueing theory, more precisely, by means of one of its rapidly growing sections, namely, priority system theory. Let's note by the way that the approach based on the queueing theory was successfully used by a number of authors to analyze process evolution in complex systems such as information and computation ones. See, for example, monographs [1,2]. The remarkable examples of such analysis are work [3], investigating a class of dynamic load balancing algorithms which uses probabilities of task transference in a $M|M|1$ queueing system, and article [4] which uses $M|G|1$ queueing system to examine scheduling procedures in a single-processor real-time computer system[1].

Note, that both in articles mentioned above, and in a number of others, requests in the input flow are assumed homogeneous. It seems to us that in many cases heterogeneity (preemptive character) of the request flow is more natural and that it describes more adequately actual processes taking place in a transputer network.

In some other cases, introducing various preemptive service strategies, one can succeed in redistribution of arriving tasks, which results in raising the efficiency of computing process and in optimization of some its parameters.

We assume, that the input task flow divides into an arbitrary number r, of homogeneous task classes. It should be noted, that there always exists a possibility to divide a Poisson flow into several flows applying on of the screening operations to it (see the monograph [5]). The resulting flows are also Poisson ones, while their priorities could be found in the case the parameters of the original flow is known.

The service time B_i by a host i, i=1,...,r, is assumed to be a stochastic value with its distribution function $B_i(x)$ of an arbitrary kind. Homogeneous request classes obtained as a result of division of the original request flow are called below urgency classes or priority classes. A value meaning the level of urgency or the priority is assigned to each class. We assume that the classes are enumerated in a descending order of their priorities. To find the order in which requests belonging to different classes are chosen for service, some priority rules are then defined , they establish a measure of preemption of requests belonging

[1] We would like to remind, that notation $\alpha|\beta|\gamma$, introduced by Kendall to describe briefly a queueing system, means the following: α is input flow characteristic, β is service characteristic, while γ is the number of processors.$\alpha = M$ means that the input flow is supposed to be a Poisson one; $\beta = M$ means that the service time has exponential distribution; $\beta = G$ means that the service time has an arbitrary distribution; $\gamma = 1$ means that the system has a single processor handling the requests in sequence.

to one class with respect to requests of another one. The order of choosing the requests belonging to the same class is defined also. For more details about the question see, for example, monographs [6], [7] or other literature on the priority systems theory.

In already mentioned Kendall notation this queuing system is designated as $M_r|G_r|1$. In the case $r = 1$ (a single homogeneous input flow or a single priority class), the queueing system would be, naturally, of $M|G|1$ kind. The more particular case, when $G = M$, i.e. it is known that the service time is distributed exponentially, the queuing system would be $M|M|1$.

Let us return to the $M_r|G_r|1$ queueing system and make some additional, and, from our point of view, essential assumptions. It is known (see [8], pp.. 84, 85), that the initial step of loading a transputer network is to reset it, as a result every processor in the network switches itself to the state of loading through one of its communication channels. As a result of the reset operation each transputer waits for data arrival via any one of its communication channels. Therefore we assume that at each node there exist one another stochastic process besides the process of service of arriving requests. This additional stochastic process models loss in time for so-called switching, or changing the orientation to the respective priority class. In different situations the concept of switching or orientation can be treated in different ways depending on the formalization of either actual process in the system. For example, there can exist losses in time to transfer and analyses the information amongst load balancing programs, while message routing in the network is performed. We assume that this losses are random variables designated C_{ij}, having their distribution functions $C_{ij}(x), i, j = 1, ..., r; i \neq j$. This notation will be used in the whole article. Indices i,j mean switching of the processor to the tasks of j priority class, provided that the previous task was from i class. The condition $i \neq j$ means that time losses to switch inside the same priority class are equal to zero. That is, time losses exist only when the priority class of requests changes. The fact whether this change is from a lower priority class to a greater one, or vice versa, is unimportant. To simplify mathematical notation in the text below, in the case $r \geq 2$ we assume that distribution functions $C_{ij}(x)$ depend only on index j. That is, we assume, that random variables $C_{ij} = C_j$, and, respectively, their distribution functions $C_{ij}(x) = C_j(x)$. It means, that the time loss to switch the processor from i priority class to j one depends only on j.

Note, that there exist no restrictions about both the kind of distribution functions $C_{ij}(x)$, and the kind of distribution functions $B_i(x)$. This permits us to use complemented in the way described above $M_r|G_r|1$ queueing system in a wide range of modelling.

And another one addition. We assume that as soon as the period of activity of the processor (its busy state time interval) completes, its current switch state (its orientation) is immediately lost. The reset operation occurs instantly. At the beginning of the next period of activity the service starts only after random time interval C_j^0, depending only on the priority class index $j, j = 1, ..., r$. This value may not be obligatory equal to C_j^0. However, although the general case can be examined, we assume that $C_j^0 = C_j \neq 0$, thus slightly simplifying the general model.

In this article the problem of detailed examination of each algorithm belonging to the class introduced above is not considered. Also is not considered the problem of program implementation of above-said algorithms. Although an example of such implementation (in Turbo-Pascal language using object-oriented methodology) is given in one of the subsequent parts of the article.

At this stage of our research we pursue a more global aim, namely, to use analytical studies in the field of a theory of so-called priority systems with orientation [9], [10] and the respective software [11] to create algorithms modelling the node load conditions, and balancing this load, and prove their validity. In each particular case these algorithms implement specific analytical formulae. For example, algorithms controlling the system in its stationary load state implement the necessary stationary condition of semimarkov processes of service. Each algorithm of this class depends on the parameters of input flows, on their priorities, on the given preemption rules to chose the tasks from different priority classes, on time losses to switch to the priority classes, on the distribution of service time, etc. Let us also note, that each algorithm needs to use non-trivial tools both from the field of modern mathematics (such as numerical solution of recurrent functional equations, numerical inversion of Laplace-Stieltjes transformations etc.), and from the field of computer science (problem-oriented approach to the software implementation, fast algorithms of Laplace transformation and of Laplace-Stieltjes transformation etc.) We concern the problems listed above, each of them may be a subject for independent and detailed investigation, only as it is necessary. Mathematical proofs of the statements made above are also not produced here.

2. ELEMENTARY HOMOGENEOUS CASE

Before considering the main results of the work let us analyses two particular classes of the general $M_r|G_r|1$ queueing system with time losses for switching the priority classes. Sections 2 and 3 consider the homogeneous case (without priorities). Namely, in this section the case of $r = 1, C_1 = C_1^0 = 0$ is considered, while in section 3 the case of $r = 1, C_1^0 \neq 0$, $C_1 = 0$ is considered.

Let us examine the following situation: to the processor arrives a homogeneous task flow to be served. In the case the processor is busy servicing any task, a new task arrived in this period of time is queued for service. Let us assume that tasks arrive at random time moments, and that the service time is a random variable. The served task leaves the processor and is not queued. Immediately after the processor is released of the current task service, the next task from the queue is taken for service, and the service process continues. In the case when in the moment the processor is released there is no task in the queue, the processor is idle. The idle state longs until a new task arrives and is immediately taken for service. Let us examine the busy period (period of activity), meaning by this term the period of time starting in the moment a new task arrives at the processor in idle state, and completing in the moment the processor is released, there is no new task to serve, and new idle period starts.

In the situation described above of interest for us are queue length fluctuation and capacity of the processor to serve the input flow. It is obvious that busy period length is a random variable.

An example of some "usual" pattern of busy periods can be represented by the following diagram:

Where  is busy period

 is idle period

The word "usual" taken into quotes is to be made more exact. There could exist the following three typical cases of the busy period pattern:

1. There are no requests. The idle period is infinite. The busy period length is equal to zero.

2. Request arrival rate is greater than service rate. In this case, on the contrary, busy period is (or tends to be) infinite.

3. There exist a balance of task arrival and task service rates. The length of queue (the number of requests waiting the service) varies, i.e. sometimes increases, sometimes decreases. There may exist moments when the length of queue equals to zero. In this case the busy period length is a finite value.

The purpose of this section, and the subsequent ones, is to obtain analytical formulae, on the basis of which algorithms controlling or modelling the system for the three cases listed above. Let us note, that for the system considered in this section, it could be made without any difficulty. The corresponding formulae are simple, quite natural, and usual for the specialist in the field of queueing theory. We pay our attention to them only because that the maters, considered in both this section and the next one, may be regarded as an introduction to a more general and more complex heterogeneous case.

For the purposes of mathematical analysis of the system under consideration one needs to formalize the assertions made in the beginning of this section. Let us assume that the input task flow is a Poisson one with the parameter a. Task service time is assumed to be a random variable with its distribution function $B(x)$. Let us denote by Π the length of busy period.

Let us also denote the distribution function of the stochastic variable Π by $\Pi(x) = P\{\Pi < x\}$, and by $\pi(s)$ and $\beta(s)$ Laplace-Stieltjes transforms of the functions $B(x)$ and $\Pi(x)$ respectively, i.e.

$$\pi(s) = \int_0^\infty e^{-st} d\Pi(t), \beta(s) = \int_0^\infty e^{-st} dB(t),$$

by β_1 and π_1 first moments of this functions respectively:

$$\pi_1 = \int_0^\infty t d\Pi(t), \quad \beta_1 = \int_0^\infty t dB(t)$$

It is known, that the following functional equation:

$$\pi(s) = \beta(s + a - a\pi(s)) \tag{2.1}$$

determines the unique function $\pi(s)$, being Laplace-Stieltjes transform of distribution function $\Pi(t)$, which is analytical when

$$Re \quad s > 0.$$

Distribution function $\Pi(t)$ is a proper one when $a\beta_1 \leq 1$ and an improper one when $a\beta_1 > 1$.

Moreover, in the case

$$\rho = a\beta_1 < 1, \tag{2.2}$$

then $\pi_1 < \infty$, and

$$\pi_1 = \frac{\beta_1}{1 - a\beta_1} \tag{2.3}$$

In the case $a\beta_1 > 1$ the fact that $\Pi(t)$ is an improper distribution function, i.e. $\Pi(+\infty) < 1$, consists in a condition $1 - \Pi(+\infty) > 0$, that is there exist a positive probability that the busy period could be as long as one wants. Thus, in the case condition (2.2) does not hold, the queue length grows infinitely, and the processor is incapable to serve all the arriving requests.

The exact solution for $\Pi(t)$ could be obtained using equation (2.1), although, unfortunately, analytical solution of this equation does not exist except for one or two particular cases. However a numerical approach to the problem was developed. It enables one to solve rather efficiently the equation (2.1) and to obtain the value of $\Pi(s)$ with satisfactory accuracy.

Inverting, again numerically, the equation

$$\pi(s) = \int_0^\infty e^{-st} d\Pi(t), \tag{2.4}$$

it is possible to find and $\Pi(t)$ too. We have developed and investigated effective numerical algorithms solving equations (2.1) and (2.4).

Let us point out the method to obtain (2.3) using (2.1). It can be proved that

$$\pi_1 = -\pi'(0), \tag{2.5}$$

which permits one to obtain easily first moments of distributions if there exist the correspondent functional equations like (2.1).

3. HOMOGENEOUS CASE WITH NON-ZERO LOSSES.

The difference between the queuing system used in this section and the previous one is that each time before busy period starts and the processor is ready to begin the service, an additional segment of time of random length C_1^0 must expire. Its distribution function we denote by $\Gamma(t) = P\{C_1^0 \leq x\}$, while Laplace-Stieltjes transform of this function by $\gamma(s)$. Let's assume $\gamma_1 = \int_0^\infty td\Gamma(t) < \infty$.

Now let us adduce formulae analogous to the (2.1) to (2.3) from the previous section using the same notation. Through $\Pi^R(t)$ we designate the distribution function of the busy period length, while through $\pi^R(s)$ its Laplace-Stieltjes transform. It can be proved, that

$$\pi^R(s) = \gamma(s + a - a\pi(s))\pi(s), \tag{3.1}$$

$$\pi(s) = \beta(s + a - a\pi(s))$$

If

$$\rho = a\beta_1 < 1, \tag{3.2}$$

then

$$\pi_1^R = \frac{\gamma_1 + \beta_1}{1 - a\beta_1}, \tag{3.3}$$

where π_1^R is the average busy period length,

$$\pi_1^R = \int_0^\infty td\Pi^R(t)$$

Let's note, that condition (3.2) is the same as condition (2.2) for the previous queuing system.

4. GENERAL HETEROGENEOUS CASE.

In this section a general basic algorithm is described. It deals with a multi-dimensional (multi-class) task flow, arbitrary service time distribution functions for each priority class, and arbitrary switching time losses distribution functions for each class. The basis of this algorithm is a multidimensional analogue of formula (2.2), deduced as a result of study of a general $M_r|G_r|1$ queuing system with time losses for task class switching.

Let us designate by a_i the parameter of an i-th input Poisson flow, where $i = 1, 2, ..., r, r \geq 2;$;$B_i(x)$ let be the distribution function of service time length for requests from i-th priority class, and $C_j(x)$ be the distribution function of time losses for switching the calculation process to the class j, $j = 1, ..., r$, $i \neq j$. Input flows corresponding to the priority classes are assumed to be enumerated in a descending order of their priorities.

The priority is assumed to be absolute as it concerns both service process and switching. This means the following: if during the service of a particular task another task with a greater priority arrives, then the service process is interrupted; just the same if during the switching period to the particular task another task with a greater priority arrives, the process of switching is interrupted. Various strategies about how the interrupted processes are managed afterwards could be considered.

Let's introduce some more notations: let $\Pi_k(x)$ be the distribution function of busy period length for all the priorities including the k-th one and the greater ones, and $\Pi(x)$ be the distribution function of busy period length. Note [9], that $\Pi_r(x) = \Pi(x)$. Furthermore, let's assume that

$$\sigma_k = a_1 + ... + a_k, \sigma = \sigma_r, \sigma_0 = 0;$$

$$\beta_i(s) = \int_0^\infty e^{-sx} dB_i(x), c_j(s), \pi(s), \pi_k(s)$$

are Laplace-Stieltjes transforms of distribution functions $B_i(x), C_j(x)$, $\Pi(x), \Pi_k(x)$; respectively; $\beta_{k1}, c_{k1}, \pi_{k1}, ...$ are first moments of functions $B_k(x)$, $C_k(x)$, $\Pi_k(x), ..., k = 1, ..., r$ respectively.

The condition of stationary load for the processor is

$$\rho_k = \sum_{i=1}^{k} a_i b_i < 1, \tag{4.1}$$

where

$$b_1 = \frac{\beta_{11} + c_{11}}{1 + a_1 c_{11}}$$

$$b_i = \Phi_1 \dots \Phi_{i-1} \frac{1}{\sigma_{i-1}} [\frac{1}{\beta_i(\sigma_{i-1})} - 1](1 + \sigma_{i-1} c_{i1}),$$

$$\Phi_1 = 1, \Phi_i = 1 + [\sigma_i - \sigma_{i-1}\pi_{i-1}(a_i)]c_{i1}, i = 2, ..., k.$$

Functions $\pi_{i-1}(a_i)$ are found using the following system of recurrent equations:

$$\sigma_k \pi_k(s) = \sigma_{k-1}\pi_{k-1}(s+a_k) + \sigma_{k-1}\{\pi_{k-1}(s+a_k[1-\bar{\pi}_k(s)])-$$

$$-\pi_{k-1}(s+a_k)\}\nu_k(s+a_k[1-\bar{\pi}_k(s)]) + a_k\pi_{kk}(s), \tag{4.2}$$

$$\pi_{kk}(s) = \nu_k(s+a_k[1-\bar{\pi}_k(s)])\bar{\pi}_k(s),$$

$$\bar{\pi}_k(s) = h_k(s+a_k[1-\bar{\pi}_k(s)]),$$

where

$$h_k(s) = \beta_k(s+\sigma_{k-1})\{1-\frac{\sigma_{k-1}}{s+\sigma_{k-1}}[1-\beta_k(s+\sigma_{k-1})]\pi_{k-1}(s)\nu_k(s)\}^{-1}, \tag{4.3}$$

$$\nu_k(s) = c_k(s+\sigma[1-\pi_{k-1}(s)]). \tag{4.4}$$

Here for purposes of definiteness it is assumed, that an interrupted (in consequence of a task with a greater priority arrival) service of a particular task is started afterwards from its very beginning, while interrupted switching starts form the interruption point.

In the case condition (4.1) holds, using the procedure (2.5) we can find from equation (4.2) to (4.4) average values not only for the busy period length, but for a number of other functions.

Namely, if $\rho_k < 1$

$$\sigma_k \pi_{k1} = \frac{\Phi_2 \dots \Phi_k + \rho_{k-1}}{1-\rho_k}, \tag{4.5}$$

$$\bar{\pi}_{k1} = \frac{b_k}{1-\rho_k}, \tag{4.6}$$

$$h_{k1} = \frac{b_k}{1-\rho_{k-1}}, \tag{4.7}$$

$$\nu_{k1} = \frac{\Phi_2 \dots \Phi_k}{1-\rho_{k-1}} c_{k1}. \tag{4.8}$$

For example, (4.7) gives us average total value of the time during which a query belonging to priority class k stays at the processor, (4.8) is an average value of the full time lost for switching to the priority class k.

A particular case of heterogeneous algorithm could be obtained in the case one assumes, that switching occurs instantly (i.e. takes no time). In this case $c_{i1} = 0$, $\phi_i = 1$.

In particular, from formulae (4.5) to (4.8) it follows, that

$$\sigma_k \pi_{k1} = \frac{\rho_k}{1-\rho_k}, h_{k1} = \frac{b_k}{1-\rho_{k-1}},$$

which coincides with results obtained by B.V.Gnedenco et all..[6] for this type of absolute priority. To obtain (4.1) we used for this type of absolute priority. To obtain (4.1) we used Smith's theorem [12] about regenerative process. Start moments of busy periods serve as regeneration points. The regenerative cycle length is the sum of a busy period length and the consequent idle period length. Because these random variables are mutually independent, and the latter has exponential distribution, the distribution function of their sum, i.e. of the regenerative cycle length, is absolutely continuous.

In accordance with one of the Smith's theorem assertions if an average value of regenerative cycle length is bounded, then process $m(t)$ is ergodic. As far as the boundedness of regenerative cycle length is equivalent to the boundedness of average busy period length, the ergodicity condition is equivalent to condition (4.1).

5. PROGRAM IMPLEMENTATION.

Computational problems associated with program implementation of condition (4.1) test are far more complex than for implementation of its homogeneous analogue. To find ρ_r, and also π_{r1} itself, one must calculate Φ_i, hence, must calculate $\pi_{i-1}(a_i)$. $\pi_{i-1}(a_i)$ is Laplace-Stieltjes transform of a distribution function of so-called Π_{i-1} period in the point $s = a_i$. $\pi_i(a_i), i = 1, ..., r$, are found from recurrent equations (4.2) to (4.4). Let's assume that $a_i, B_i(x), C_i(x)$ (hence $\beta_i(s)$ and $c_i(s)$) are given. Then the calculation of $\pi_{i-1}(a_i)$, as the mentioned equations indicate, even for exponential distribution functions $B_i(x)$ and $C_i(x)$, is a nontrivial problem.

To solve functional equations of the kind

$$\bar{\pi}_i(s) = h_i(s + a_i[1 - \bar{\pi}_i(s)])$$

the following iterative process is created:

$$\bar{\pi}_i(s)^n = h_i(s + a_i[1 - \bar{\pi}_i(s)^{n-1}])$$

Calculations stop if

$$[\bar{\pi}_i(s)^n - \bar{\pi}_i(s)^{n-1}] \leq \epsilon,$$

where $\epsilon > 0$ is a defined in advance small number, i.e. the measure of proximity of $\bar{\pi}_i(s)^n$ to $\bar{\pi}_i(s)$.

In the capacity of $\bar{\pi}_i(s)$ its approximation $\bar{\pi}_i(s)^n$, calculated with an accuracy up to ϵ, is taken. The validity of this calculation procedure follows from the following inequality:

$$0 \leq \bar{\pi}_i(s)^n \leq \bar{\pi}_i(s)^{n+1} \leq 1$$

and the fact, that $\bar{\pi}_i(s)^n$ is an essentially monotonic function. The proof is carried out by mathematical induction method, just like as in work [6] when studying the equation

$$\pi(s) = \beta(s + a - a\pi(s))$$

We have proposed some effective algorithms to solve systems of functional equations of a kind (4.2) to (4.4), including those which use fast Laplace-Stieltjes transform. Analysis of the state of a node in the transputer system and calculation of the parameters named above is implemented using object-oriented programming technique, and is based on an approach, which was developed by us earlier, and is intended for general purposes of preemptive systems modelling. Being based on previously elaborated object class system, which is included into PRIOR software package, a new object class TAverMode was developed, which inherits from Tpk object class (implementing Laplace-Stieltjes transformation of busy period length distribution function). The fields pointing to an array containing the results of calculations, i.e. the parameters listed above, and to a method, i.e. a procedure filling the above mentioned array, are added. At each step with $k \geq 2$, where k is one of the chosen input flow numbers,ρ_k is calculated and the stationary condition $\rho_k < 1$ of the system is tested. In the case $\rho_k \geq 1$ calculations stop, otherwise parameter values $p_{k1}, p_{kk1}, \bar{p}_{k1}$, are calculated, for which purpose the procedure included as a method in Tpk object class is used, calculating busy period parameters in the terms of Laplace-Stieltjes transformation.

The result of the calculation are stored in the memory array used to fill the collection which is further used to put out them to the screen.

Tables 1 to 6 show some particular results of modelling with various distribution laws for service and switching times. Table 1 and 2, for example, show us the situation when priority class number $k = 7$ gives rise to an infinite growth of queue length, and as a result of this growth the processor enters into an infinite busy period.

It is interesting, that the maximal time losses are associated with switching to the priority class $k = 2$, while the minimal losses are associated with switching to the class $k = 5$. It is a typical situation corresponding to the case 2 described in section 2.

The results of modelling shown at the tables 3 to 6, on the contrary, represent the stationary state of the transputer system. In addition one can examine also average load values by each priority class.

Regime:Reset Queue		Ser.Disc: Restart Service			Or.Disc: Restart Orient		
Exp	0.26500	Exp	0.600		Erlang	1.1450	3.0000
Exp	0.05600	Erlang	3.200	4.00	Exp	0.5000	
Exp	0.00300	Exp	2.200		Exp	2.5000	
Exp	0.00200	Uniform	5.000	7.00	Uniform	0.0025	1.0025
Exp	0.02800	Erlang	2.230	3.00	Exp	7.9800	
Exp	0.00500	Exp	1.220		Exp	2.6540	
Exp	0.02300	Uniform	0.500	3.00	Erlang	1.1540	5.0000

k	$pk1$	$pkk1$	$\bar{p}k1$	$hk1$	$\nu k1$	rok
2	29.30982	22.85307	10.47937	6.58347	7.77355	0.79339
3	31.29819	5.70703	3.20689	3.17432	2.47474	0.79549
4	43.51979	191.48955	186.84366	134.19094	3.33667	0.85312
5	69.18400	25.57848	23.73772	14.25983	1.10579	0.91177
6	75.83931	20.87833	14.85556	13.85322	5.61644	0.91772
7	Non stationary regime					

Table 1. Non stationary regime

Regime:Reset Queue		Ser.Disc: Restart Service			Or.Disc: Restart Orient		
Exp	0.26500	Exp	4.000		Erlang	1.1450	3.0000
Exp	0.05600	Erlang	3.200	4.00	Exp	0.5000	
Exp	0.10000	Erlang	5.876	6.00	Exp	2.5000	
Exp	0.00200	Uniform	1.654	2.99	Uniform	0.0025	1.0025
Exp	0.02800	Erlang	2.230	3.00	Exp	7.9800	
Exp	0.00500	Exp	1.220		Exp	2.6540	
Exp	0.02300	Uniform	0.500	3.00	Erlang	1.1540	5.0000

k	$pk1$	$pkk1$	$\bar{p}k1$	$hk1$	$\nu k1$	rok
2	11.92795	10.63573	5.05401	3.93186	4.34240	0.57160
3	16.36866	7.61924	5.84985	3.69079	1.11634	0.72971
4	18.61242	26.51427	23.80706	22.67351	2.57830	0.74258
5	23.34047	12.18611	11.35098	8.61340	0.63372	0.80467
6	24.72271	9.30997	6.68442	6.47368	2.54278	0.81082
7	Non stationary regime					

Table 2. Non stationary regime

Regime:Reset Queue		Ser.Disc: Restart Service			Or.Disc: Restart Orient		
Exp	0.26500	Exp	10.00		Uniform	0.9000	2.6400
Exp	0.05600	Exp	6.00		Exp	0.5000	
Exp	0.05000	Erlang	5.87	6.00	Erlang	0.5000	3.0000
Exp	0.00200	Uniform	2.70	3.00	Uniform	0.2000	1.0025
Exp	0.02800	Erlang	2.23	3.00	Exp	7.9810	
Exp	0.00500	Exp	4.20		Exp	3.1240	
Exp	0.00600	Uniform	1.10	2.50	Erlang	2.9000	2.0000

k	$pk1$	$pkk1$	$\bar{p}k1$	$hk1$	$\nu k1$	rok
2	7.21616	3.99723	0.39382	0.38525	3.52502	0.35212
3	20.26932	53.91628	17.65119	9.37616	19.26372	0.65585
4	24.79678	49.74352	45.02740	41.13753	4.30870	0.68558
5	35.29413	19.23540	17.88354	11.91649	0.90079	0.79049
6	36.90954	6.52607	2.97654	2.93401	3.49881	0.79348
7	49.20578	60.74054	49.80815	37.62647	8.25863	0.84399

Table 3. Stationary regime

Regime:Reset Queue		Ser.Disc: Restart Service			Or.Disc: Restart Orient		
Exp	0.26544	Exp	10.00		Uniform	0.9300	2.6400
Exp	0.05647	Exp	6.00		Exp	0.5100	
Exp	0.05000	Erlang	5.90	6.00	Erlang	0.5800	3.0000
Exp	0.00210	Uniform	2.70	3.00	Uniform	0.2200	1.0025
Exp	0.02800	Erlang	2.20	3.00	Exp	7.9800	
Exp	0.00487	Exp	4.20		Exp	3.1200	
Exp	0.00650	Uniform	1.10	2.50	Erlang	2.9000	2.0000

k	$pk1$	$pkk1$	$\bar{p}k1$	$hk1$	$\nu k1$	rok
2	7.20937	3.97680	0.37678	0.36893	3.52502	0.35151
3	16.29295	39.47194	10.30758	6.80198	19.24559	0.57206
4	18.11097	15.29768	11.74809	11.46523	3.46412	0.58236
5	20.25692	7.48209	6.67827	5.62622	0.67719	0.64815
6	20.76798	3.69982	1.60399	1.59155	2.07959	0.65088
7	22.42898	16.85331	11.61951	10.80355	4.86626	0.67540

Table 4. Stationary regime

Regime:Reset Queue		Ser.Disc: Restart Service			Or.Disc: Restart Orient		
Exp	0.02540	Exp	5.690		Uniform	0.9000	2.6400
Exp	0.07540	Exp	6.000		Exp	0.5000	
Exp	0.10000	Erlang	5.880	6.00	Erlang	0.5000	3.0000
Exp	0.00023	Uniform	2.700	3.00	Uniform	0.2000	1.0025
Exp	0.09860	Erlang	2.230	3.00	Exp	7.9800	
Exp	0.00487	Exp	4.200		Exp	3.1200	
Exp	0.00650	Uniform	1.100	2.50	Erlang	2.9000	2.0000
Exp	0.00210	Exp	3.500		Exp	0.5600	

k	$pk1$	$pkk1$	$\bar{p}k1$	$hk1$	$\nu k1$	rok
2	12.72425	2.65095	0.18641	0.18383	2.43038	0.06049
3	15.27691	14.58728	2.95135	2.27880	8.98434	0.27459
4	16.78455	14.86592	12.97126	12.93200	1.88893	0.27678
5	20.73773	8.80940	8.13320	4.51360	0.37526	0.59864
6	23.33923	3.23483	1.45177	1.44158	1.77055	0.60146
7	25.99345	22.54230	17.99806	16.11304	4.06831	0.64320
8	26.80001	14.35468	2.86044	2.84336	11.42561	0.64533

Table 5. Stationary regime

Regime:Reset Queue		Ser.Disc: Restart Service			Or.Disc: Restart Orient		
Exp	0.0254	Exp	5.698		Uniform	0.90	2.6400
Exp	0.0754	Exp	6.00		Exp	0.50	
Exp	0.1000	Erlang	5.876	6.00	Erlang	0.50	3.0000
Exp	0.0002	Uniform	2.700	3.00	Uniform	0.20	1.0025
Exp	0.0986	Erlang	2.230	3.00	Exp	7.98	
Exp	0.0048	Exp	4.200		Exp	3.12	
Exp	0.0650	Uniform	1.100	2.50	Erlang	2.90	2.0000
Exp	0.0021	Exp	3.500		Exp	0.56	

k	$pk1$	$pkk1$	$\bar{p}k1$	$hk1$	$\nu k1$	rok
2	12.72350	2.65001	0.18561	0.18305	2.43038	0.06044
3	14.84996	13.90181	2.59063	2.05759	8.98380	0.25376
4	16.15433	8.94508	7.10584	7.09404	1.83619	0.25500
5	16.61359	5.62037	5.07385	3.38193	0.36428	0.50343
6	18.18123	2.53180	1.09326	1.08747	1.43092	0.50606
7	19.03177	11.40276	7.95447	7.56341	3.27876	0.53034
8	19.41844	10.66564	1.98415	1.97592	8.64547	0.53229

Table 6. Stationary regime

References

[1] KLEINROCK A. (1979) Computer systems with queues (Russian). - Moscow, Mir.

[2] BASHARIN G.P. ET AL. (1989) Queue analysis in computer networks (Russian). - Moscow, Nauka.

[3] D.J.EVANS AND W.U.K.BUTT. (1993) Dynamic load balancing using task-transfer probabilities. - Parallel Computing, 19, p.897-916.

[4] BOGUNOVIC N. (1987) Processes scheduling procedure for a class of a real-time computer systems. - IEEE Trans. Ind. Electron., v.34, No 1, p.29-34.

[5] KLIMOV G.P. (1966) Stochastic queueing systems (Russian). - Moscow, Nauka.

[6] GNEDENKO B.V. ET AL. (1973) Priority queueing systems (Russian). - Moscow, MSU.

[7] JASWOLL N.K. (1973) Queues with priorities (Russian). - Moscow, Mir.

[8] (1993) Transputers. Architecture and software (Russian). - Moscow, Radio i svyaz.

[9] KLIMOV G.P.AND MISKOY G.K. (1979) Priority queueing systems with orientation (Russian). - Moscow, MSU.

[10] MISKOY G.K. (1989) Priority queueing systems with orientation, their software and some applications (Russian). Ph. D. thesis. - Kiev.

[11] MISKOY G.K. (1990) Software for priority service systems with orientation (Russian). - Kishinev, Shtiintsa.

[12] KOKS D. AND SMITH V. (1967) Regeneration theory (Russian). - Moscow, Sovietskoye Radio.

STATISTICAL ENDPOINT MONITORING OF BREAST CANCER CLINICAL TRIALS: A REVIEW AND CASE STUDY OF NSABP PROTOCOL B-14

JOHN BRYANT, JAMES J. DIGNAM, H. SAMUEL WIEAND

National Surgical Adjuvant Breast and Bowel Project (NSABP) and

Departments of Biostatistics (JB, JJD, HSW) and Statistics (JB)

University of Pittsburgh, Pittsburgh, PA, USA

SUMMARY

We review methods of interim analysis which are appropriate for monitoring Phase III clinical trials of adjuvant therapy for breast cancer. Since the results of a clinical trial will impact clinical practise only if there is agreement throughout the clinical research community that its findings are compelling, we argue that early stopping decisions should explicitly consider the degree to which the early results are sufficiently strong to achieve such a concensus of opinion. Bayesian methods are ideally suited to address this question, since they may be used to evaluate the sensitivity of posterior conclusions over a range of prior beliefs regarding treatment effectiveness. These considerations are illustrated in a case study of NSABP Protocol B-14, which evaluated tamoxifen in the treatment of early stage breast cancer. An extension of the study re-randomized tamoxifen patients who were disease-free at 5 years to an additional 5 years of tamoxifen or placebo, and was terminated early when interim analyses indicated no advantage for the additional tamoxifen. This closure has been controversial, and the question of optimal tamoxifen duration is still a source of debate. We consider interim monitoring of the B-14 study using both frequentist and Bayesian methods.

Key Words and Phrases: Interim Analysis; Early Stopping; Phase III Clinical Trials; Clinical Consensus; Confirmatory Clinical Trials; Bayesian Methods; Stochastic Curtailment

1. INTRODUCTION

In this paper, we discuss interim monitoring of Phase III clinical trials in early stage breast cancer. The rationale for study monitoring is reviewed and the structure of typical group sequential monitoring rules is briefly summarized. We then describe the statistical rationale which prompted early closure of a breast cancer clinical trial (NSABP B-14), and consider alternative approaches to assessing early termination in the face of evidence against a benefit. We argue that Bayesian methods are valuable in assessing the degree to which current results may be expected to be interpreted as compelling by the research community, by allowing the incorporation of a range of prior beliefs regarding treatment effectiveness, and evaluating the sensitivity of posterior conclusions over this range.

2. INTERIM MONITORING OF BREAST CANCER CLINICAL TRIALS

2.1 Issues and Rationale for Clinical Trials Monitoring. The monitoring of some aspects of an adjuvant therapy breast cancer clinical trial, including accrual, adverse events, protocol adherence, and patient compliance, is essentially continuous. In contrast, analyses of efficacy endpoints are scheduled at intervals prescribed in the study protocol, which also includes formal monitoring rules governing the possibility of early termination. Historically, the incorporation of early stopping rules into U.S. cooperative group trials was not primarily intended to improve efficiency, but rather to prohibit premature stopping caused by uncontrolled multiple inspections of the outcome data (Simon, 1994). Other factors not explicitly accounted for in formal stopping rules also argue for the adoption of a conservative approach. These include (i) the need to adequately address secondary endpoints and to compare treatment effects within key patient subsets; (ii) the desire for long-term follow-up to determine whether early treatment differences will attenuate with time; and (iii) possible biases associated with lags in data submission or differential follow-up. It should also be considered that in order to impact clinical practice, the results of a trial must be broadly accepted within the clinical research community; given the wide range of prior opinion liable to exist within that community, this is unlikely to occur if study results are perceived as marginal.

Most common monitoring rules are based on group sequential boundaries, which we briefly describe here. A more complete review is found in the excellent article of Jennison and Turnbull (1990) and in the Proceedings of the Workshop on Early Stopping Rules In Cancer

Clinical Trials (*Statistics In Medicine* Vol. 13 No. 13/14, 1994). Other methods will be discussed in Section 3 in the context of the B-14 trial.

2.2 Group Sequential Hypothesis Tests. In a clinical trial designed to compare treatments A and B with respect to a survival endpoint, treatment differences can be summarized by the parameter $\delta = \ln(\lambda_A(t)/\lambda_B(t))$, the logarithm of the ratio of instantaneous hazard rates $\lambda_A(t)$ and $\lambda_B(t)$. It is assumed here that this ratio is time-independent, in which case it is well known that the logrank test (Mantel, 1966; Peto and Peto, 1972) provides a fully efficient test of H_o: $\delta = 0$.

Suppose that K-1 interim analyses are scheduled, so that the final analysis will be the Kth, and that analysis k is scheduled to occur following observation of the D_kth death, k=1,2,...,K. Let L_k represent the kth logrank statistic; Assuming equal allocation of patients to either treatment arm, the variance of L_k is $V_k \approx D_k/4$. This approximation is accurate when either $|\delta|$ is not extreme or when D_k is moderately small relative to total accrual, two conditions that typically obtain in early stage breast cancer trials. The ratios $t_k = V_k/V_K \approx D_k/D_K$, k=1,2,...,K are referred to as "information time", since they represent the fraction of total information to be observed prior to each interim analysis. Define $Z_k = L_k/\sqrt{V_k}$, k=1,2,...,K. A group sequential test of H_o: $\delta = 0$ against a two-sided alternative hypothesis $H_A: \delta \neq 0$ takes the following form: *Let b_k, k=1,2,...,K, be specified critical values. At the kth analysis, if $|Z_k| \le b_k$ continue to the next analysis; otherwise stop and reject H_o. At analysis K, if $|Z_k| \le b_k$ accept H_o; otherwise reject H_o.*

This rule can equivalently be based on the statistics $W_k = L_k/\sqrt{V_K} = Z_k \bullet \sqrt{t_k}$, k=1,2,...,K, rather than the Z_k, with corresponding critical values for W_k related to those for Z_k by $B_k = b_k \bullet \sqrt{t_k}$. Under appropriate conditions (roughly, sequential entry, randomized treatment assignment, loss-to-follow-up independent of entry time and treatment assignment) the W_k behave asymptotically like Brownian motion: Defining $\Delta t_k = t_k - t_{k-1}$ and $\eta = \delta \bullet \sqrt{D_K}/2$, the increments $W_k - W_{k-1}$ are approximately uncorrelated normal random variables with means $\eta \bullet \Delta t_k$ and variances Δt_k (Tsiatis (1981, 1982), Gail et al, 1982). This important result permits the extension of sequential methods based on evolving sums of independent normal variates to the survival setting, and in particular, the direct application of the recursive integration scheme of Armitage, McPherson and Rowe to compute the density $f_k(w; \eta) = d\Pr\{\tau \ge k, W_k \le w; \eta\}/dw$, where τ represents the number of the terminal analysis: Letting $\phi\{\bullet\}$ represent the standard normal density, $f_1(w;\eta) = (1/\sqrt{t_1}) \bullet \phi\{(w-\eta \bullet t_1)/\sqrt{t_1}\}$, and for k=2,3,...,K

$$f_k(w;\eta) = (1/\sqrt{\Delta t_k}) \int_{-B_{k-1}}^{B_{k-1}} f_{k-1}(y;\eta) \bullet \phi\{(w-y-\eta \bullet \Delta t_k)/\sqrt{\Delta t_k}\} dy . \tag{2.1}$$

From this result all operating characteristics of the group sequential procedure (size, power, stopping probabilities, etc.) may be obtained. In practice, the critical values b_k (equivalently B_k) are chosen to maintain a desired experimentwise Type 1 error rate. Given these critical values, the duration D_K of the trial is selected to achieve power 1-β against a specified alternative H_A:$\delta=\delta_A$, by determining that value of η which yields Pr{Reject $H_o|\eta$} = 1-β, and then setting $D_K = 4 \bullet \eta^2/\delta_A^2$.

Most of the common sequential monitoring rules fit into this general framework. The best known of these are the rules proposed by Pocock (1977, 1982) and O'Brien and Fleming (1979). Both procedures assume that analysis will take place at equally spaced increments of information time, so that t_k = k/K. The Pocock boundaries are obtained by constraining the z-critical values b_k to be identical for each k, while the W-critical values for the O'Brien-Fleming procedure are constant: B_k = constant, or equivalently $b_k = \text{constant}/\sqrt{t_k}$. An advantage of the Pocock boundaries is that for procedures of matched size and adequate power (say 1-β ≥ .80), a smaller expected number of deaths need to be observed under the alternative H_A than is the case for O'Brien-Fleming boundary. On the other hand, the O' Brien-Fleming procedure requires a smaller maximum number of deaths (D_K); in fact D_K is only slightly greater than would be required had no interim analyses been performed, since for the O'Brien-Fleming procedure the final critical value b_K is only slightly larger than the 1-α/2 quantile of the standard normal distribution. For this reason, if the trial is not stopped early the final test of H_o is nearly identical to the fixed sample size test, a fact that has contributed to the popularity of the O'Brien-Fleming boundaries.

Fleming, Harrington and O'Brien (1984) proposed a class of boundaries which retain the desirable property of the O'Brien-Fleming boundary that the nominal level of the Kth analysis is nearly equal to α, while at the same time avoiding the extremely conservative nature of that procedure for small k when K>3. Wang and Tsiatis (1987) proposed a class of symmetric group sequential boundaries which include Pocock and O-Brien-Fleming boundaries as special cases.

2.3 Unpredictable Analysis Times. It is not usually possible to conduct interim analyses precisely according to a fixed schedule of numbers of deaths. DeMets and Gail (1985) showed by simulation that the impact of moderately unequal information intervals on the Type I error rate is generally negligible. To accomodate irregular scheduling, Slud and Wei (1982) specify the probabilities Pr{τ=k|H_0} directly, and determine the corresponding critical values via (2.1). At information time t_k, it is not necessary to know the values of t_j for j>k in order to compute b_k. Therefore, one can accommodate scheduling difficulties by pre-specifying the amount of type I error to be spent at each analysis, without precisely specifying the exact information time at which the analysis is to take place. Lan and DeMets (1983) proposed a similar procedure in which an "alpha-spending" function $\alpha(t)$ is specified satisfying $\alpha(0)=0$, $\alpha(1)=1$, $\alpha'(t)>0$, $0<t<1$. If the kth interim analysis occurs at information time t_k, the critical value b_k is chosen so that $\Pr\{\tau=k|H_0\} = \alpha(t_k) - \alpha(t_{k-1})$.

2.4 One-sided Tests. The foregoing development may be adapted to accommodate one-sided hypothesis tests using asymmetric group sequential boundaries. The reader is referred to DeMets and Ware (1980, 1982), Whitehead (1979), Whitehead and Stratton (1983), Jennison (1987), Emerson and Fleming (1989), and Pampallona and Tsiatis (1994). Most of the common group sequential procedures are now available through one of two commercial software packages (East, Cytel Software, or Pest3, MPS Research Unit, University of Reading). Emerson (1996) has prepared a comparative review of these packages.

3. EARLY STOPPING OF A BREAST CANCER CLINICAL TRIAL: NSABP PROTOCOL B-14

3.1 Background and Design of NSABP B-14. We will describe the early closure of NSABP Protocol B-14, a randomized trial investigating the optimal duration of tamoxifen in treatment of early stage breast cancer patients (Fisher et al, 1996). We consider this decision from both frequentist and Bayesian perspectives. We argue that the application of Bayesian methods is instructive in ascertaining the degree to which trial results might be regarded as definitive throughout the clinical research community.

In 1982, the NSABP initiated Protocol B-14, a double-blind comparison of five years of tamoxifen (10 mg b.i.d.) with placebo in patients having estrogen receptor-positive tumors and no axillary node involvement. The first report of findings (Fisher, et al 1989) indicated improved disease-free survival (DFS, defined as time to either breast cancer recurrence, contralateral breast cancer or other new primary cancer, or death from any cause) (percent event-free at four years = 83% vs. 77%, $p<.00001$). Subsequent follow-up (Fisher, et al 1996) confirmed this benefit (DFS at 10 years, 69% vs. 57%, $p<.00001$), and also showed a significant survival advantage (percent alive at 10 years, 80% vs. 76%, $p=.02$).

As the first results became evident, the question arose as to whether more than five years of tamoxifen might offer increased protection. In April 1987 a second randomization was initiated. Patients who had received tamoxifen and were event-free through five years were re-randomized to either continue tamoxifen for an additional 5 years or to receive placebo. 1172 patients were re-randomized between April 1987 and December 1993. The definitive endpoint for the trial was DFS. To provide for a .05 level one-sided test with a power of at least 0.85 under the assumed alternative of a 40% reduction in failure rate, a total of 115 events were specified prior to definitive analysis. Four interim analyses were scheduled at approximately equal increments of information time. Stopping boundaries were obtained using the method of Fleming, Harrington, and O'Brien (1984) at the two-sided 0.10 level. Confidential interim endpoint analyses were to be compiled by the study statistician and presented to the independent Data Monitoring Committee (DMC) of the NSABP.

3.2. Results at Interim Analyses. At the first interim analysis, based on all data received as of September 30, 1993, more events had occurred in the tamoxifen group (28 events) than among those receiving placebo (18 events). There had been 6 deaths on the placebo arm, and 9 among tamoxifen patients. By the second interim analysis (data received as of September 30, 1994), there were 24 events on the placebo arm and 43 on the tamoxifen arm (relative risk = 0.57, nominal 2p=.03), with 10 deaths on the placebo arm and 19 among patients receiving tamoxifen. While there was concern regarding the possibility of a less favorable outcome for patients continuing tamoxifen, we recommended that the trial be continued to the next scheduled interim analysis because the early stopping criterion was not achieved (2α =.0030) and follow-up for the majority of patients was relatively short (mean 3.75 years). Additional statistical considerations discussed below also influenced the decision to continue.

Figure 1 shows DFS at the third interim analysis (data received as of June 30, 1995). At this time there were 32 events on the placebo arm and 56 on the treatment arm (relative risk = 0.59). Four-year DFS was 92% for patients on placebo and 86% for patients on tamoxifen. The boundary for early termination (2α=0.0035) was not crossed (2p = 0.015). However, calculations showed that even if the remaining 27 events (of 115) all occurred on the placebo arm, the logrank statistic would not approach significance. The imbalance in deaths also persisted (13 placebo arm, 23 tamoxifen, 2p=.11). Additionally, unpublished data were made available to the NSABP by the leadership of the Scottish Trial (Stewart et al, 1996), which reinforced the pattern seen in B-14. Following this interim analysis, a recommendation to terminate the trial was made by NSABP statisticians to the DMC, who concurred and informed the Group Chair. A brief announcement was prepared and disseminated from the National Cancer Institute in November, 1995, and a detailed presentation of findings has been published (Fisher et al, 1996).

3.3 Additional Statistical Evaluation of Interim Analysis Results. A common predictive approach to early stopping is based on the concepts of conditional power and stochastic curtailment (Lan, Simon, and Halperin 1982; Halperin, Lan, Ware et al, 1982, Davis and Hardy 1991). Define the *conditional power* function $\gamma = Pr(S \in R \mid D,H)$, where S represents a test statistic to be computed at the end of the trial, R is the rejection region of the test, D represents the current data, and H denotes either the null hypothesis H_o or an alternative hypothesis H_A. If this conditional probability is sufficiently large under H_o, one may decide to stop and immediately reject H_o. On the other hand, if this probability is sufficiently small under a `realistic' alternative hypothesis H_A, we may decide that continuation is futile because H_o will ultimately not be rejected regardless of further observations. Often, conditional power is evaluated over a range of alternatives.

At the second interim analysis, we computed the conditional probability of rejecting the null hypothesis at the scheduled final analysis (115 events), given the current data. Results suggested that even under extremely optimistic assumptions concerning the true state of nature, the null hypothesis could almost certainly not be rejected: Even under the assumption of a 67% reduction in failures the conditional probability of eventual rejection was less than 5%. We also considered an 'extended trial', repeating conditional power calculations as if we had intended to observe a total of 229 events before final analysis (this number of events would allow for the detection of a 30% reduction in event rate with a power of 85%). Results indicated that, were the trial to continue and the underlying relative risk actually be strongly in favor of tamoxifen, then the null hypothesis could possibly be rejected (Figure 2). This consideration did influence our decision to continue.

At the third interim analysis, there was no longer any possibility that a significant result favoring tamoxifen could be achieved at the 115th event. For the extended trial allowing follow-up to 229 events, Figure 2 shows that conditional power was about 15% under the planning alternative of 40% reduction in relative risk, and was only 50% under the assumption of a two-fold benefit. At this time, we also considered an early stopping rule proposed by Wieand, Schroeder, and O'Fallon (1994). In its simplest form, the rule calls for performing an interim analyses when one half of the required events have taken place. At that time, if the event rate on the experimental arm exceeds that on the control arm, then termination should be considered. If this rule is superimposed on Fleming-Harrington-O'Brien boundaries (replace the lower boundary $-b_k$ with 0 for any information time $t_k \geq \frac{1}{2}$) and the result is viewed as an asymmetric procedure testing a one-sided hypothesis, there is almost no change in operating characteristics. To illustrate, suppose the lower boundaries at the third, fourth and final analysis of the B-14 monitoring plan were replaced with zeros. Then the (upper) level of significance is reduced from

.0501 to .0496, and the power under the alternative of a 40% reduction in event rate is reduced from .8613 to .8596.

By the third interim analysis, considerably more events had occurred on the treatment arm than on the control arm. Had a conservative "futility" rule such as that described above been incorporated into the monitoring plan, it would have suggested termination. This consideration influenced our decision to terminate at the third analysis.

3.4 Bayesian Approaches to Interim Analysis of the B-14 Study. As discussed, the approaches taken in considering the early termination of the B-14 study were frequentist. However, Bayesian methodology is well suited to problems in trial monitoring (Freedman and Spiegelhalter, 1983; Kass and Greenhouse, 1989; Spiegelhalter et al, 1994; Freedman et al, 1994; Greenhouse and Wasserman, 1994; Berry and Stangl, 1996; Parmar, Ungerleider and Simon, 1996). In this section we apply Bayesian methods and assess whether we might have arrived at dissimilar conclusions had we used this approach.

Bayesian analyses have often been criticized on grounds of subjectivity in specifying prior information. However, the goal of any clinical trial is ultimately to influence clinical practice, and to do so its results must be sufficiently strong to prove compelling to a community of clinical researchers whose prior opinions are diverse. Thus, the Bayesian paradigm is of value in considering the early termination of a study, precisely because it does allow for the contemplation of diverse prior opinions, and provides a mechanism for judging whether observed results are strong enough to overcome this diversity. If the termination of a study is considered, an analysis of the robustness of conclusions over a range of priors thought to resemble the a priori beliefs of reasonable members of the clinical research community should provide insight into the impact that the trial might be expected to exert on clinical practice.

We define δ as in Section 2, and adopt the assumptions of that Section. Denote the partial maximum likelihood estimate of the log hazard ratio as $\hat{\delta}$. Asymptotically, δ has a normal likelihood with mean $\hat{\delta}$ and variance 4/m, where m is the total number of events observed. We assume a normal prior distribution with specified mean δ_p and variance σ_p^2. The values of δ_p and σ_p^2 may be determined to reflect an individual's prior level of enthusiasm regarding a proposed regimen, and these parameters may be altered to reflect varying degrees of enthusiasm. The notion of "skeptical" and "optimistic" prior distributions is discussed by Kass and Greenhouse (1989), and others (Freedman et al, 1994; Spiegelhalter et al, 1994; Parmar et al, 1996). It is suggested that a skeptical but reasonable member of the cancer clinical community may adopt a prior for δ which is centered at 0, since relatively few new regimens have led to material improvements in recurrence or survival. Nevertheless, if a trial is well designed, its designers will have specified a planning alternative for δ, $\delta = \delta_A$, say, which they feel is both clinically meaningful and relatively probable. If the skeptic is reasonably open-minded, he or she would probably be willing to admit some probability that this effect could be achieved, perhaps on the order of 5%. Therefore, one might model the skeptical prior with mean $\delta_p = 0$ and standard deviation $\sigma_p = \delta_A/1.645$. Similarly, one might consider the trial designers as being among the most optimistic of its proponents, but even they would be compelled to admit as much as a 5% chance that the proposed regimen will have no effect, i.e. that $\delta \leq 0$. It may therefore be reasonable to model an optimist's prior by setting $\delta_p = \delta_A$ and $\sigma_p = \delta_A/1.645$.

For a given prior distribution and the observed likelihood, a posterior density can be computed for δ, and one may then determine the probability that the effect is in some specified range of interest, say $\delta > 0$, indicating presence of a benefit, or $\delta \geq \delta_{ALT} > 0$, corresponding to some clinically relevant effect size. The posterior distribution for δ is normal with mean $(n_0\delta_p + m\hat{\delta})/(n_0+m)$ and variance $4/(n_0+m)$, where $n_0 = 4/\sigma_p^2$.

Figure 3 shows the likelihood for the B-14 data at the third interim analysis, having mean ln(0.586) = -0.534 and standard deviation 0.213. Also shown is an "optimistic" prior distribution centered at $\delta_p = \delta_A = 0.511$, corresponding to a 40% reduction in failures for patients continuing on tamoxifen relative to those stopping at five years. The prior standard deviation is $\sigma_p = \delta_A/1.645 = 0.311$. The resulting posterior distribution has mean -0.199 and standard deviation 0.176 (also shown). From this distribution one can determine that the posterior probability that $\delta > 0$ is $1 - \Phi(0.199/0.176) = 0.13$, where $\Phi(\bullet)$ is the standard normal distribution function. To the degree that this prior distribution represents that of a clinical researcher who was initially very optimistic, these calculations suggest that even in the face of the negative trial findings as of the third interim analysis, this individual would still assign a small but non-negligible probability to the possibility that continued tamoxifen has some benefit. On the other hand this individual would now assign essentially no probability ($\approx 3 \times 10^{-5}$) to the possibility that the benefit is as large as a 40% reduction in risk.

To formulate a predictive distribution, let m be the number of events observed thus far, and let n the number of additional events to be observed. Denote by $\hat{\delta}_n$ the log relative risk which maximizes that portion of the partial likelihood corresponding to failures m+1, m+2,...,m+n. The predictive distribution of $\hat{\delta}_n$ is approximately normal with the same mean as the posterior distribution and variance $\sigma^2_{pred} = 4/(n_0+m) + 4/n$. If the trial were to be extended to allow for a total of 229 events, or 141 events beyond the third interim analysis, we obtain μ_{pred} = -0.199 and σ_{pred} = 0.243. The predictive probability of a significant treatment comparison following the 229th event is determined as follows: If $\hat{\delta}_{m+n}$ denotes the estimated log relative risk based on all the data, then the ultimate result will be significant at the .05 level if $\hat{\delta}_{m+n} > 1.645 \bullet \sqrt{(4/229)} = 0.217$. Since approximately $\hat{\delta}_{m+n} \approx (88\hat{\delta} + 141\hat{\delta}_n)/229$, this event requires that $\hat{\delta}_n > 0.686$. The predictive probability of this occurrence is $1 - \Phi(\{0.686 + 0.199\}/0.243) \approx 0.0001$.

We considered a family of priors, identical to the prior of Fig. 3 in spread (σ_p = 0.311) but having mean δ_p varying between 0 (skeptical prior) and 0.511 (optimistic prior). Results are summarized in Table 1, which also includes the noninformative prior. For each tabulated prior, Table 1 shows the percent reduction in failures most likely under that prior, the prior probability of benefit for continued tamoxifen, the posterior probability of benefit given the results of the third interim analysis, and the posterior 95% credible interval.

To summarize the accumulation of evidence over time, we computed the posterior distribution of the log hazard ratio at each of the three interim analyses as well as at the time of publication (data received as of December 31, 1995) and at one year subsequent to publication (December 31, 1996), starting with the optimistic prior described above (Fig. 4). At the first interim analysis, the posterior probability of a benefit for continued tamoxifen was 0.57, but by the third interm analysis, when the study was terminated, this probability had diminished to 0.13. At the time of publication the posterior probability of benefit was 0.05, and one year later was 0.04. At that point in time 135 total events had occurred, 85 among patients who had received continued tamoxifen, and 50 among patients rerandomized to placebo (relative risk = 0.60, nominal 2p=0.002). There were 36 deaths among tamoxifen patients and 17 among control patients (nominal 2p=0.01).

4. DISCUSSION

Prior to the B-14 results, the most common expectation among cancer researchers was that continuation of tamoxifen beyond five years would prove efficacious. Thus the closure was received with surprise, and several commentaries have reflected a sentiment that it was premature (Peto 1996, Swain 1996, Current Trials Working Party 1996). However, the analyses described in Section 3 support our decision to stop the trial. Although the pre-specified significance level for early stopping was not achieved at the third interim analysis, it was apparent that had the trial had been completed as planned, it could not have led to any result that would have supported the continued use of tamoxifen. Even if reporting were to have been delayed for several additional years, stochastic curtailment and predictive power calculations indicated little chance of a dramatic reversal of the patterns seen up to the third interim analysis. Since the majority of patients on the trial were still receiving their re-randomized treatment, we believed it was not justifiable to continue.

The Bayesian perspective allows for the incorporation of an individual's prior expectation into the analysis of interim data, and in the B-14 study lent some insight into why challenges to the early stopping decision have been mounted. Given the results of the third interim analysis, under an optimistic but not unreasonable prior distribution, the data were not sufficiently strong to completely rule out the possibility of moderate benefit and thereby would not be convincing to some individuals. However, the Bayesian perspective also showed that this same reasonable optimist would believe the chances to be very small that continued follow-up would eventually lead to a positive conclusion. Thus, even an advocate of continued testing might argue that B-14 should have been closed and reported, if for no other reason than to make way for a confirmatory trial in which participants could be adequately consented.

Concurrently with the release of our results, the Scottish trial of five years versus continued tamoxifen was concluding (Stewart et al, 1996). At five years, approximately 72% of those continuing tamoxifen were event-free, compared to 78% of those who discontinued. The conclusion of this study was that a benefit from continuing tamoxifen appeared unlikely. Also, a study from the Eastern Cooperative Oncology Group (ECOG) appeared (Tormey et al, 1996) shortly after the publication of the B-14 findings. In this study, 194 node-positive women who had been treated by one year of chemotherapy followed by five years of tamoxifen were randomized to either observation or continuation of tamoxifen. Overall, 85% of patients who continued were alive and relapse-free at five years compared to 73% of patients who discontinued. The ECOG study authors suggested that these findings should not necessarily be considered inconsistent with those of the B-14 study. They speculated that patients with positive nodes may benefit from continued tamoxifen while node-negative patients do not, or that the chemotherapy that these patients received plays a role in the relative benefit of long-term tamoxifen.

REFERENCES

Armitage P, McPherson CK, Rowe BC. (1969). Repeated significance tests on accumulating data. *JRSS*. Series A 132, 235-244.

Berry DA, Stangl DK, editors. (1996). *Bayesian Biostatistics*. New York: Marcel Dekker.

Current Trials Working Party of the Cancer Research Campaign Breast Cancer Trials Group. (1996). Preliminary results from the Cancer Research Campaign trial evaluating tamoxifen duration in women aged fifty years or older with breast cancer. *J Natl Cancer Inst*. 88, 1834-1839.

Davis BR, Hardy RJ. (1991). Upper bounds for type I and type II error rates in conditional power calculations. *Communications in Statistics - Theory and Methods*. 19(10), 3571-3584.

DeMets DL, Ware JH. (1980). Group sequential methods for clinical trials with a one-sided hypothesis. *Biometrika*. 67, 651-660.

DeMets D.L., Ware JH. (1982). Asymmetric group sequential boundaries for monitoring clinical trials. *Biometrika*. 69, 661-663.

DeMets D.L., Gail MH. (1985). Use of logrank tests and group sequential methods at fixed calendar times. *Biometrics*. 41, 1039-1044.

Emerson S.S., Fleming T.R. (1989). Symmetric group sequential test designs. *Biometrics*. 45, 905-923.

Emerson S.S. (1996). Statistical packages for group sequential methods. *The American Statistician*. 50(2), 183-192.

Fisher B, Dignam J, Bryant J, et al. (1996). Five versus more than five years of tamoxifen therapy for breast cancer patients with negative lymph nodes and estrogen receptor-positive tumors. *J Natl Cancer Inst*. 88, 1529-1542.

Fisher B, Costantino J, Redmond C, et al. (1989). A randomized trial evaluating tamoxifen in the treatment of patients with node-negative breast cancer who have estrogen-receptor positive tumors. *N Engl J Med*. 320, 479-484.

Fleming TR, Harrington DP, O'Brien PC. (1984). Designs for group sequential tests. *Controlled Clinical Trials*. 5, 348-361.

Freedman LS, Spiegelhalter DJ. (1983). The assessment of subjective opinion and its use in relation to stopping rules for clinical trials. *The Statistician*. 32, 153-160.

Freedman LS, Spiegelhalter DJ, Parmar MKB. (1994). The what, why and how of Bayesian clinical trials monitoring. *Stat in Med*. 13, 1371-1383.

Gail MH, DeMets DL, Slud EV. (1982). Simulation studies on increments of the two-sample logrank score test for survival time data, with application to group sequential boundaries. In *Survival Analysis*, Crowley J and Johnson RA, (eds). Institute of Mathematical Statistics. Lecture Notes - Monograph Series, Vol. 2. Hayward, CA, 287-301.

Greenhouse J, Wasserman L. (1994). Robust Bayesian methods for monitoring clinical trials. *Statistics in Medicine*. 14, 1379-1391.

Halperin M, Lan KKG, Ware JH, et al. (1982). An aid to data monitoring in long-term clinical trials. *Controlled Clinical Trials*. 3, 311-323.

Jennison C. (1987). Efficient group sequential tests with unpredictable group sizes. *Biometrika* 74(1), 155-165.

Jennison C, Turnbull BW. (1990). Statistical approaches to interim monitoring of medical trials: A review and commentary. *Statistical Science*. 5(3), 299-317.

Kass R, Greenhouse J. (1989). Comment on " Investigating therapies of potentially great benefit: ECMO" by J. Ware. *Statistical Science*. 4, 310-317.

Lan KKG, Simon R, Halperin M. (1982). Stochastically curtailed tests in long-term clinical trials. *Communications in Statistics - Sequential Analysis*. 1(3), 207-219.

Lan KKG, DeMets DL. (1983). Discrete sequential boundaries for clinical trials. *Biometrika* 70, 659-663.

Mantel N. (1966). Evaluation of survival data and two new rank order statistics arising in its consideration. *Cancer Chemotherapy Reports*. 50, 163-170.

O'Brien PC, Fleming TR. (1979). A multiple testing procedure for clinical trials. *Biometrics*. 35, 549-556.

Pampallona S, Tsiatis AA. (1994). Group sequential designs for one-sided and two-sided hypothesis testing with provision for early stopping in favor of the null hypothesis. *Journal of Statistical Planning and Inference*. 42, 19-35.

Parmar MKB, Ungerleider RS, Simon R. (1996). Assessing whether to perform a confirmatory randomized clinical trial. *J Natl Cancer Inst*. 88, 1645-1651.

Peto R and Peto J. (1972). Asymptotically efficient rank invariant test procedures (with discussion). *JRSS*. Series A 135, 185-207.

Peto R. (1996). Five years of tamoxifen - Or more? *J Natl Cancer Inst*. 88, 1791-1793.

Pocock SJ. (1977). Group sequential methods in the design and analysis of clinical trials. *Biometrika*. 64(2), 191-199.

Pocock SJ. (1982). Interim analyses for randomized clinical trials: The group sequential approach. *Biometrics*. 38, 153-162.

Simon R. (1994). Some practical aspects of the interim monitoring of clinical trials. *Stat in Med*. 13, 1401-1409.

Slud E, Wei LJ. (1982). Two-sample repeated significance tests based on the modified wilcoxon statistic. *JASA*. 77(380), 862-868.

Spiegelhalter DJ, Freedman LS, Blackburn PR. (1986). Monitoring clinical trials: conditional or predictive power? *Controlled Clinical Trials*. 7, 8-17.

Spiegelhalter DJ, Freedman LS, Parmar MKB. (1994). Bayesian approaches to randomized trials (with discussion). *J R Stat Soc. (A)* 157, 357-416.

Stewart HJ, Forrest AP, Everington D. (1996). Randomized trial of five years of adjuvant tamoxifen with continuous therapy for operable breast cancer. The Scottish Cancer Trials Breast Group. *British Journal of Cancer*.74, 797-799.

Swain SM. (1996). Tamoxifen: the long and short of it. *J Natl Cancer Inst*. 88, 1510-1512.

Tormey DC, Gray R, Falkson HC. (1996). Postchemotherapy adjuvant tamoxifen therapy beyond five years in patients with lymph node positive breast cancer. *J Natl Cancer Inst.* 88, 1828-1833.

Tsiatis AA. (1981). The asymptotic joint distribution of the efficient scores test for the proportional hazards model calculated over time. *Biometrika.* 68, 311-315.

Tsiatis AA. (1982). Repeated significance testing for a general class of statistics used in censored survival analysis. *JASA.* 77(380), 855-861.

Wang SK, Tsiatis AA. (1987). Approximately optimal one-parameter boundaries for group sequential trials. *Biometrics.* 43, 193-199.

Whitehead J. Jones, D. (1979). The analysis of sequential clinical trials. *Biometrika.* 66, 443-452.

Whitehead J, Stratton I. (1983). Group sequential clinical trials with triangular continuation regions. *Biometrics.* 39, 227-236.

Wieand S, Schroeder G, O'Fallon JR. (1994). Stopping when the experimental regimen does not appear to help. *Stat in Med.* 13, 1453-1458.

Table 1: Prior and Posterior Probability of a Benefit for Continued Tamoxifen, Given Data at Third Interim Analysis and a Range of Prior Distributions

Prior Mean δ_p	Most Probable Relative Risk Based on Prior	Prior Pr{δ>0}	Posterior Pr{δ>0}	95% Credible Interval For Relative Risk
0.511	0.60	0.95	0.13	(0.58,1.16)
0.357	0.70	0.87	0.08	(0.55,1.10)
0.223	0.80	0.76	0.05	(0.53,1.05)
0.105	0.90	0.63	0.03	(0.51,1.02)
0.000	1.00	0.50	0.02	(0.49,0.98)
Diffuse Prior	----	0.50	0.006	(0.39,0.89)

At the third interim analysis, there were 32 failures on the placebo arm and 56 failures on the treatment arm. The estimated relative risk was 0.586, corresponding to a log relative risk = -.534. The prior standard deviation = $-\ln(.6)/z_{.05}$.

Figure Legends

Fig. 1: NSABP B-14: Summary of disease-free survival at third interim analysis; events include local, regional and distant treatment failures, incidence of contralateral breast cancer, other second primary cancers and deaths prior to recurrence. Analysis is based on all 1166 patients with follow-up who were event-free at five years.

Fig. 2: Conditional probability of finding a significant effect, if analysis were deferred to the 229th event. Probabilities are conditional on the results of the second and third interim analysis, and are graphed as a function of the assumed relative risk. The solid line is based on equation (1) of Davis and Hardy (1991); the dashed line is based on a binomial calculation following from a Poisson occurrence assumption.

Fig. 3: Prior distribution, likelihood and posterior distribution of the logged relative risk following the third interim analysis. An "optimistic" prior is assumed, under which the most probable treatment effect is a 40% reduction in risk, with only a 5% prior probability that the treatment provides no benefit. The prior is assumed to be normal.

Fig. 4: Posterior distribution of the log hazard ratio over successive analyses. At publication, there were 36 events among placebo patients and 66 on the tamoxifen arm (2p=.003); there were 14 deaths among the patients receiving placebo and 25 for those receiving tamoxifen (2p=.08). Through December 31, 1996, there were 50 events among placebo patients and 85 events on the tamoxifen arm (relative risk = 0.60, 2p=.002; estimated 5-year DFS: 90% for placebo, 84% for tamoxifen); there were 17 deaths for placebo patients and 36 for tamoxifen patients (estimated 5-year survival: 96% for placebo, 92% for tamoxifen, 2p=.01).

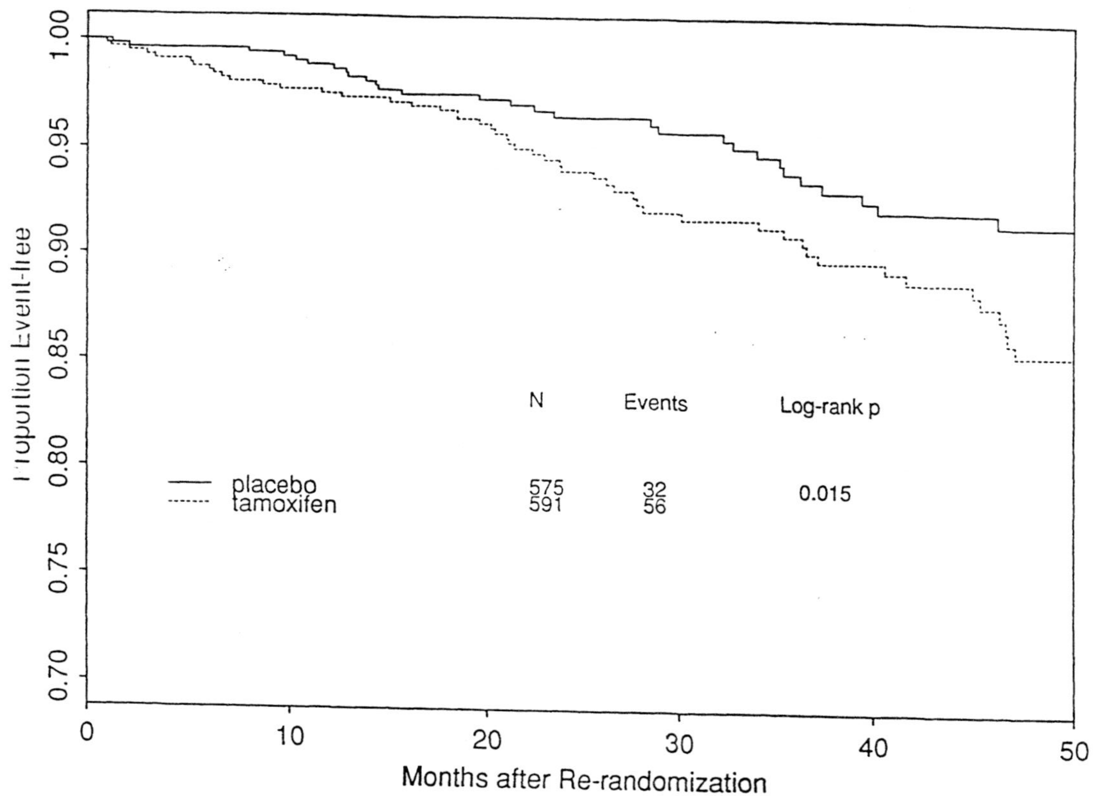

Fig. 1: NSABP B-14: Summary of disease-free survival at third interim analysis; events include local, regional and distant treatment failures, incidence of contralateral breast cancer, other second primary cancers and deaths prior to recurrence. Analysis is based on all 1166 patients with follow-up who were event-free at five years.

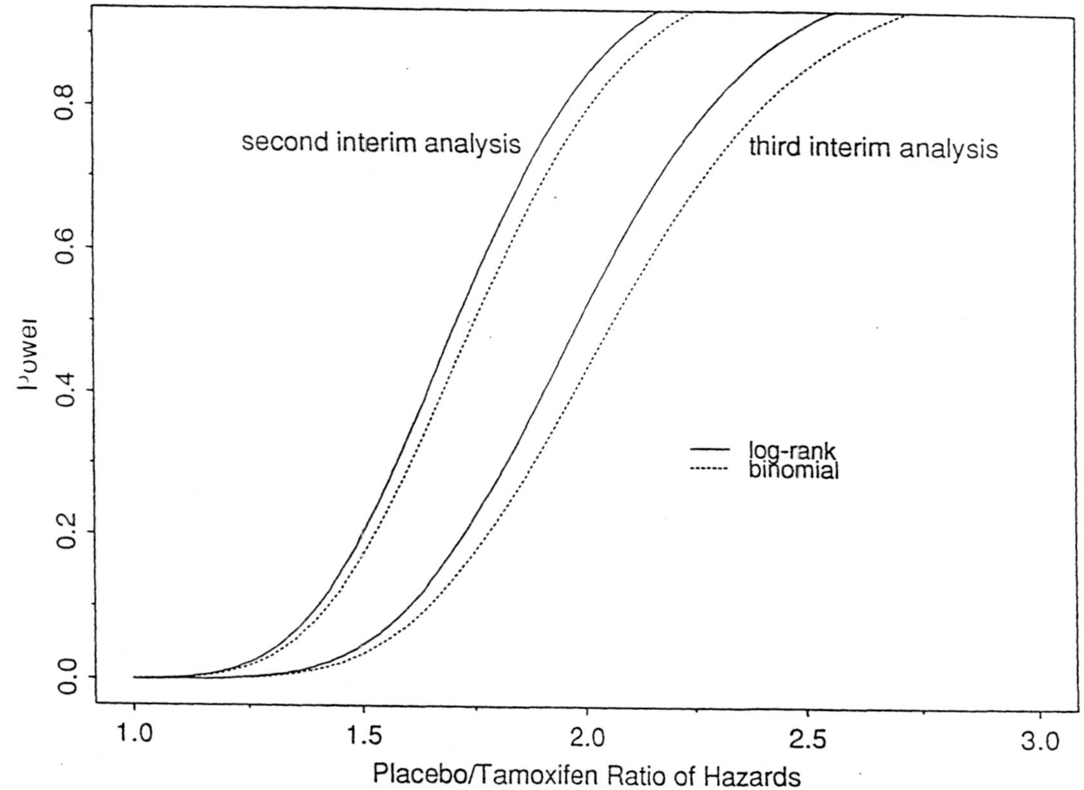

Fig. 2: Conditional probability of finding a significant effect, if analysis were deferred to the 229th event. Probabilities are conditional on the results of the second and third interim analysis, and are graphed as a function of the assumed relative risk. The solid line is based on equation (1) of Davis and Hardy (1991); the dashed line is based on a binomial calculation following from a Poisson occurrence assumption.

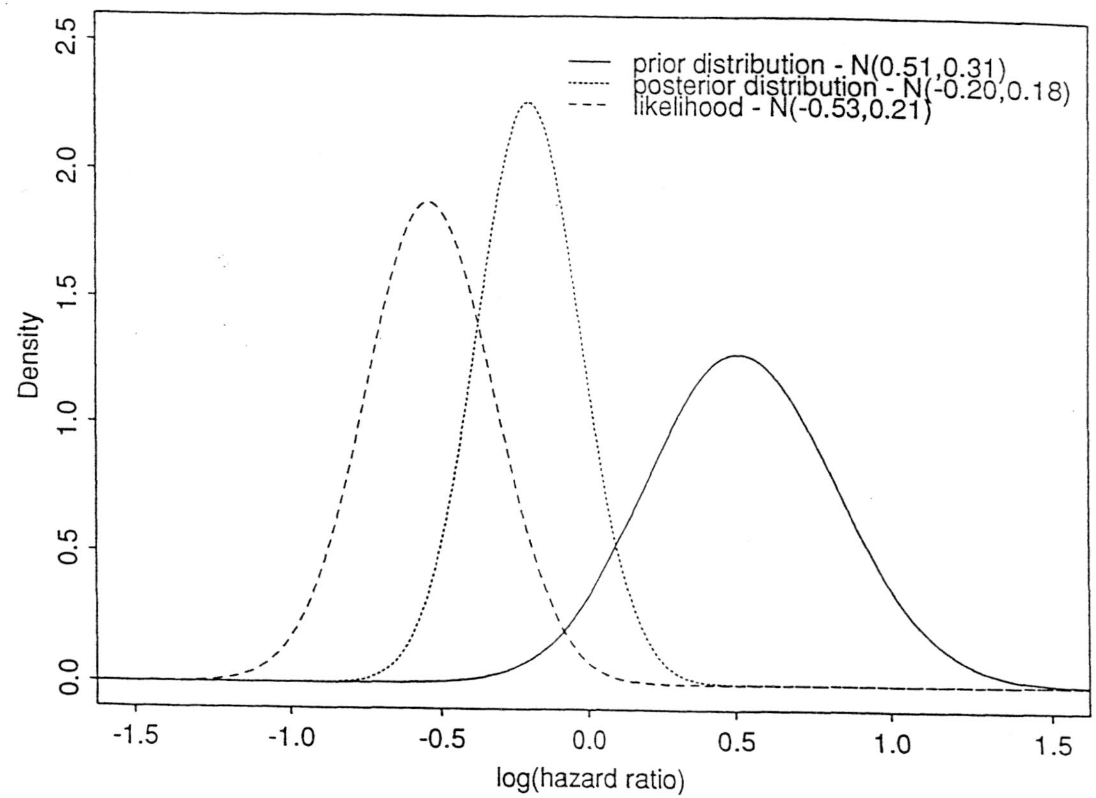

Fig. 3: Prior distribution, likelihood and posterior distribution of the logged relative risk following the third interim analysis. An "optimistic" prior is assumed, under which the most probable treatment effect is a 40% reduction in risk, with only a 5% prior probability that the treatment provides no benefit. The prior is assumed to be normal.

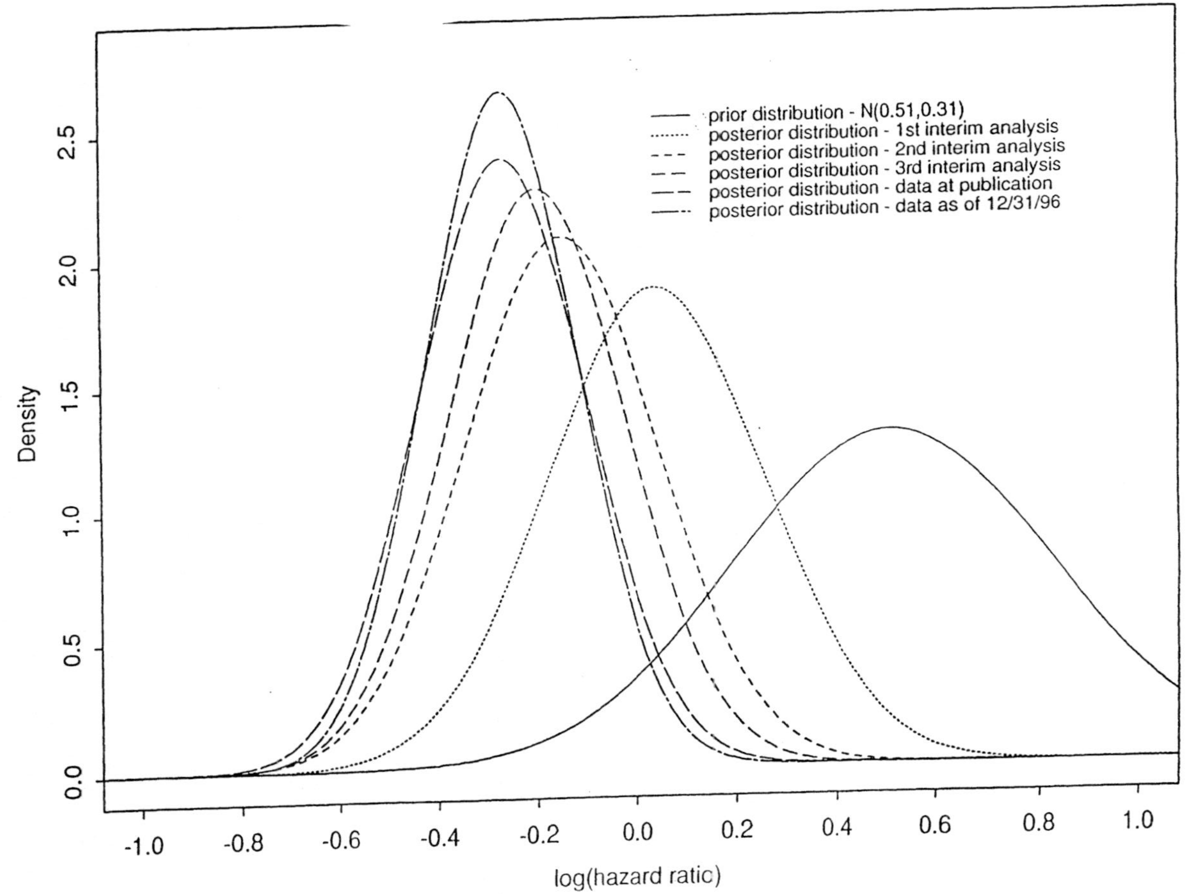

Fig. 4: Posterior distribution of the log hazard ratio over successive analyses. At publication, there were 36 events among placebo patients and 66 on the tamoxifen arm (2p--.003); there were 14 deaths among the patients receiving placebo and 25 for those receiving tamoxifen (2p=.08). Through December 31, 1996, there were 50 events among placebo patients and 85 events on the tamoxifen arm (relative risk -- 0.60, 2p=.002; estimated 5-year DFS: 90% for placebo, 84% for tamoxifen); there were 17 deaths for placebo patients and 36 for tamoxifen patients (estimated 5-year survival: 96% for placebo, 92% for tamoxifen, 2p--.01).

A Note on Quasilikelihood Estimation for GARCH Models

S. Kim [1] and S.D. Lee [2]

Abstract

Quasilikelihood estimation has been a very useful tool for estimating the parameter of interest. when the underlying distribution is not fully spesified except for the mean and the variance structure. The original idea of quasilikelihood estimation was proposed by Wedderburn (1974) and this has been widdly used in generalized linear models. On the other hand, Godambe (1960, 1985) proposed the optimal estimating functions for iid case and for discrete time stochastic processes. We are mainly concerned with the asymptotic normality of the estimator which is a consistent solution of the quasilikelihood estimating equation. Simulation results are presented to investigate the properties of the quasi-score estimators for the GARCH model by comparing with the ML estimators.

keywords: Quasilikelihood Estimation, Estimating Functions, Asymptotic Normality, GARCH models, Maximum likelihood Estimation.

1. Introduction

Quasilikelihood estimation was originally developed by Wedderburn (1974) in connection with generalized linear models. Godambe (1960) proposed an optimality criterion for general estimation functions. Godambe (1985) has further developed the theory of optimal estimating functions for discrete time stochastic processes. Several authors have studied the relationship between quasilikelihood estimation and optimal estimating functions. See the excellent review paper of Desmond (1997). In this paper, we will present the optimal estimating function as a generalization of Wedderburn's (1974) quasilikelihood function. And we will prove the asymptotic normality of the estimator obtained as a consistent solution of the quasilikelihood estimating equation. The maximum likelihood(ML) estimating function is also given to have the ML estimator as a solution of the ML estimating equation. In the last section, we will present some simulation results to compare the asymptotic variances of the quasilikelihood estimators and the ML estimators for the GARCH model.

[1]Dept. of statistics Chung-Ang University, Seoul, Korea
[2]Dept. of statistics, Chungbuk National University, Cheongju, Korea

2. Quasilikelihood Estimating Functions

Let X_t, $t \geq 1$, be a time series model satisfying a p-th order autoregressive equation

$$X_t = \theta^T \underline{X}_{t-1} + \nu_t^{1/2}(\theta) e_t \tag{2.1}$$

where e_t is an unobserved sequence of identically and independently distributed (iid) random variables with mean 0 and variance 1. $\underline{X}_{t-1} = (X_{t-1}, \cdots, X_{t-p})$ and θ is a $p \times 1$ vector, $\nu_t(\theta) = Var(X_t|F_{t-1})$, and F_{t-1} is a σ-field generated by $X_{t-1}, \cdots, X_{t-p}, p \geq 1$.

Consider the GARCH(p,q) model

$$\begin{aligned} &E(\epsilon_t|F_{t-1}) = 0, \\ &Var(\epsilon_t|F_{t-1}) = \nu_t(\theta) \\ &= \alpha_0 + \sum_{i=1}^{q} \alpha_i \epsilon_{t-i}^2 + \sum_{j=1}^{p} \beta_j \nu_{t-j}(\theta), \\ &\text{where} \quad \epsilon_t = \nu_t^{1/2}(\theta) e_t. \end{aligned} \tag{2.2}$$

In this paper, we only consider the GARCH(1,1) model for simplicity. The GARCH(1,1) model is as follow.

$$\begin{aligned} &E(\epsilon_t|F_{t-1}) = 0, \\ &\nu_t(\theta) = \alpha_0 + \alpha_1 \epsilon_{t-1}^2 + \beta_1 \nu_{t-1}(\theta). \end{aligned} \tag{2.3}$$

The necessary and sufficient conditions for the stationarity of the GARCH(1,1) model are

$$\alpha_0 > 0, \alpha_1 \geq 0, \beta_1 \geq 0 \;\; and \;\; \alpha_1 + \beta_1 < 1 \tag{2.4}$$

(Bollerslev (1986)).

Godambe (1985) showed that the optimal estimating function in the class of unbiased estimating functions of the type

$$g(X_t; \theta) = \sum_{i=1}^{n} W_t(\theta)\{X_t - \mu(\underline{X}_{t-1}; \theta)\} \tag{2.5}$$

is given by

$$g^*(X_t; \theta) = \sum_{i=1}^{n} W_t^*(\theta)\{X_t - \mu(\underline{X}_{t-1}; \theta)\} \tag{2.6}$$

where

$$\begin{aligned} &W^*(\theta) = \tfrac{dH(\underline{X}_{t-1};\theta)}{d\theta}[Var(X_t|F_{t-1})]^{-1}, \\ &\mu(\underline{X}_{t-1}; \theta) = E(X_t|F_{t-1}). \end{aligned}$$

Tne next definition illustrate a more general representation of the quasilikelihood estimation.

Definition 2.1

The quasilikelihood estimating function is defined by

$$S_n(\theta) = \sum_{i=1}^{n}[X_t - \mu_t(\underline{X}_{t-1};\theta)]\nu_t^{-1}(\theta)\frac{d\mu_t(\underline{X}_{t-1};\theta)}{d\theta} \tag{2.7}$$

where

$$\mu_t(\underline{X}_{t-1};\theta) = E(X_t|F_{t-1}),\ \nu_t(\theta) = Var(X_t|F_{t-1}),\ \underline{X}_{t-1} = (X_{t-1},\cdots,X_{t-p}),\ p \geq 1.$$

Godambe's (1985) optimal estimating function was based on finite samples. In this section we will establish the asymptotic normality of a consistent solution of the equation $S_n(\theta) = 0$ in (2.7).

2.1. The Asymptotic Normality

In order to investigate the asymptotic normality of the estimator which satisfied $S_n(\theta) = 0$ in (2.7), we need the following regularity conditions.

Conditions

1. X_t is a stationary and ergodic process with $EX_t^2 < \infty$.

2. For any θ^* with $\theta^* \xrightarrow{p} \theta$,

   $\frac{1}{n}\sum_{i=1}^{n}[\nu_t^{-1}(\theta)\frac{d\mu_t(\underline{X}_{t-1};\theta)}{d\theta}\frac{d\mu_t(\underline{X}_{t-1};\theta)^T}{d\theta}]|_{\theta=\theta^*} \xrightarrow{d} F(\theta)\ as\ n \to\infty,$ where
   $F(\theta) = E[\nu_t^{-1}(\theta)\frac{d\mu_t(\underline{X}_{t-1};\theta)}{d\theta}\frac{d\mu_t(\underline{X}_{t-1};\theta)^T}{d\theta}]$ and $F(\theta)$ is assumed to be a $p \times p$ positive definite matrix.

3. For any θ^* with $\theta^* \xrightarrow{p} \theta$, $\frac{1}{n}\sum_{i=1}^{n}[h_t(\theta^*) - h_t(\theta)] \xrightarrow{p} 0$ as $n\to\infty$,

$$h_t(\theta) = [X_t - \mu_t(\underline{X}_{t-1};\theta)]\frac{d}{d\theta}[\nu_t^{-1}(\theta)\frac{d\mu_t(\underline{X}_{t-1};\theta)}{d\theta}].$$

Theorem 2.1.

Suppose that there exists a consistent estimator $\hat{\theta}_n$ such that $S_n(\hat{\theta}_n) = 0$. Then, under regularity conditions 1 to 3, we have

$$\sqrt{n}(\hat{\theta}_n) \xrightarrow{d} N(0,\ F^{-1}(\theta)) \tag{2.8}$$

where

$$F(\theta) = E[\nu_t^{-1}(\theta)\frac{d\mu_t(\underline{X}_{t-1};\theta)}{d\theta}\frac{d\mu_t(\underline{X}_{t-1};\theta)^T}{d\theta}].$$

3. Maximum Likelihood Estimator for Nonlinear Time Series models

Suppose that $X_t, t \geq 1$ are observations from a nonlinear time series model. Assume that X_t has a conditional normal density such as

$$X_t|F_t - 1N(\mu_t(\underline{X}_{t-1};\theta),\ \mu_t(\theta)) \tag{3.1}$$

where

$$\mu_t(\underline{X}_{t-1};\theta) = E(X_t|F_{t-1}),\ \ \nu_t(\theta) = Var(X_t|F_{t-1}),$$
$$F_{t-1} = \sigma(X_{t-1},\cdots,X_{t-p}),\ \ p \geq 1,$$

and

$$\theta = (\theta_1, \theta_2, \cdots, \theta_p)^T.$$

Therefore, a likelihood function in this case is

$$L(\theta) = (\frac{1}{2\pi})^{\frac{n}{2}}[\prod_{t=1}^{n} \nu_t^{-\frac{1}{2}}(\theta)] \exp[-\frac{1}{2}\sum_{i=1}^{n} \frac{\{X_t - \mu_t(\underline{X}_{t-1};\theta)\}^2}{\nu_t(\theta)}]. \tag{3.2}$$

The above likelihood yields the following maximum likelihood (ML) estimating equation

$$S_n^*(\theta) = 0,$$

where

$$S_n^*(\theta) = S_n(\theta) - \frac{1}{2}\sum_{i=1}^{n}[\frac{\nu_t(\theta) - \{X_t - \mu_t(\underline{X}_{t-1};\theta)\}^2}{\nu_t^2(\theta)}]\frac{d\nu_t(\theta)}{d\theta}. \tag{3.3}$$

and S_n^* is the quasilikelihood estimating function in (2.7). Under standard regularity conditions a consistent solution of $S_n^*(\theta) = 0$ in (3.3) will asymptotically normal with variance-covariance matrix $F_{ML}^{-1}(\theta)$ where $F_{ML}(\theta)$ is the Fisher information matrix.

4. Simulation Results

A simulation study was carried out to compare the sample variances of the maximum likelihood (ML) estimator and the proposed quasilikelihood (QL) estimator of the AR(1) model with the GARCH(1,1) error. We consider 3 different error distributions, i.e., standard normal, maxture of two normal, i.e., $\alpha N(0,1) + (1-\alpha)N(0,\ \sigma^2)$, where $\sigma^2 > 1$ and $a < \alpha < 1$ and the standard double exponential distribution. The sample sizes of n=200 and 400 are used. The ML estimator is the soultion of (3.3) and denoted by $\hat{\theta}_{ML}$.

Let $\hat{\theta}_{QL}$ be the solution of the estimating estimating equation in (2.7) under the GARCH(1,1) error. We compare the limiting variances of the QL estimators with the ML estimators of θ by simulation. The asymptotic relative efficiency (ARE) of $\hat{\theta}_{QL}$ with respect to $\hat{\theta}_{ML}$ is given by

$$ARE = Var(\hat{\theta}_{QL})/Var(\theta_{ML}) \tag{4.1}$$

In the tables, the simulation results show that the AREs are littel bit bigger than 1 under the standard normal error distribution. This means that the ML estimators better than the QL estimators under the normal assumption. Under the heavy-tailed error distributions, the AREs are smaller than 1 especially shen σ^2 is getting larger in case of the contaminated normal cases. This implies that the QL estimators perform quite well under the heavy-tailed cases and work better the ML estimators as well.

Table 1 : AREs of ther QL Estimators for a GARCH(1,1) model with
$\theta = 0.1, \alpha_0 = 0.3, \alpha_1 = 0.2, \beta_1 = 0.2$

		standard normal	Standard Double exponential (1)	Contiminated normal		
				$\alpha = 10(10\%)$		
				Variance		
				5	10	20
n	200	1.0203	0.9279	0.7736	0.7474	0.7282
	400	1.0143	0.9299	0.7551	0.7405	0.7186

Table 2 : AREs of ther QL Estimators for a GARCH(1,1) model with
$\theta = 0.3, \alpha_0 = 0.3, \alpha_1 = 0.2, \beta_1 = 0.2$

		standard normal	Standard Double exponential (1)	Contiminated normal		
				$\alpha = 10(10\%)$		
				Variance		
				5	10	20
n	200	1.0265	0.9377	0.7701	0.7475	0.7447
	400	1.0183	0.9034	0.7682	0.7338	0.7235

Table 1 : AREs of ther QL Estimators for a GARCH(1,1) model with
$\theta = 0.5, \alpha_0 = 0.3, \alpha_1 = 0.2, \beta_1 = 0.2$

		standard normal	Standard Double exponential (1)	Contiminated normal		
				$\alpha = 10(10\%)$		
				Variance		
				5	10	20
n	200	1.0254	0.9243	0.7762	0.7747	0.7402
	400	1.0200	0.9208	0.7661	0.7378	0.7181

Table 1 : AREs of ther QL Estimators for a GARCH(1,1) model with
$\theta = 0.7, \alpha_0 = 0.3, \alpha_1 = 0.2, \beta_1 = 0.2$

		standard normal	Standard Double exponential (1)	Contiminated normal		
				$\alpha = 10(10\%)$		
				Variance		
				5	10	20
n	200	1.0232	0.9335	0.7672	0.7324	0.7211
	400	1.0103	0.9327	0.7553	0.7405	0.7109

References

[1] Basawa, I. V. (1997). Inference in stochastic processes. Technical Report, Department of Statistics, University of Georgia, Athens.

[2] Bollerslev, T. (1986). Generalized Autoregressive Conditional Heteroskedasticity, Journal of Econometrics, 31, 307-327.

[3] Dosmond, A. F. (1997). Optimal estimating functions, quasilikelihood and statistical modelling, Journal of Statistical Planning and inference 60, 77-121.

[4] Engle, R. F. (1982). Autoregressive Conditional Heteroskedasticity with estimates of the variance of U.K. inflation, Econometrica, 50, 987-1007.

[5] Godambe, V.P. (1960). An optimum property of regular maximum likelihood estimation. Annals of mathematical statistics, 31, 1208-1211.

[6] Godambe, V.P. (1985). The foundations of finite sample estimation in stochastic processes, Biometrika. 72, 419-428.

[7] Hall, P. and Heyde, C. C. (1980). Martingale Limit Theory and Applications. London : Academic Press.

[8] Kim, S. (1998). Inference for nonlinear time series models via estimating functions. Ph.D, dissertation, Univ. of Georgia.

[9] Tavaneswaran, A. and Abraham, B. (1988). Estimation for non-linear time series models using estimating equations. Journal of Time Series Analysis. 9. 99-108.

[10] Wedderburn, R. P. M. (1974). Quasilikelihood functions, generalized linear models and the Gauss-Newton method. Biometrika 61, 439-447.

SOME INFERENTIAL PROBLEMS OF TYPE II DISTRIBUTION BASED ON RECORDS

M. Ahsanullah
Department of Management Sciences
Rider University
Lawrenceville, NJ 08648-3099

Dinesh S. Bhoj
Department of Mathematical Sciences
Rutgers University
Camden, NJ 08102

ABSTRACT

Several distributional properties of record values from type II extreme value distribution are discussed. The problem of the estimation of the location and scale parameters is studied. Predictors of the sth record value based on the first m (m < s) record values are derived.

Key Words: Records, Best Linear Unbiased Estimator, Invariant Estimator, Prediction of Records

1. INTRODUCTION

A random variable (rv) is said to have Type II extreme value distribution if its cumulative distribution function (cdf), F(x), is of the following form:

$$F(x) = \exp\left(-\left(\frac{x-\mu}{\sigma}\right)^{-\delta}\right), \quad x > \mu, \sigma > 0, \delta > 0,$$

$$= 0, \text{ otherwise.} \qquad (1.1)$$

We will write $X \in TP_2(\mu,\sigma,\delta)$ if X has the cdf as given in (1.1). Suppose $\{U_n, n \geq 1\}$ is a sequence of independent and identically distributed (i.i.d.) random variables and $M_n = \max\{U_1,\ldots,U_n\}$ with distribution function F. It is well known that for suitable constants α_n and β_n, if $(M_n-\alpha_n)/\beta_n$ converges to a non degenerate distribution as $n \to \infty$, then it will converge to one of the three types extreme value distributions:

(i) Type I (Gumbel) $F(x) = \exp(-e^{-x})$, $-\infty < x < \infty$,
(ii) Type II (Frechet) $F(x) = \exp(-x^{-\delta})$, $x > 0$, $\delta > 0$,
(iii) Type III (Weibull) $F(x) = \exp(-(-x)^{-\delta})$, $x < 0$, $\delta > 0$.

Extreme value distributions have been used in the analysis of data concerning floods, extreme sea levels and air pollution problems; for details see Gumbel (1958), Horwitz (1980) and Jenkinson (1955). For detailed discussions for various distributional properties of extreme value distributions, see Galambos (1987) and Reiss (1989). Type II (Frechet) is the limiting distribution of $U_{n,n}$ when F is Pareto, Cauchy etc.

2. RECORD VALUES AND SOME DISTRIBUTIONAL PROPERTIES

In general, let $X_1, X_2, \ldots$ be a sequence of independent and identically distributed (i.i.d.) rvs with cdf F(x) and the corresponding probability density function (pdf) f(x). Set $Y_n = \min\{X_1, \ldots, X_n\}$ for $n > 1$. We say that X_j is a lower record value of $\{X_n\}$ if $Y_j < Y_{j-1}$. By definition, X_1 is a lower record value. The indices at which the lower record values occur are called the lower record value times { L(m)}. We define $\{ L(n), n \geq 1\}$ as : $L(1) = 1$, and for $m \geq 1$, $L(m) = \min_j\{j \mid j > L(m-1), X_j < X_{L(m-1)}\}$.

For a given set of n observations, let $X_{1,n} \leq \ldots \leq X_{n,n}$ be the associated order statistics. Suppose that $P\{ a_n (X_{n,n} - b_n) \leq x \} \to F(x)$ as $n \to \infty$ for some suitable constants a_n and b_n. Then it is known (see Leadbetter *et al.* 1983 p.33) that

$$P\{a_n(X_{n-m,n} - b_n) \leq x\} \xrightarrow{d} F(x) \sum_{s=0}^{m} (1/s!)\{-\ln F(x)\}^s \quad \text{where } F(x) > 0$$

(and zero when $F(x) = 0$), $m \geq 0$.

It will follow from Lemma 2.1 below that this is the distribution of the (m+1)th lower record value from a distribution F(x). For a given set of n observations we can obtain m records if m is not too large and $m < n$. One can transform from lower records to upper records by replacing the original sequence of rvs by $-X_j$, $j \geq 1$ or (if $P\{X_i > 0\}=1$) by $\{X_i^{-1}, i \geq 1\}$. The upper records of these sequences will correspond to the lower records of the original sequence. Thus it suffices to confine our attention to just lower record values.

Record values were first discussed by Chandler (1952). In many practical situations one finds that the only available data are the records; records involving sporting endeavors provide a well-known example. Feller (1966) gave some examples of record values in the context of gambling problems. For various distributional properties of upper or lower record values see Ahsanullah (1995) and Nevzorov (1988).

Lemma 2.1.

$$P(X_{L(m)} \le x) = \int_{-\infty}^{x} \frac{\{H(u)\}^{m-1}}{\Gamma(m)}\, dF(u), \text{where } H(u) = -\ln F(u),\ 0 < F(u) < 1, \tag{2.1}$$

and the joint pdf of $X_{L(r)}$ and $X_{L(s)}$ is

$$f_{r,s}(x,y) = \frac{(H(x))^{r-1}}{(r-1)!} \frac{[H(y)-H(x)]^{s-r-1}}{(s-r-1)!} h(x) f(y), \text{ for } s > r \text{ and } -\infty < y < x < \infty, \tag{2.2}$$

where $h(u) = \dfrac{f(u)}{F(u)}$.

For proof see Ahsanullah (1995) and Nagaraja (1988).

Lemma 2.2 Let $\{ Y_i,\ i = 1,2,\ldots \}$ be a sequence of independent and identically distributed random variables with $Y_i \in TP_2(0,1,,\delta)$ and $Y_{L(1)}$, $Y_{L(2)}$, .. be the corresponding lower record values of the sequence. Then for $k < n\delta$ and $1 \le m < n$, $r < m\delta$, $r + s < n\delta$. $\mu_n^k = \dfrac{\Gamma(n-\frac{k}{\delta})}{\Gamma(n)}$, and

$$\mu_{m,n}^{r,s} = \frac{\Gamma(m-\frac{r}{\delta})\Gamma(n-\frac{r+s}{\delta})}{\Gamma(m)\Gamma(n-\frac{r}{\delta})}, \text{where } \mu_n^k = E(Y_{L(n)}^k) \text{ and } \mu_{m,n}^{r,s} = E(Y_{L(m)}^r\, Y_{L(n)}^s)$$

Proof.

$$E(Y_{L(N)}^k) = \int_0^\infty \frac{u^{-\frac{k}{\delta}} u^{n-1} e^{-x}}{\Gamma(n)} du = \frac{\Gamma(n-\frac{k}{\delta})}{\Gamma(n)} \text{ and for } m < n,$$

$E(Y_{L(m)}^r\ Y_{L(n)}^s) =$

$$\int_0^\infty \int_y^\infty \frac{x^r x^{-\delta(m-1)}}{\Gamma(m)\Gamma(n-m)} y^s \left[y^{-\delta} - x^{-\delta}\right]^{n-m-1} \delta x^{-\delta-1}\ y^{-\delta-1} e^{-y^\delta}\, dy\, dx$$

$$= \int_0^\infty \frac{B(n-m, m-\frac{r}{\delta})}{\Gamma(m)\Gamma(n-m)} y^{r+s-(n-1)\delta})\ y^{-\delta-1} e^{-y^\delta}\, dy = \frac{\Gamma(m-\frac{r}{\delta})\Gamma(n-\frac{r+s}{\delta})}{\Gamma(m)\Gamma(n-\frac{r}{\delta})}.$$

Let $Y_{L(n)} = \dfrac{X_{L(n)} - \mu}{\sigma}$, then $X \in TP_2(\mu,\sigma,\delta)$ and we have

$$E(X_{L(n)}) = \mu + \sigma \frac{\Gamma(n-\frac{1}{\delta})}{\Gamma(n)}, \quad Var(X_{L(n)}) = \sigma^2\{\frac{\Gamma(n-\frac{2}{\delta})}{\Gamma(n)} - (\frac{\Gamma(n-\frac{1}{\delta})}{\Gamma(n)})^2\}$$

and

$$Cov(X_{L(m)} X_{L(n)}) = \sigma^2 \{\frac{\Gamma(m-\frac{1}{\delta})}{\Gamma(m)}[\frac{\Gamma(n-\frac{2}{\delta})}{\Gamma(n-\frac{1}{\delta})} - \frac{\Gamma(n-\frac{1}{\delta})}{\Gamma(n)}]\}.$$

We can rewrite the covariance expression as $Cov(X_{L(m)} X_{L(n)}) = \sigma^2$ $a_m b_n$, where $a_m = \frac{\Gamma(m-\frac{1}{\delta})}{\Gamma(m)}$ and $b_n = \frac{\Gamma(n-\frac{2}{\delta})}{\Gamma(n-\frac{1}{\delta})} - \frac{\Gamma(n-\frac{1}{\delta})}{\Gamma(n)}, 1 \le m \le n$.

(a) Minimum Variance Linear Unbiased Estimators (MVLUE)

Suppose $\delta \neq 0$. Then the minimum variance linear unbiased estimators of $\hat{\mu}$ and $\hat{\sigma}$ of μ and σ are respectively

$$\hat{\mu} = w_{11} X_{L(1)} + w_{12} X_{L(2)} + \cdots + w_{1m} X_{L(m)}$$

and

$$\hat{\sigma} = w_{21} X_{L(1)} + w_{22} X_{L(2)} + \cdots + w_{2m} X_{L(m)}$$

where

$$w_{1k} = \frac{a_m e_{1k}}{D_0 b_m}, \quad w_{2k} = -\frac{e_{1k}}{D_0 b_m} = -\frac{w_{1k}}{a_m}, \quad k = 1, 2, .., m-1$$

$$w_{1m} = \frac{1}{D_0}\left[\frac{a_m(b_{m-1} - b_m)}{b_m^2} \cdot \frac{1}{a_m b_{m-1} - a_{m-1} b_m} - \frac{1}{b_m^2}\right]$$

$$w_{2m} = \frac{1}{D_0}\left[-\frac{b_{m-1} - b_m}{b_m^2} \cdot \frac{1}{a_m b_{m-1} - a_{m-1} b_m} + \frac{D_1}{b_m}\right] = \frac{1}{a_m} - \frac{w_{1m}}{a_m}. \quad (2.3)$$

Proof: Let $V_1 = \frac{1}{\sqrt{a_1 b_1}} X_{L(1)}$,

$$V_k = \sqrt{\frac{b_{k-1}}{b_k}\{a_k b_{k-1} - a_{k-1} b_k\}^{-1}}\left[X_{L(k)} - \frac{b_k}{b_{k-1}} X_{L(k-1)}\right],$$

$k = 2, \ldots, m$. Then we can write

$V = AX$, $X' = (X_{L(1)}, X_{L(2)}, \ldots, X_{L(m)})$, where

$$A=\begin{pmatrix} a_{11} & a_{12} & \cdots & a_{1m} \\ a_{21} & a_{22} & \cdots & a_{2m} \\ \cdot\cdot & \cdots & \cdots & \cdots \\ a_{m1} & a_{m2} & \cdots & a_{mm} \end{pmatrix}, \qquad a_{11}=\frac{1}{\sqrt{a_1 b_1}}, a_{1i}=0, i>1$$

$$a_{k\,k-1}=-\sqrt{\frac{b_k}{b_{k-1}}}\frac{1}{\sqrt{a_k b_{k-1}-a_{k-1}b_k}},\ a_{kk}=\sqrt{\frac{b_{k-1}}{b_k}}\frac{1}{\sqrt{a_k b_{k-1}-a_{k-1}b_k}},\ a_{k,k+i}=0.i>0$$

$a_{k,k-i}=0, i>1, k=2,3,\ldots,m.$

In terms of matrix notation, we have $E(X)=\mu\,1+\sigma\,a=B\theta$, where $a'=(a_1,a_2,\cdots,a_m)$, $a_i=E\left(\frac{X_{L(i)}-\mu}{\sigma}\right)$, $i=1,2,\ldots,m$, $1'=(1,1,\ldots,1)$

$$B=\begin{pmatrix} 1 & a_1 \\ 1 & a_2 \\ \cdot & \cdot \\ 1 & a_m \end{pmatrix},$$ $\theta'=(\mu,\sigma)$. Thus $E(V)=AB\theta=C\theta$, say, where

$$C=\begin{pmatrix} c_{11} & c_{12} \\ c_{21} & c_{22} \\ \cdot & \cdot \\ c_{m1} & c_{m2} \end{pmatrix}, c_{11}=\frac{1}{\sqrt{a_1 b_1}}, c_{12}=\sqrt{\frac{a_1}{b_1}}, c_{1i}=0, i>2$$

$$c_{k1}=\frac{b_{k-1}-b_k}{\sqrt{b_{k-1}b_k}}\frac{1}{\sqrt{a_k b_{k-1}-a_{k-1}b_k}}, c_{k2}=\frac{\sqrt{(a_k b_{k-1}-a_{k-1}b_k)}}{\sqrt{b_k b_{k-1}}}, c_{ki}=0, i>2,\ .$$

$Var(V)=\sigma^2 I$, where I is an identity matrix of order $m\times m$.
Using the ordinary least squares, we get the minimum variance unbiased estimator of θ as

$$\hat{\theta}=(C'C)^{-1}C'V=(C'C)^{-1}C'AX=WX \tag{2.4}$$

Let D_0 = the determinant of $C'C$. then $D_0=\frac{a_m}{b_m}D_1-\frac{1}{b_m^2}$ and $D_1=\sum_{i=1}^{m}c_{i1}^2$.

Write $C'A=E=\begin{pmatrix} e_{11} & e_{12} & \cdot & e_{1m} \\ e_{21} & e_{22} & \cdot & e_{2m} \end{pmatrix}$, with

$e_{11}=\frac{a_2-a_1}{a_1(a_2 b_1-a_1 b_2)}$, $e_{21}=0$, $e_{2k}=0$, $k=2,\ldots,m-1$.

$$e_{1k} = \frac{b_{k-1}-b_k}{b_k} \cdot \frac{1}{a_k b_{k-1} - a_{k-1} b_k} - \frac{b_k - b_{k+1}}{b_k} \cdot \frac{1}{a_{k+1} b_k - a_k b_{k+1}}, \ k = 2,3,\ldots, m-1,$$

$$e_{1m} = c_{m1}\, a_{mm} = \frac{b_{m-1}-b_m}{b_m} \cdot \frac{1}{a_m\, b_{m-1} - a_{m-1}\, b_m} \text{ and } e_{2m} = c_{m2}\, a_{mm} = \frac{1}{b_m}.$$

Let $W=(C'C)^{-1} E$, then $W = \frac{1}{D_0}\begin{pmatrix} \frac{a_m}{b_m} & -\frac{1}{b_m} \\ -\frac{1}{b_m} & D_1 \end{pmatrix}\begin{pmatrix} e_{11} & e_{12} & \cdots & e_{1m} \\ 0 & 0 & \cdots & \frac{1}{b_m} \end{pmatrix}$

Writing $W = \begin{pmatrix} w_{11} & w_{12} & \cdots & w_{1m} \\ w_{21} & w_{22} & w_{23} & w_{2m} \end{pmatrix}$, we have

$\hat{\mu} = w_{11} X_{L(1)} + w_{12} X_{L(2)} + \cdots + w_{1m} X_{L(m)}$ and

$\hat{\sigma} = w_{21} X_{L(1)} + w_{22} X_{L(2)} + \cdots + w_{2m} X_{L(m)}$

where w_{1k} and $w_{2k}, k = 1,2,\ldots, m$ are given in (2.3). The variances and covariance of the estimators $\hat{\mu}$, $\hat{\sigma}$ are respectively

$\mathrm{Var}(\hat{\mu}) = \frac{a_m}{b_m D_0}\sigma^2$, $\mathrm{Var}(\hat{\sigma}) = \frac{D_1}{D_0}\sigma^2$ and

$\mathrm{Cov}(\hat{\mu},\hat{\sigma}) = -\frac{\sigma^2}{b_m D_0}$.

Table 1 gives the values of w_{1i} *and* w_{2i} for $\delta = 2.5$, 3.0, 4.0 and 5.0. For given values of m and δ, $w_{1i} < 0$, $i = 1,2,.., m-1$ and $w_{1m} > 0$. We observe that the weight attached to the last record value is much larger than the previous record values $X_{L(i)}$, $i = 1,2\ldots, m-1$. In the case of w_{2i}, we note that $w_{2i} > 0$ for $i = 1,2,\ldots, m-1$ and $w_{2m} < 0$. Also the absolute value of w_{2m} is much larger than w_{2i} for $i = 1,2,\ldots, m-1$. For example for $\delta = 2.5$ and $m = 6$, we have

$$\hat{\mu} = -0.0376 r_1 - 0.4386 r_2 - 0.7309 r_3 - 0.9967 r_4 - 1.2468 r_5 + 4.4505 r_6$$

and

$$\hat{\sigma} = 0.0733 r_1 + 0.8550 r_2 + 1.4249 r_3 + 1.9431 r_4 + 2.4306 r_5 - 6.7268 r_6.$$

Table 3 gives the variances of $\hat{\mu},\hat{\sigma}$ and $Cov(\hat{\mu},\hat{\sigma})$. For $\delta = 2.5$ and m = 6, we have

$$Var(\hat{\mu})=0.1150\sigma^2, Var(\hat{\sigma})=0.4690\sigma^2 \text{ and } Cov(\hat{\mu},\hat{\sigma})=-0.2233\sigma^2$$

(b) Best Invariant Estimates (BLIE)

Suppose $\delta \neq 0$. Then the best linear invariant(best in the sense of minimum mean squared error and invariant with respect to the location parameter μ) estimators (BLIE) $\tilde{\mu},\tilde{\sigma}$ of μ and σ are respectively

$$\tilde{\mu} = \hat{\mu} - d_m\hat{\sigma} \text{ and } \tilde{\sigma} = g_m\hat{\sigma}, \text{ where } d_m = \frac{E_{12}}{1+E_{22}} \text{ and } g_m = \frac{1}{1+E_{22}},$$

where $E_{11} = \frac{a_m}{b_m D_0}$, $E_{12} = -\frac{1}{b_m D_0}$ and $E_{22} = \frac{D_1}{D_0 b_m}$.

The mean squares errors (MSE) of $\tilde{\mu}$ *and* $\tilde{\sigma}$ are :

$$MSE(\tilde{\mu}) = \sigma^2 \{ E_{11} - E_{12}^2 (1+E_{22})^{-1} \},$$

$$MSE(\tilde{\sigma}) = \sigma^2 E_{22} (1+E_{22})^{-1},$$

$$E(\tilde{\mu}-\mu)(\tilde{\sigma}-\sigma) = \sigma^2 E_{12} (1+E_{22})^{-1}$$

Table 2 gives values of d_m and g_m for $\delta = 2.5, 3.0$, 4.0 and 5.0. As m increases, g_m increases while d_m decreases. Table 3 presents the MSE of $\tilde{\mu}$ *and* $\tilde{\sigma}$. For fixed δ, as m increases, as expected, $Var(\tilde{\mu}), Var(\tilde{\sigma}), MSE(\tilde{\mu})$, and $MSE(\tilde{\sigma})$ decrease. However the differences $Var(\hat{\mu})$ - $MSE(\tilde{\mu})$, and $Var(\hat{\sigma}) - MSE(\tilde{\sigma})$, decrease as m increases. For m = 6 and $\delta = 2.5$, we have $d_m = 0.1527$ and $g_m = 0.6807$. Thus $\tilde{\mu} = \hat{\mu} - 0.1587\hat{\sigma}$ and $\tilde{\sigma} = 0.6807\hat{\sigma}$. The mean squared errors (MSE) are $MSE(\tilde{\sigma}) = 0.3193\sigma^2$ and $MSE(\tilde{\mu}) = 0.0808\sigma^2$.

(c) Maximum Likelihood Estimates (MLE)

Suppose $\delta \neq 0$. Then we can write the log likelihood function ln L based on the record values $r_1, r_2, \ldots, r_m$ as

$$\ln L=\sum_{i=1}^{m-1}\ln\{\frac{f(r_i)}{F(r_i)}\}+\ln f(r_m)$$

$$=m\ln\delta+m\delta\ln\sigma-(\delta+1)\sum_{i=1}^{m-1}\ln(r_i-\mu)-\sigma^{\delta}(r_m-\mu)^{-\delta}$$

Differentiating ln L with respect to σ and μ respectively and equating the equations to zero, we get

$\frac{m\delta}{\sigma}=\delta\sigma^{\delta-1}(r_m-\mu)^{-\delta}$ and $\sum_{i=1}^{m}\frac{\delta+1}{r_i-\mu}=\delta\sigma^{\delta}(r_m-\mu)^{-(\delta+1)}$. On further simplification, we obtain

$\hat{\mu}_{ml}=r_m-\frac{\hat{\sigma}_{ml}}{m^{1/\delta}}$ and $\hat{\sigma}_{ml}=\frac{\delta}{\delta+1}m^{\frac{\delta}{\delta+1}}\sum_{I=1}^{m}\frac{1}{r_i-\mu}$, where $\hat{\mu}_{ml}$ and $\hat{\sigma}_{ml}$ are the maximum likelihood estimators of μ and σ. Solving the above two equations, we obtain the maximum likelihood estimators $\hat{\mu}_{ml}$ and $\hat{\sigma}_{ml}$ For m = 2,

$$\hat{\mu}_{ml}=\frac{(1-\delta)r_1+(1+\delta)r_2}{2} \text{ and } \hat{\sigma}_{ml}=2^{\frac{1-\delta}{2}}(1-\delta)(r_2-r_1).$$

Remarks. If δ is unknown, then we can estimate μ, σ and δ by the maximum likelihood method. Differentiating ln L with respect to μ, σ and δ, and equating the equations to zero, we get the following equations:

$\sigma-\frac{\delta}{\delta+1}m^{\frac{1+\delta}{\delta}}(\sum_{i=1}^{m}\frac{1}{r_i-\mu})^{-1}=0,\quad \mu+\sigma m^{-\frac{1}{\delta}}-r_m=0$ and

$\frac{m}{\delta}-\sum_{i=1}^{m}\ln(r_i-\mu)+m\ln(r_m-\mu)=0$. These equations have to be solved by iteration.

3. Prediction of $X_{L(s)}$ (s > m)

We will present here predictors of the s th lower record value based on the first m record values for s > m. Let K' = ($K_1, K_2, \ldots, K_m$), where $\sigma^2 K_i=Cov\left(V_i, X_{L(s)}\right), i=1,\ldots,m$ and $a_s=\sigma^{-1}E\left(X_{L(s)}-\mu\right)$.

The best linear unbiased predictor of $X_{L(s)}$ is

$$\hat{X}_{L(s)}=\hat{\mu}+\hat{\sigma}a_s+K'Z^{-1}(V-C\hat{\theta}),$$

$\sigma^2 K_i = \mathrm{Cov}\left(V_i, X_{L(s)}\right), i = 1,\ldots,m$ and $a_s = \sigma^{-1}E\left(X_{L(s)} - \mu\right)$.

The best linear unbiased predictor of $X_{L(s)}$ is

$$\hat{X}_{L(s)} = \hat{\mu} + \hat{\sigma} a_s + K'Z^{-1}(V - C\hat{\theta}),$$

where $\sigma^2 Z^{-1}$ = variance covariance matrix of V. It can easily be shown that $K'Z^{-1}(V - C\hat{\theta}) = 0$. Thus $\hat{X}_{L(s)} = \hat{\mu} + \hat{\sigma} a_s$ and $\mathrm{Var}(\hat{X}_{L(s)}) = \mathrm{Var}(\hat{\mu}) + (a_s)^2 \mathrm{Var}(\hat{\sigma}) + 2a_s \mathrm{Cov}(\hat{\mu}, \hat{\sigma})$. The best linear invariant predictor of $X_{L(s)}$ is

$$\tilde{X}_{L(s)} = \hat{X}_{L(s)} - \frac{\tilde{c}_{12}}{1+\tilde{c}_{22}}\hat{\sigma} = \hat{\mu} + (a_s - \frac{\tilde{c}_{12}}{1+\tilde{c}_{22}})\hat{\sigma}\text{ , where}$$

$$\tilde{c}_{22}\sigma^2 = \mathrm{Var}(\hat{\sigma}) = \sigma^2 \frac{D_1}{D_0}$$

and

$$\tilde{c}_{12}\sigma^2 = \mu_n^k = E(Y_{L(n)}^k) \text{ and } \mu_{m,n}^{r,s} = E(Y_{L(m)}^r\, Y_{L(n)}^s)[\hat{\sigma}, (1 - \frac{b_s}{b_m})\hat{\mu} + (a_s - \frac{b_s}{b_m} a_m)\hat{\sigma}]$$

REFERENCES

Ahsanullah, M. (1995). Record Statistics. Nova Science Publications, Inc. Commack, NY.

Arnold, B. C., N. Balakrishnan and H. N . Nagaraja (1992). A First Course in Order Statistics, John Wiley & Sons, New York.

Chandler, K. M. (1952). The distribution and frequency of record values. J. R. Statist. Soc. B 14, 220-228.

Feller, W. (1966). An Introduction to Probability Theory and its Applications. Vol. II, Wiley, New York.

Galambos, J. (1987). The Asymptotic Theory of Extreme Order Statistics. Robert E. Krieger Publishing Co. Malaban, Florida.

Goldberger, A.S. (1962). Best linear unbiased prediction in the generalized linear regression model. JASA 57, 369-379.

Gumbel, E. J. (1958). Statistics of Extremes. Columbia Univ. Press, New York.

Horwitz, J. (1980). Extreme values from a non stationary stochastic process: an application to air quality analysis (with discussion) . Technometrics 22,469-482.

Jenkinson, A.F. (1955). The frequency distribution of the annual maximum

(or minimum) values of meteorological elements. Quart. J. Meter. Soc. 87,158-171.

Leadbetter, M.R., Lindgreen, G. and Rootzen, H (1983). Extremes and Related Properties of Random Sequences and Series, Springer-Verlag, New York, N.Y.

Nagaraja, H. N. (1988) Record values and related statistics -- a review, Commun. Statist. Theory Meth. 17, 2223-2238.

Nevzorov, V. B. (1988). Records. Theo. Prob. Apl. 32, 201-228.

Reiss, R.D. (1989). Approximate Distributions of Order Statistics. Springer - Verlag, New York, NY.

Table 1: Coefficients of the MVLUE of $\hat{\mu}$ and $\hat{\sigma}$

$\delta = 2.5$

m/ i	1	2	3	4	5	6	7	8	9	10	11	12	13	14	15
2	-1.5000	2.5000													
2	1.6788	-1.6788													
3	-0.2500	-2.9167	4.1667												
3	0.3497	4.0804	-4.4302												
4	-0.1046	-1.2209	-2.0349	4.3605											
4	0.1689	1.9708	3.2847	-5.4245											
5	-0.0584	-0.6810	-1.1350	-1.5478	4.4222										
5	0.1047	1.2215	2.0357	2.7760	-6.1379										
6	-0.0376	-0.4386	-0.7309	-0.9967	-1.2468	4.4505									
6	0.0733	0.8550	1.4249	1.9431	2.4306	-6.7268									
7	-0.0264	-0.3080	-0.5134	-0.7000	-0.8757	-1.0413	4.4648								
7	0.0551	0.6434	1.0723	1.4622	1.8291	2.1751	-7.2372								
8	-0.0197	-0.2293	-0.3822	-0.5212	-0.6520	-0.7754	-0.8904	4.4702							
8	0.0435	0.5081	0.8468	1.1547	1.4444	1.7177	1.9726	-7.6877							
9	-0.0152	-0.1778	-0.2964	-0.4042	-0.5056	-0.6013	-0.6905	-0.7994	4.4904						
9	0.0355	0.4147	0.6912	0.9425	1.1790	1.4021	1.6102	1.8641	-8.1395						
10	-0.0122	-0.1425	-0.2375	-0.3239	-0.4051	-0.4818	-0.5533	-0.6405	-0.6750	4.4718					
10	0.0298	0.3478	0.5796	0.7904	0.9887	1.1757	1.3502	1.5631	1.6473	-8.4726					
11	-0.0100	-0.1170	-0.1951	-0.2660	-0.3327	-0.3957	-0.4544	-0.5260	-0.5544	-0.6174	4.4687				
11	0.0255	0.2975	0.4958	0.6761	0.8458	1.0058	1.1551	1.3372	1.4093	1.5695	-8.8178				
12	-0.0084	-0.0978	-0.1629	-0.2222	-0.2779	-0.3305	-0.3796	-0.4394	-0.4631	-0.5157	-0.6299	4.5275			
12	0.0221	0.2579	0.4298	0.5861	0.7332	0.8719	1.0013	1.1592	1.2217	1.3605	1.6617	-9.3056			
13	-0.0071	-0.0831	-0.1385	-0.1888	-0.2362	-0.2809	-0.3226	-0.3734	-0.3936	-0.4383	-0.5353	-0.5061	4.5040		
13	0.0194	0.2267	0.3779	0.5153	0.6446	0.7666	0.8803	1.0191	1.0740	1.1961	1.4609	1.3813	-9.5624		
14	-0.0061	-0.0717	-0.1196	-0.1631	-0.2040	-0.2426	-0.2786	-0.3225	-0.3399	-0.3785	-0.4623	-0.4371	-0.4097	4.4355	
14	0.0173	0.2020	0.3367	0.4591	0.5743	0.6830	0.7843	0.9080	0.9569	1.0657	1.3016	1.2306	1.1536	-9.6732	
15	-0.0054	-0.0626	-0.1043	-0.1422	-0.1779	-0.2116	-0.2430	-0.2813	-0.2964	-0.3301	-0.4032	-0.3812	-0.3574	-0.4765	4.4732
15	0.0155	0.1814	0.3023	0.4122	0.5157	0.6132	0.7042	0.8153	0.8592	0.9569	1.1687	1.1050	1.0358	1.3812	-10.0668

Table 1: *(continued)*

$\delta = 3.0$

m/ i	1	2	3	4	5	6	7	8	9	10	11	12	13	14	15
2	-2.0000	3.0000													
2	2.2155	-2.2156													
3	-0.5000	-3.0000	4.5000												
3	0.6646	3.9879	-4.6525												
4	-0.2353	-1.4118	-2.1175	4.7646											
4	0.3519	2.1113	3.1666	-5.6298											
5	-0.1400	-0.8400	-1.2599	-1.6202	4.8601										
5	0.2284	1.3704	2.0554	2.6431	-6.2974										
6	-0.0942	-0.5652	-0.8478	-1.0902	-1.3090	4.9064									
6	0.1647	0.9880	1.4818	1.9056	2.2880	-6.8282									
7	-0.0684	-0.4103	-0.6154	-0.7914	-0.9502	-1.0979	4.9336								
7	0.1266	0.7594	1.1390	1.4647	1.7586	2.0320	-7.2801								
8	-0.0523	-0.3138	-0.4706	-0.6052	-0.7267	-0.8396	-0.9328	4.9410							
8	0.1016	0.6098	0.9145	1.1761	1.4121	1.6316	1.8127	-7.6585							
9	-0.0415	-0.2489	-0.3733	-0.4800	-0.5763	-0.6659	-0.7398	-0.8392	4.9650						
9	0.0841	0.5046	0.7569	0.9733	1.1687	1.3503	1.5002	1.7018	-8.0400						
10	-0.0339	-0.2033	-0.3050	-0.3922	-0.4709	-0.5441	-0.6045	-0.6857	-0.7023	4.9417					
10	0.0714	0.4282	0.6422	0.8258	0.9915	1.1457	1.2728	1.4438	1.4789	-8.3003					
11	-0.0283	-0.1696	-0.2544	-0.3271	-0.3927	-0.4538	-0.5042	-0.5719	-0.5858	-0.6897	4.9774				
11	0.0616	0.3694	0.5541	0.7125	0.8555	0.9885	1.0983	1.2458	1.2761	1.5024	-8.6644				
12	-0.0239	-0.1431	-0.2147	-0.2761	-0.3315	-0.3830	-0.4255	-0.4827	-0.4944	-0.5821	-0.7899	5.1467			
12	0.0536	0.3215	0.4823	0.6202	0.7446	0.8604	0.9559	1.0843	1.1106	1.3076	1.7744	-9.3155			
13	-0.0206	-0.1239	-0.1858	-0.2389	-0.2869	-0.3315	-0.3683	-0.4177	-0.4279	-0.5038	-0.6836	-0.2550	4.8439		
13	0.0477	0.2862	0.4293	0.5521	0.6629	0.7659	0.8509	0.9652	0.9887	1.1640	1.5796	0.5892	-8.8819		
14	-0.0180	-0.1079	-0.1619	-0.2082	-0.2500	-0.2888	-0.3209	-0.3640	-0.3728	-0.4389	-0.5957	-0.2222	-0.6685	5.0177	
14	0.0427	0.2560	0.3839	0.4937	0.5928	0.6849	0.7609	0.8632	0.8841	1.0409	1.4126	0.5269	1.5853	-9.5279	
15	-0.0159	-0.0952	-0.1427	-0.1836	-0.2204	-0.2547	-0.2829	-0.3209	-0.3287	-0.3870	-0.5252	-0.1959	-0.5894	-0.4263	4.9688
15	0.0385	0.2312	0.3467	0.4459	0.5354	0.6186	0.6873	0.7796	0.7986	0.9402	1.2758	0.4759	1.4319	1.0357	-9.6413

Table 1: *(continued)*

δ = 4.0

m/ i	1	2	3	4	5	6	7	8	9	10	11	12	13	14	15
2	-3.0000	4.0000													
2	3.2642	-3.2643													
3	-1.0000	-3.3333	5.3333												
3	1.2435	4.1450	-5.3885												
4	-0.5294	-1.7647	-2.3532	5.6472											
4	0.7182	2.3938	3.1922	-6.3042											
5	-0.3383	-1.1277	-1.5039	-1.8052	5.7751										
5	0.4895	1.6318	2.1761	2.6121	-6.9095										
6	-0.2396	-0.7985	-1.0648	-1.2782	-1.4576	5.8386									
6	0.3649	1.2162	1.6219	1.9469	2.2201	-7.3699									
7	-0.1809	-0.6029	-0.8040	-0.9651	-1.1005	-1.2258	5.8792								
7	0.2875	0.9582	1.2778	1.5339	1.7491	1.9483	-7.7549								
8	-0.1428	-0.4759	-0.6346	-0.7618	-0.8687	-0.9676	-1.0484	5.8997							
8	0.2353	0.7844	1.0460	1.2556	1.4318	1.5948	1.7280	-8.0758							
9	-0.1164	-0.3880	-0.5174	-0.6211	-0.7083	-0.7889	-0.8548	-0.9138	5.9088						
9	0.1981	0.6602	0.8804	1.0568	1.2050	1.3422	1.4544	1.5547	-8.3518						
10	-0.0972	-0.3239	-0.4319	-0.5184	-0.5912	-0.6585	-0.7135	-0.7627	-0.8555	5.9525					
10	0.1700	0.5667	0.7558	0.9072	1.0345	1.1523	1.2486	1.3347	1.4972	-8.6671					
11	-0.0826	-0.2754	-0.3672	-0.4408	-0.5026	-0.5599	-0.6066	-0.6485	-0.7274	-0.7779	5.9888				
11	0.1483	0.4942	0.6591	0.7911	0.9022	1.0049	1.0888	1.1639	1.3057	1.3962	-8.9545				
12	-0.0712	-0.2372	-0.3163	-0.3797	-0.4330	-0.4823	-0.5226	-0.5586	-0.6267	-0.6701	-0.7929	6.0906			
12	0.1307	0.4357	0.5810	0.6974	0.7953	0.8858	0.9598	1.0260	1.1510	1.2308	1.4562	-9.3498			
13	-0.0627	-0.2091	-0.2788	-0.3346	-0.3816	-0.4251	-0.4606	-0.4923	-0.5523	-0.5906	-0.6987	-0.2148	5.7011		
13	0.1176	0.3921	0.5229	0.6277	0.7158	0.7973	0.8639	0.9235	1.0359	1.1078	1.3106	0.4029	-8.8181		
14	-0.0552	-0.1839	-0.2453	-0.2944	-0.3358	-0.3740	-0.4052	-0.4332	-0.4859	-0.5196	-0.6148	-0.1890	-1.1061	6.2425	
14	0.1055	0.3518	0.4691	0.5631	0.6422	0.7153	0.7750	0.8285	0.9294	0.9938	1.1758	0.3614	2.1154	-10.0265	
15	-0.0496	-0.1655	-0.2207	-0.2649	-0.3021	-0.3365	-0.3646	-0.3897	-0.4372	-0.4675	-0.5531	-0.1700	-0.9951	0.0932	5.6232
15	0.0967	0.3222	0.4297	0.5158	0.5882	0.6552	0.7099	0.7589	0.8513	0.9103	1.0770	0.3311	1.9377	-0.1815	-9.0027

Table 1: *(continued)*

δ = 5.0

m/ i	1	2	3	4	5	6	7	8	9	10	11	12	13	14	15
2	-4.0001	5.0001													
2	4.2948	-4.2949													
3	-1.4999	-3.7503	6.2502												
3	1.7893	4.4740	-6.2634												
4	-0.8421	-2.1056	-2.6307	6.5784											
4	1.0764	2.6913	3.3625	-7.1302											
5	-0.5590	-1.3978	-1.7464	-2.0197	6.7229										
5	0.7521	1.8807	2.3496	2.7175	-7.7000										
6	-0.4072	-1.0180	-1.2719	-1.4710	-1.6243	6.7924									
6	0.5706	1.4268	1.7826	2.0617	2.2765	-8.1182									
7	-0.3142	-0.7856	-0.9815	-1.1351	-1.2534	-1.3803	6.8501								
7	0.4555	1.1390	1.4230	1.6458	1.8172	2.0012	-8.4817								
8	-0.2526	-0.6317	-0.7892	-0.9128	-1.0079	-1.1099	-1.1531	6.8572							
8	0.3771	0.9428	1.1779	1.3623	1.5042	1.6565	1.7210	-8.7417							
9	-0.2087	-0.5219	-0.6520	-0.7541	-0.8327	-0.9170	-0.9526	-1.1112	6.9502						
9	0.3195	0.7989	0.9981	1.1543	1.2746	1.4036	1.4583	1.7010	-9.1084						
10	-0.1771	-0.4427	-0.5531	-0.6397	-0.7064	-0.7779	-0.8082	-0.9427	-0.7764	6.8242					
10	0.2772	0.6931	0.8660	1.0015	1.1059	1.2178	1.2652	1.4758	1.2155	-9.1182					
11	-0.1526	-0.3816	-0.4768	-0.5514	-0.6089	-0.6705	-0.6966	-0.8125	-0.6692	-0.8849	6.9050				
11	0.2438	0.6096	0.7616	0.8809	0.9727	1.0711	1.1128	1.2980	1.0691	1.4136	-9.4333				
12	-0.1332	-0.3329	-0.4160	-0.4811	-0.5312	-0.5850	-0.6077	-0.7089	-0.5838	-0.7720	-0.8645	7.0161			
12	0.2166	0.5417	0.6768	0.7827	0.8643	0.9518	0.9888	1.1534	0.9500	1.2561	1.4066	-9.7889			
13	-0.1172	-0.2930	-0.3660	-0.4233	-0.4674	-0.5147	-0.5348	-0.6238	-0.5138	-0.6793	-0.7607	-0.9061	7.2000		
13	0.1939	0.4848	0.6056	0.7005	0.7734	0.8517	0.8849	1.0322	0.8501	1.1240	1.2587	1.4993	-10.2591		
14	-0.1046	-0.2616	-0.3269	-0.3781	-0.4175	-0.4597	-0.4776	-0.5571	-0.4588	-0.6067	-0.6794	-0.8092	-0.4139	6.9510	
14	0.1759	0.4397	0.5493	0.6353	0.7015	0.7726	0.8026	0.9362	0.7711	1.0195	1.1417	1.3599	0.6955	-10.0009	
15	-0.0951	-0.2378	-0.2971	-0.3436	-0.3794	-0.4178	-0.4341	-0.5064	-0.4170	-0.5514	-0.6175	-0.7355	-0.3762	0.0277	6.3815
15	0.1622	0.4054	0.5065	0.5858	0.6469	0.7124	0.7401	0.8633	0.7110	0.9401	1.0528	1.2539	0.6413	-0.0472	-9.1747

Table 2. Coefficients d_m and g_m for $\tilde{\mu}$ and $\tilde{\sigma}$

δ	$\backslash m$	2	3	4	5	6	7	8	9	10	11	12	13	14	15
2.5	d_m	0.7446	0.4156	0.2740	0.1985	0.1527	0.1224	0.1011	0.0854	0.0735	0.0643	0.0567	0.0506	0.0456	0.0414
	g_m	0.1449	0.3883	0.5299	0.6194	0.6807	0.7251	0.7587	0.7851	0.8062	0.8235	0.8382	0.8506	0.8610	0.8701
3.0	d_m	0.6771	0.3983	0.2708	0.2005	0.1567	0.1273	0.1063	0.0907	0.0788	0.0693	0.0613	0.0552	0.0499	0.0455
	g_m	0.2282	0.4474	0.5746	0.6553	0.7106	0.7508	0.7812	0.8051	0.8241	0.8399	0.8538	0.8644	0.8741	0.8824
4.0	d_m	0.6127	0.3784	0.2660	0.2018	0.1608	0.1327	0.1123	0.0970	0.0849	0.0753	0.0673	0.0612	0.0555	0.0512
	g_m	0.3177	0.5151	0.6270	0.6977	0.7461	0.7813	0.8079	0.8287	0.8456	0.8596	0.8715	0.8807	0.8897	0.8964
5.0	d_m	0.5821	0.3676	0.2629	0.2020	0.1628	0.1354	0.1155	0.1002	0.0885	0.0790	0.0711	0.0643	0.0588	0.0546
	g_m	0.3641	0.5519	0.6560	0.7215	0.7661	0.7986	0.8230	0.8425	0.8577	0.8704	0.8812	0.8907	0.8984	0.9044

Table 3: Variances and Covariance of MVLUE and MSE of $\tilde{\mu}$ and $\tilde{\sigma}$

$\delta = 2.5$

m	$V(\hat{\mu})$	$V(\hat{\sigma})$	$Cov(\hat{\mu},\hat{\sigma})$	$MSE(\tilde{\mu})$	$MSE(\tilde{\sigma})$
2	4.5909	5.9004	-5.1380	0.7651	0.8551
3	0.7651	1.5756	-1.0704	0.3203	0.6117
4	0.3203	0.8872	-0.5170	0.1787	0.4701
5	0.1787	0.6144	-0.3204	0.1151	0.3806
6	0.1150	0.4690	-0.2243	0.0808	0.3193
7	0.0808	0.3791	-0.1688	0.0601	0.2749
8	0.0602	0.3180	-0.1333	0.0467	0.2413
9	0.0467	0.2736	-0.1088	0.0374	0.2149
10	0.0374	0.2404	-0.0912	0.0307	0.1938
11	0.0307	0.2143	-0.0780	0.0257	0.1765
12	0.0256	0.1930	-0.0677	0.0218	0.1618
13	0.0218	0.1756	-0.0595	0.0188	0.1494
14	0.0188	0.1615	-0.0530	0.0164	0.1390
15	0.0164	0.1493	-0.0476	0.0144	0.1299

$\delta = 5.0$

m	$V(\hat{\mu})$	$V(\hat{\sigma})$	$Cov(\hat{\mu},\hat{\sigma})$	$MSE(\tilde{\mu})$	$MSE(\tilde{\sigma})$
2	1.4892	1.7468	-1.5989	0.5585	0.6359
3	0.5584	0.8120	-0.6662	0.3135	0.4481
4	0.3135	0.5243	-0.4007	0.2082	0.3440
5	0.2081	0.3861	-0.2800	0.1516	0.2785
6	0.1516	0.3053	-0.2124	0.1170	0.2339
7	0.1170	0.2522	-0.1696	0.0940	0.2014
8	0.0941	0.2150	-0.1404	0.0778	0.1770
9	0.0777	0.1869	-0.1190	0.0658	0.1575
10	0.0659	0.1659	-0.1032	0.0568	0.1423
11	0.0568	0.1489	-0.0908	0.0496	0.1296
12	0.0496	0.1348	-0.0807	0.0438	0.1188
13	0.0436	0.1227	-0.0722	0.0390	0.1093
14	0.0390	0.1130	-0.0655	0.0351	0.1016
15	0.0354	0.1057	-0.0604	0.0321	0.0956

AN AUTOREGRESSIVE DISAGGREGATION METHOD

by ERIN M. HODGESS
Department of Computer and Mathematical Sciences,
University of Houston–Downtown
and
WILLIAM W.S. WEI
Department of Statistics, Temple University

Abstract. We propose a procedure to disaggregate autoregressive integrated moving average ARIMA(p, d, r) models. We show how to calculate the autocovariances of a disaggregate autoregressive integrated process from those of the aggregate ARIMA process, and demonstrate that their relationship is unique. The disaggregate series is then derived from the autoregressive integrated model.

Keywords. Disaggregation; Aggregation; ARIMA process.

1. INTRODUCTION

We receive much of our data in aggregate form. In the U.S. economic data alone, all of the National Income and Product Account data are available on an annual and quarterly basis only. With such aggregation comes a loss of information. We will demonstrate that some of this information can be re-captured through the process of disaggregation.

Consider the following basic series, x_t, which follows the ARIMA(p, d, q) model:

$$\phi_p(B)(1 - B)^d x_t = \theta_q(B)a_t, \tag{1.1}$$

where B is the backshift operator such that $B^j x_t = x_{t-j}$, and the a_t series is a Gaussian white noise series with mean zero and variance σ_a^2. We assume that the roots of the $\phi_p(B)$ real-valued polynomial are outside of the unit circle to ensure stationarity. Similarly, we assume that the roots of the $\theta_q(B)$ real-valued polynomial are outside of the unit circle to ensure invertibility. Finally, we assume that these two polynomials have no roots in common. There are nm observations in the basic series. Notice that we do not have this basic series available; rather we will reconstruct this series via disaggregation.

The temporal mth order aggregate series, Y_T, is constructed as:

$$Y_T \quad = \sum_{t=m(T-1)+1}^{mT} x_t = (1 + B + \cdots + B^{m-1})x_{mT}. \tag{1.2}$$

Stram and Wei (1986a) show that the aggregate series follows the ARIMA(p, d, r) model:

$$\alpha_p(\mathcal{B})(1 - \mathcal{B})^d Y_T \quad = \quad \beta_r(\mathcal{B})C_T, \tag{1.3}$$

where $\mathcal{B}$ is the backshift operator on the aggregate model, and the C_T is a Gaussian white noise series with mean zero, variance σ_C^2, and $r = \left[p + d + 1 + \frac{(q-p-d-1)}{m}\right]$. There are n observations in the aggregate series.

Stram and Wei (1986b) produced a generalized least squares procedures for the disaggregation of time series. We will use a method similar to theirs, and we need to present additional notation. We will let $\mathbf{x} = \{x_1, x_2, \ldots, x_{nm}\}'$ be our disaggregate vector which has length nm. Next, we let $w_t = (1 - B)^d x_t$, and the differenced disaggregate vector becomes $\mathbf{w} = \{w_{d+1}, w_{d+2}, \ldots, w_{nm}\}'$. This $\mathbf{w}$ vector is a stationary process with a variance matrix $\mathbf{V}_w$. We follow that logic for the aggregate series; i.e., $\mathbf{Y} = \{Y_1, Y_2, \ldots, Y_n\}'$ and its differenced vector, $\mathbf{U} = \{U_{d+1}, U_{d+2}, \ldots, U_n\}'$, where $U_T = (1 - \mathcal{B})^d Y_T$, and $\mathbf{U}$ follows a stationary process with a variance matrix of $\mathbf{V}_U$. We can express the relationship between the original and differenced series as:

$$\mathbf{w} = \mathbf{\Delta}_{nm}^d \mathbf{x} \quad \text{and} \quad \mathbf{U} = \mathbf{\Delta}_n^d \mathbf{Y}, \tag{1.4}$$

where

$$\mathbf{\Delta}_k^d \quad = \begin{pmatrix} \delta_0 & \delta_1 & \ldots & \delta_d & 0 & \ldots & \ldots & 0 \\ 0 & \delta_0 & \delta_1 & \ldots & \delta_d & & & \\ & & \ddots & \ddots & & \ddots & & \\ & & & \ddots & \ddots & & \ddots & 0 \\ 0 & \ldots & \ldots & 0 & \delta_0 & \delta_1 & \ldots & \delta_d \end{pmatrix}, \tag{1.5}$$

is a $(k - d) \times k$ matrix and the δ_i are the coefficients of B^i in $(B - 1)^d$. Also, $U_T = (1 + B + \cdots + B^{m-1})^{d+1} w_{mT}$ for $T = d + 1, \ldots, n$, and thus $\mathbf{U} = \mathbf{C}^d \mathbf{w}$, where $\mathbf{C}^d$ is an $(n-d) \times (nm-d)$ matrix such that:

$$\mathbf{C}^d = \begin{pmatrix} \mathbf{c}' & \underset{(n-d-1)m}{\mathbf{0}'} & & & \\ \underset{m}{\mathbf{0}'} & \mathbf{c}' & \underset{(n-d-2)m}{\mathbf{0}'} & & \\ \underset{2m}{\mathbf{0}'} & & \mathbf{c}' & \underset{(n-d-3)m}{\mathbf{0}'} & \\ \vdots & \vdots & & & \\ \underset{(n-d-1)m}{\mathbf{0}'} & & & & \mathbf{c}' \end{pmatrix}. \tag{1.6}$$

The $\underset{k}{\mathbf{0}'}$ is a $1 \times k$ vector of zeros and the $\mathbf{c}' = \left(c_0, c_1, \ldots, c_{(d+1)(m-1)}\right)$ is a $1 \times ((d+1)(m-1)+1)$ row vector of c_i such that the c_i are the coefficients of B^i in the polynomial $(1 + B + \cdots + B^{m-1})^{d+1}$.

Stram and Wei (1986b) developed the following disaggregation procedure for a zero mean Gaussian process with an assumed $\mathbf{V}_w$:

$$\hat{\mathbf{w}} = \mathbf{V}_w \left(\mathbf{C}^d\right)' \mathbf{V}_U^{-1} \mathbf{\Delta}_n^d \mathbf{Y}, \tag{1.7}$$

$$\hat{\mathbf{x}} = \left(\frac{\mathbf{\Delta}_{nm}^d}{0 \mid \mathbf{I}_d \otimes \mathbf{J}_m}\right)^{-1} \left(\frac{\mathbf{V}_w \left(\mathbf{C}^d\right)' \mathbf{V}_U^{-1} \mathbf{\Delta}_n^d}{0 \mid \mathbf{I}_d}\right) \mathbf{Y}, \tag{1.8}$$

where $\mathbf{I}_d$ is the $d \times d$ identity matrix, $\mathbf{J}_m$ is a $1 \times m$ vector of ones, and $\otimes$ represents the Kronecker product. Wei and Stram (1990) further developed this approach by demonstrating a method to calculate $\hat{\mathbf{V}}_w$. Hodgess (1995) and Hodgess and Wei (1997) proposed a different method for the $\hat{\mathbf{V}}_w$ calculation. We will continue in this same vein and not only show a new method of $\hat{\mathbf{V}}_w$ calculation, but will produce a disaggregate ARI(p, d) series as well.

For historical reference, we show the development of these models:

Source	Aggregate Order	Disaggregate Order
Wei and Stram (1990)	ARIMA(p, d, r)	ARIMA(p, d, r^{**}))
Hodgess and Wei (1997)	ARIMA(p, d, r)	ARIMA(p, d, r)
New Method	ARIMA(p, d, r)	ARI(p, d)

Note: r^{**} refers to $\max(r, (p + d + 1))$, as found in Wei and Stram (1990), and p, d, and r are the orders of the fitted aggregate time series model.

2. ISSUES IN AGGREGATION AND DISAGGREGATION

Wei and Stram (1990) have shown that a disaggregate ARI(p, d) model will become an ARIMA(p, d, r) model upon aggregation. Therefore, we propose that an ARIMA(p, d, r) model can be successfully disaggregated to an ARI(p, d) model.

However, several disaggregate models will aggregate to ARIMA(p, d, r) models. Hence, we must determine under which circumstances the aggregate models can be disaggregated to ARI(p, d) models.

To illustrate this, suppose we have access to the original basic series, as seen in Equation (1.1). We performed an mth order aggregation. For no hidden periodicity, Wei (1978) and Wei and Stram (1990) have shown that the order of the aggregate moving average polynomial is $r = \left[p + d + 1 + \frac{(q-p-d-1)}{m}\right]$. Also, Wei and Stram (1990) have shown that the relationship between the basic autoregressive polynomial and the aggregated autoregressive polynomial can be constructed as:

$$\begin{aligned}
\phi_p(B) &= 1 - \phi_1 B - \cdots - \phi_p B^p, \\
&= \prod_{i=1}^{p}(1 - r_i B), \\
\alpha_p(\mathcal{B}) &= \prod_{i=1}^{p}(1 - r_i^m \mathcal{B}),
\end{aligned}$$

where the r_i are the roots of the $\phi_p(B)$. There are no explicit closed form solutions for the values of the moving average parameters and the σ_C^2 in the aggregate model, in terms of the disaggregate model.

For example, consider the AR(1) theoretical model:

$$z_t = \phi z_{t-1} + a_t,$$

where a_t is a Gaussian white noise process with mean zero and $\sigma_a^2 = 1$, and $|\phi| < 1$. As we are dealing with an AR(1) model, it is easily shown that the aggregate model has the order ARMA(1, 1), for all $m \geq 2$. This follows, since

$$\begin{aligned}
r &= \left[p + d + 1 + \frac{q - p - d - 1}{m}\right], \\
&= \left[1 + 0 + 1 + \frac{0 - 1 - 0 - 1}{m}\right], \\
&= \left[2 - \frac{2}{m}\right].
\end{aligned}$$

Hence, $r = 1$ for all values of m.

Let $\gamma_U(k)$ and $\gamma_w(k)$ be the autocovariance functions for $U_T = (1 - \mathcal{B})^d Y_T$ and $w_t = (1 - B)^d x_t$, respectively. Wei and Stram (1986a) show that the autocovariances of the U_T amd w_t are related as:

$$\gamma_U(k) = (1 + B + \cdots + B^{m-1})^{2(d+1)} \gamma_w(mk + (m-1)(d+1)). \qquad (2.1)$$

To see whether every aggregate ARMA(1,1) model can be disaggregated to an AR(1) model, for various values of ϕ and m, we calculated autocovariances of the given disaggregate AR(1) model. We then derived the autocovariances of the aggregate ARMA(1,1) model using Equation (2.1). We then calculated numerically the values of β and σ_C^2. Table 1 gave us some insight as to the range of the theoretical values of the aggregate moving average coefficient and the variance. This information helped us determine which ARMA(1,1) models can be successfully disaggregated to AR(1) models. For example, Table 1 clearly shows that the moving average parameter β of the aggregate model has a lower bound. In fact, Teles and Wei (1998) show that the lower bound is approximately equal to -0.268. Thus, if we have an aggregate model with $\beta = -0.5$, then there is no disaggregate AR(1) model corresponding to this aggregate ARMA(1,1) model. Hence, autoregressive disaggregation cannot be used to disaggregate this series. If we ignored this information, and still tried to disaggregate the model, we would find that Equation (2.1) would not hold for all values of k. Hence, we will use Equation (2.1) as our verification tool for disaggregation of more complex models.

Table 1: Moving Average Parameters of an Aggregate ARMA(1,1) Model

	$m = 2$		$m = 3$		$m = 4$	
ϕ	β	σ_C^2	β	σ_C^2	β	σ_C^2
−0.90	0.86	1.04	0.00	2.50	0.74	1.98
−0.80	0.73	1.10	0.02	2.18	0.54	2.01
−0.70	0.61	1.16	0.03	2.00	0.38	2.06
−0.60	0.49	1.23	0.05	1.92	0.26	2.16
−0.50	0.38	1.31	0.07	1.93	0.18	2.30
−0.40	0.28	1.41	0.07	2.01	0.12	2.39
−0.30	0.20	1.52	0.07	2.16	0.08	2.74
−0.20	0.12	1.66	0.05	2.37	0.05	3.06
−0.10	0.06	1.81	0.03	2.64	0.02	3.47
0.00	0.00	2.00	0.00	3.00	0.00	4.00
0.10	−0.05	2.21	−0.04	3.45	−0.03	4.69
0.20	−0.08	2.46	−0.07	4.02	−0.06	5.58
0.30	−0.11	2.75	−0.11	4.73	−0.09	6.76
0.40	−0.13	3.07	−0.14	5.62	−0.12	8.35
0.50	−0.15	3.43	−0.17	6.75	−0.16	10.49
0.60	−0.16	3.83	−0.19	8.16	−0.19	13.42
0.70	−0.16	4.27	−0.20	9.93	−0.21	17.48
0.80	−0.17	4.74	−0.21	12.12	−0.23	23.11
0.90	−0.17	5.27	−0.22	14.83	−0.24	30.87

3. DISAGGREGATION TO ARI MODELS

Suppose that all we have is the aggregate series. We will fit ARIMA(p, d, r) models to the aggregate series, but consider that the disaggregate models are ARI(p, d) models.

Theorem 3.1 *Given the nonconstant stochastic univariate aggregate process:*

$$\alpha_p(\mathcal{B})(1-\mathcal{B})^d Y_T \;=\; \beta_r(\mathcal{B})C_T$$

where $\alpha_p(\mathcal{B}) = 1 - \alpha_1\mathcal{B} - \cdots - \alpha_p\mathcal{B}^p = \prod_{i=1}^{p}(1-\delta_i\mathcal{B})$, *and* $\mathbf{C}_T$ *is a white noise series with mean zero and variance* σ_C^2. *The* $\beta_r(\mathcal{B}) = 1 - \beta_1\mathcal{B} - \cdots - \beta_r\mathcal{B}^r$, *and the* α *and* β *polynomials have no roots in common. Assume that there is no hidden periodicity, and that* $\alpha_p(\mathcal{B})$ *has real roots. Let the disaggregate process be the following ARI(p, d) model,*

$$\phi_p(B)(1-B)^d x_t \;=\; a_t, \tag{3.1}$$

where $\phi_p(B) = \prod_{i=1}^{p}(1-\delta_i^{\frac{1}{m}}) = (1-\phi_1 B - \cdots - \phi_p B^p)$, *and* a_t *is a white noise process with mean zero and variance* σ_a^2. *Then the autocovariances* $\gamma_w(k)$ *of* $w_t = (1-B)^d x_t$ *can be derived from the known autocovariances* $\gamma_U(k)$ *of* $U_T = (1-\mathcal{B})^d Y_T$ *from the aggregate model.*

Proof: By the equation from Stram and Wei (1986a) which denotes the relationship between the $\gamma_w(k)$ and the $\gamma_U(k)$:

$$\gamma_U(k) \;=\; (1+B+\cdots+B^{m-1})^{2(d+1)}\gamma_w(mk+(m-1)(d+1)), \tag{3.2}$$

where B operates on $\gamma_w(j)$ such that $B\gamma_w(j) = \gamma_w(j-1)$, we can see that $\gamma_U(k)$ is a linear combination of the autocovariances $\gamma_w(j)$ for $j = mk-(m-1)(d+1)$ to $j = mk+(m-1)(d+1)$. As Wei and Stram (1990) show, we can write:

$$\begin{pmatrix} \gamma_U(0) \\ \gamma_U(1) \\ \vdots \\ \gamma_U(k) \end{pmatrix} = \mathbf{A}\begin{pmatrix} \gamma_w(-(m-1)(d+1)) \\ \vdots \\ \gamma_w(0) \\ \vdots \\ \gamma_w(mk+(m-1)(d+1)) \end{pmatrix}, \tag{3.3}$$

$$\text{where } \mathbf{A} = \begin{pmatrix} \mathbf{a}' & \underset{km}{\mathbf{0}'} & \\ \underset{m}{\mathbf{0}'} & \mathbf{a}' & \underset{(k-1)m}{\mathbf{0}'} \\ \vdots & & \\ \underset{km}{\mathbf{0}'} & & \mathbf{a}' \end{pmatrix}, \tag{3.4}$$

and $\mathbf{a}'$ is a $\times(2(m-1)(d+1)+1)$ vector of a_i, where the a_i are the coefficients of B^i in the polynomial $(1+B+\cdots+B^{m-1})^{2(d+1)}$. In the univariate case, for all positive integers i, we can write $\gamma_w(-i)=\gamma_w(i)$, and thus reduce the order of the $\mathbf{A}$ matrix. We can combine the columns of $\mathbf{A}$ which correspond to $\gamma_w(-i)$ with those of $\gamma_w(i)$ by addition. We will now write the vector of autocovariances of a series Z_t from lag 0 to lag n as $\mathbf{\Gamma}(Z,n)$. We can produce the following equation, as seen in Wei and Stram (1990):

$$\mathbf{\Gamma}(U,k) = \mathbf{A}_m^d(l)\mathbf{\Gamma}(w, mk+(m-1)(d+1)), \tag{3.5}$$

where $\mathbf{A}_m^d(l)$ with $l=mk+(m-1)(d+1)$ as the compressed $(k+1)\times l$ coefficient matrix.

We know that for an ARIMA(p,d,r) model, the autocovariances $\gamma_U(k)$ can be written as:

$$\gamma_U(k) = \sum_{i=1}^{p}\alpha_i\gamma_U(k-i), \text{ for } k>\max((p-1),r), \tag{3.6}$$

or the $\gamma_U(k)$ can be written as a linear combination of $\gamma_U(j)$ for $j=0,1,\ldots,\max((p-1),r)$. Since our disaggregate model follows an ARI(p,d) model, we can write:

$$\gamma_w(k) = \phi_p(B)\gamma_w(k-p), \text{ for } k\geq p. \tag{3.7}$$

So all autocovariances $\gamma_w(k)$ can also be written as a linear combination of $\gamma_w(j)$, for $j=0,1,\ldots,(p-1)$. Hence we write:

$$\begin{pmatrix}\gamma_w(0)\\ \gamma_w(1)\\ \vdots\\ \vdots\\ \gamma_w(l)\end{pmatrix} = \mathbf{G}\begin{pmatrix}\gamma_w(0)\\ \gamma_w(1)\\ \vdots\\ \gamma_w(p-1)\end{pmatrix}, \tag{3.8}$$

$$\text{where } \mathbf{G} = \left(\frac{\mathbf{I}_p}{\mathbf{H}}\right). \tag{3.9}$$

The $l=m(p-1)+(m-1)(d+1)$, and the $\mathbf{H}$ matrix is of the order $(l-p)\times p$, and contains the coefficients of the linear relationship between the autocovariances. We can combine the previous statements in this fashion:

$$\mathbf{\Gamma}(U,(p-1)) = \mathbf{A}_m^d\mathbf{G}\mathbf{\Gamma}(w,(p-1)). \tag{3.10}$$

Let $\mathbf{B} = \mathbf{A}_m^d\mathbf{G}$. In order to prove a unique relationship, we need to show that $\mathbf{B}$ is nonsingular. We will follow a similar proof to that found in Hodgess (1995). Assume that $\mathbf{B}$ is singular. Hence,

$$\begin{aligned} \mathbf{\Gamma}(U,(p-1)) &= \mathbf{B}\mathbf{\Gamma}(w,(p-1)), \\ &= \mathbf{0}. \end{aligned}$$

The only case in which this is true is when $\mathbf{U}$ is a constant process, which contradicts our original assumption. Hence $\mathbf{B}$ is nonsingular. Hence,

$$\mathbf{\Gamma}(w,(p-1)) = \left(\mathbf{A}_m^d\mathbf{G}\right)^{-1}\mathbf{\Gamma}(U,(p-1)). \ \blacksquare \qquad (3.11)$$

4. STEPS OF DISAGGREGATION

From Section 2, we have seen that the ARI(p,d) basic series will aggregate to an ARIMA(p,d,r) series. But we also have seen that the MA(r) polynomial is subject to certain restrictions in order to use our new autoregressive disaggregation method; i.e., both (3.2) and (3.10) must be satisfied for all k.

For any given aggregate time series, the steps of our autoregressive disaggregation method are:

1. Fit the best ARIMA(p,d,r) model to the aggregate data. This will give us the values for the $\alpha_p(\mathcal{B})$, and $\beta_r(\mathcal{B})$ polynomials, along with σ_C^2.
2. Determine the appropriate order of the disaggregate ARI(p,d) model.
3. Calculate the $\hat{\mathbf{V}}_w$ matrix. Verify that the relationship between the aggregate and disaggregate autocovariances holds for all values of k, as in Equations (3.2) and (3.10). If this relationship does not hold, then the model cannot be disaggregated using the autoregressive disaggregation method.
4. If the check in Step 3 passes, we then calculate the disaggregate series, using Equations (1.7) and (1.8).

5. COMPARISONS: A SIMULATION STUDY

We produced a study that would compare the results of the proposed new method against those of the lower bound method found in Hodgess (1995), and the Wei and Stram method found in Wei and Stram (1990). We simulated a series with 372 data points. The first 360 data points represented 30 years of monthly data, while the remaining 12 points will be used as comparison to forecast data. This simulated series followed an AR(1) model, with various values of ϕ and $\sigma_a^2 = 1$. We then aggregated the series with $m = 3$, and used that aggregate series for our disaggregation process. Hence, $nm = 360$, $n = 120$, and $m = 3$.

The steps in our autoregressive disaggregation method are:

1. Fit the best ARIMA(p, d, r) model to the aggregate data to get the $\alpha(\mathcal{B})$ polynomial, the $\beta(\mathcal{B})$ polynomial, σ_c^2, and the $\mathbf{V}_U$ matrix. We use the standard model identification techniques, such as the sample autocorrelation function, the sample partial autocorrelation function, and the Akaike Information criterion.

2. Select the appropriate ARI(p, d) disaggregate model.

3. Calculate the $\hat{\mathbf{V}}_w$ matrix, using Equations (3.2) and (3.10). Determine if all covariance restrictions are met.

4. If the previous requirements hold, then use Equations (1.7) and (1.8) to produce the disaggregate series.

We calculated values for parameter estimation, in-sample mean square error(IMSE). We also used the derived disaggregation model to forecast 12 future disaggregate values and calculated forecast MSE(FMSE). IMSE and FMSE are defined as:

$$\begin{aligned} \text{IMSE} &= \sum_{i=1}^{360} \frac{(x_t - \hat{x}_t)^2}{360}, \quad \text{and} \\ \text{FMSE} &= \sum_{i=361}^{372} \frac{(x_t - \hat{x}^F)^2}{12}, \end{aligned}$$

where x_t is the original series, $\hat{x}_t$ is the disaggregate value, $\hat{x}_t^F$ is the disaggregate forecast value.

The values seen in the next tables were averages of 1000 replications for the Lower Bound and Wei and Stram methods. The values in the autoregressive disaggregation method reflected only the series that could be successfully disaggregated. The values in parentheses were the standard errors.

Table 2: Comparison of Parameter Estimation

ϕ	AR	LB	W/S
−0.90	−0.90(0.00)	−0.89(0.00)	−0.89(0.00)
−0.80	−0.80(0.00)	−0.76(0.01)	−0.77(0.01)
−0.70	−0.70(0.00)	−0.55(0.01)	−0.62(0.01)
−0.60	−0.62(0.01)	−0.31(0.02)	−0.47(0.01)
−0.50	−0.53(0.01)	−0.16(0.02)	−0.38(0.02)
−0.40	−0.46(0.01)	−0.08(0.02)	−0.29(0.02)
−0.30	−0.41(0.02)	−0.04(0.02)	−0.23(0.02)
−0.20	−0.39(0.02)	0.00(0.02)	−0.17(0.02)
−0.10	−0.26(0.04)	0.01(0.02)	−0.10(0.02)
0.00	−0.08(0.07)	0.03(0.02)	0.00(0.02)
0.10	0.04(0.06)	0.04(0.02)	0.11(0.02)
0.20	0.16(0.05)	0.05(0.02)	0.23(0.02)
0.30	0.35(0.02)	0.06(0.02)	0.30(0.02)
0.40	0.45(0.01)	0.11(0.02)	0.37(0.02)
0.50	0.52(0.01)	0.21(0.02)	0.40(0.02)
0.60	0.61(0.00)	0.40(0.02)	0.50(0.01)
0.70	0.70(0.00)	0.63(0.01)	0.67(0.01)
0.80	0.80(0.00)	0.78(0.00)	0.80(0.00)
0.90	0.90(0.00)	0.89(0.00)	0.89(0.00)

AR = Autoregressive Disaggregation
LB = Lower Bound
W/S = Wei and Stram

Table 3: Comparison of In-Sample MSE

ϕ	AR	LB	W/S
−0.90	0.96(0.01)	0.98(0.00)	0.98(0.00)
−0.80	1.01(0.01)	1.05(0.01)	1.04(0.01)
−0.70	1.02(0.01)	1.11(0.01)	1.07(0.01)
−0.60	1.01(0.01)	1.09(0.01)	1.05(0.01)
−0.50	0.96(0.01)	1.02(0.00)	0.99(0.00)
−0.40	0.92(0.01)	0.94(0.01)	0.93(0.00)
−0.30	0.87(0.01)	0.87(0.00)	0.86(0.00)
−0.20	0.83(0.02)	0.81(0.00)	0.81(0.00)
−0.10	0.78(0.01)	0.76(0.00)	0.75(0.00)
0.00	0.74(0.03)	0.73(0.01)	0.71(0.00)
0.10	0.67(0.02)	0.71(0.01)	0.68(0.01)
0.20	0.63(0.01)	0.74(0.01)	0.68(0.01)
0.30	0.56(0.01)	0.80(0.02)	0.70(0.02)
0.40	0.52(0.00)	0.99(0.03)	0.82(0.03)
0.50	0.47(0.00)	1.32(0.06)	1.13(0.05)
0.60	0.44(0.00)	1.50(0.08)	1.37(0.08)
0.70	0.40(0.00)	0.92(0.09)	0.92(0.09)
0.80	0.36(0.00)	0.42(0.05)	0.43(0.05)
0.90	0.33(0.00)	0.34(0.00)	0.34(0.00)

AR = Autoregressive Disaggregation
LB = Lower Bound
W/S = Wei and Stram

Table 4: Comparison of Forecast MSE

ϕ	AR	LB	W/S
−0.90	3.96(0.22)	4.25(0.14)	4.12(0.13)
−0.80	2.46(0.10)	2.73(0.07)	2.64(0.07)
−0.70	1.87(0.08)	1.97(0.04)	1.93(0.04)
−0.60	1.47(0.06)	1.58(0.03)	1.55(0.03)
−0.50	1.28(0.05)	1.35(0.02)	1.32(0.02)
−0.40	1.20(0.06)	1.20(0.02)	1.19(0.02)
−0.30	1.18(0.07)	1.11(0.02)	1.10(0.02)
−0.20	1.14(0.08)	1.06(0.01)	1.04(0.01)
−0.10	1.12(0.08)	1.03(0.01)	1.01(0.01)
0.00	1.01(0.09)	1.02(0.01)	1.00(0.01)
0.10	0.81(0.07)	1.03(0.01)	1.01(0.01)
0.20	0.95(0.07)	1.06(0.01)	1.04(0.01)
0.30	0.95(0.05)	1.13(0.02)	1.10(0.02)
0.40	1.10(0.05)	1.24(0.02)	1.20(0.02)
0.50	1.20(0.05)	1.42(0.02)	1.35(0.02)
0.60	1.47(0.06)	1.67(0.04)	1.58(0.03)
0.70	1.75(0.07)	1.98(0.05)	1.87(0.04)
0.80	2.33(0.08)	2.55(0.06)	2.43(0.06)
0.90	3.58(0.14)	3.81(0.11)	3.72(0.11)

AR = Autoregressive Disaggregation
LB = Lower Bound
W/S = Wei and Stram

6. RESULTS OF THE STUDY

There were most interesting results from this study. In all three of our investigated concepts, the autoregressive disaggregation method outperformed the other two methods for $|\phi| \geq 0.40$, as easily seen in Tables 2, 3, and 4. This indicated a strong showing for our new method. It also demonstrated that the loss of information for aggregate series in the middle range was so great that the autoregressive disaggregation method could not recover that information. There was one caveat to the autoregressive disaggregation method: many series cannot be disaggregated via this method. The percentage of successes ranged from about 35 percent in the extreme values of ϕ, to about 5 percent when ϕ was quite small. The Lower Bound method was often the middle value, with the exception of Parameter estimation, as seen in Table 2. The Wei and Stram method, however, was typically the winner when the autoregressive disaggregation method was not possible. One side note: the time for one autoregressive disaggregation simulation of 1000 replications was about 15 minutes, while the other two methods took about 40 minutes each.

We have shown that if a series can be successfully disaggregated with the autoregressive disaggregation method, this method can provide accurate and reliable estimates of all kinds.

7. CONCLUSION

We have shown by theoretical methods, and a simulation study, that an ARIMA(p, d, r) aggregate series can be disaggregated successfully to an ARI(p, d) disaggregate series. The new method performed quite well for all three of the concepts across a fairly large range. When possible, the autoregressive disaggregation method should be used, with the Wei and Stram method available as a fail-safe method. The autoregressive disaggregation method provides a parsimonious, meaningful disaggregate model, and can be used to recover much of the information lost in aggregation.

statistic. The plan of the rest of this paper is as follows. In the next section the model and the testing problem are introduced and the OLS and GLS estimators for the slope coefficients are obtained. Section 3, presents the test statistic and its distribution. The final section contains some concluding remarks.

2. THE MODEL

To begin, let's consider the regression model

$$\begin{pmatrix} Y_1 \\ Y_2 \end{pmatrix} = \begin{pmatrix} X_1 & 0 \\ 0 & X_2 \end{pmatrix} \begin{pmatrix} \beta_1 \\ \beta_2 \end{pmatrix} + \begin{pmatrix} u_1 \\ u_2 \end{pmatrix} \tag{2.1}$$

with response vectors $Y_i(i = 1, 2,)$ of n_i $(n_1 = \tau, n_2 = T - \tau)$ components, change point τ, design matrices X of size $n_i \times k$, coefficient vectors β_i and error vectors u_i of n_i components having mean vector zero and dispersion $\sigma^2 I_{n_i}(i = 1, 2)$ where I_{n_i} represents $n_i \times n_i$ unit matrix. The point of real interest is when the change point τ is unknown . Testing for stationarity in the regression model (2.1) involves testing the null hypothesis $\beta_1 = \beta_2$, $\sigma_1^2 = \sigma_2^2$ against the two sided alternative $\beta_1 \neq \beta_2$ or $\sigma_1^2 \neq \sigma_2^2$ with unknown change point τ. If τ is known the problem reduces to testing for equality of regression coefficient vectors in (2.1) with unequal error variances. Kim (1991) considered a Bayesian test for testing $\beta_1 = \beta_2$ in (2.1) when $\sigma_1^2 \neq \sigma_2^2$. He assumed u_i $(i = 1, 2)$ is n_i variates normally distributed with zero mean vector and covariance matrix $\sigma_i^2 I_{n_i}$ and such that $\beta_i \in \mathbb{R}^k$. He also developed a Bayesian test for stationarity of regresion parameters over a sequence of time series by deriving the posterior distribution of $(\beta_1 - \beta_2)$ and σ_2^2/σ_1^2 unconditionally on the change point τ. This test does not require a constant error variance assumption nor does it assume a change point fixed at the estimated point.

2.1 Inference on the Model

Note that (2.1) in its usual matrix form can be written as

$$Y = X\beta + u, \tag{2.2}$$

with error vector u of expectation zero and dispersion matrix Ω and estimate a linear contrast in $2k$ unknown regression parameters,

$$\delta = \ell'\beta. \tag{2.3}$$

Under a mild condition, without assuming normality, that the ratio σ_2^2/σ_1^2 is constant equal to ℓ (unknown). The mild condition leads to $\Omega = \sigma_1^2 v$ where

$$V = \begin{pmatrix} I_{n_1} & 0 \\ 0 & \ell I_{n_2} \end{pmatrix}.$$

Now, we could find generalized least squares (GLS) estimator of β and employ it to estimate δ, such that

$$\hat{\beta} = (X'V^{-1}X)^{-1}X'V^{-1}Y,$$
$$\hat{\delta} = \ell'\hat{\beta}$$

with $E(\hat{\delta}'|GLS) = \delta$, and $Var(\hat{\delta}|GLS) = \ell'(X'V^{-1}X)^{-1}\ell$, where

$$(X'V^{-1}X)^{-1} = \begin{pmatrix} (X_1'X_1)^{-1} & 0 \\ 0 & \ell(X_2'X_2)^{-1} \end{pmatrix},$$

and the residual matrix R is defined as,

$$R = I_\tau - V^{-1/2}X(X'V^{-1}X)^{-1}X'V^{-1/2}. \tag{2.4}$$

For ℓ partioned into ℓ_1 and ℓ_2, β partitioned into β_1 and β_2 and these are estimated independently by the ordinary least squares (OLS), then there exists an interesting relation of OLS, GLS, ℓ, β, δ, and σ. This relationship is expressed as,

$$Var(\hat{\delta}|GLS) = \ell_1'Var(\hat{\beta}_1|OLS)\ell_1/\sigma_1^2 + \ell_2'Var(\hat{\beta}_2|OLS)\ell_2/\sigma_2^2.$$

In the rest of this section we use residual matrix R which is given in (2.4) under GLS, by noting that the trace of the residual matrix, trace $(R) = (\tau - 2k)$. This refers to the degrees of freedom (d.f.), of the sum of squares of the residuals or errors. The hypothesis $H_0 : (\beta_1 = \beta_2)$ in model (2.1) or (2.2) shows that k regression parameters are to be estimated. The sum of square (SS) of residuals under $H_0(SSE(H_0))$ minus the SS of residuals (SSE) without H_0 gives the SS of departure (SSD), from $SSE(H_0) - SSE = SS$, with df., $(n - k) - (n - 2k) = k$. Under normal theory assumption, we can apply variance ratio test i.e., $F = \frac{SSD/v_1}{SSE/v_2}$, to test the significance of departure from H_0 with d.f. $v_1 = k$ and $v_2 = (n - 2k)$.

Following Dawid's (1981) notation, assume that U be a $T \times p$ random matrix. The distribution of U is called left spherical, if and only if U has the same distribution as PU for all P in $0(T)$, where $0(T)$ is the group of $T \times T$ orthogonal matrices. This is denoted as, $L(U) = L(PU)$. Under this definition the joint marginal distribution of each column array of U is left spherical on $\mathbb{R}^T$, and marginal distributions of rows are identical with mean zero and scale parameter σ. Then, one can write $L(U) = S(I_T, \Sigma)$, where I_T is an $T \times T$ identity matrix.

Theorem 1: Let $L(U) = S(I_T, \Sigma)$ be as defined above and $t(U)$ be a right invariant function of U, i.e., $t(U) = t(UA)$, where $A \in G(p)$, the group of non-singular matrices. Then the distribution of $t(U)$ is invariant and independent of Σ.

Proof: Since U is left spherical, it follows that $L(U) = L(VS)$, where V is distributed uniformly on the Stiefel manifold and S satisfies $L(S'S) = L(U'U)$ (Dawid, 1977). Hence $L(t(U)) = L(t(VS)) = L(t(V))$. Since $L(V)$ is unique the result follows. Also $t(U) = t(U\Sigma^{-1/2})$, where $\Sigma^{1/2}$ is a square root of Σ such that $\Sigma^{1/2}\left(\Sigma^{1/2}\right)' = \Sigma$.

3. THE TEST STATISTICS AND ITS DISTRIBUTION

This subsection begins by noting that the null hypothesis under - consideration (concerning no structural shift of the regression equation in (2.1) over a sample period with T observations taken sequentially) is, without reference to τ. This means that

$$\begin{aligned} \delta &= \tilde{\beta}_1 - \tilde{\beta}_2 = 0 \\ \rho &= \sigma_2^2/\sigma_1^2 = 1. \end{aligned}$$

Let $SSE_i = (Y_i - X_i\hat{\tilde{\beta}}_i)'(Y_i - X_i\hat{\tilde{\beta}}_i)$ where $\hat{\tilde{\beta}}_i$ is the ordinary least squares (OLS) estimator of $\tilde{\beta}_i$ $(i = 1, 2)$. Note that the d.f. for each SS_i is $(n - k)$. We now state the following theorem, (see Kim (1991) for a proof).

Theorem 2: Given (τ, ρ), the conditional posterior distribution of δ is

$$\pi(\delta/\tau, \rho, y, x)\rho\left\{1 + (\delta - \hat{\delta})'M(\delta - \hat{\delta})/(T - 2p)s^2\right\}^{-1/2(T-p)}$$

where $M = X_2'X_2(\rho X_1'X_1 + X_2'X_2)^{-1}\rho X_1'X_1$ and

$$s^2 = (\rho SSE_1 + SSE_2)/(T - 2k).$$

Note that the precision matrix of the posterior distribution of δ is M/s^2. Based on the above results (Theorem 2), we can obtain our F-test statistic as follows,

$$F(\delta) = Q(\delta)/ks^2 = (\delta - \hat{\delta})'M(\delta - \hat{\delta})/ks^2, \tag{3.1}$$

which is distributed, a posterior, as conditional F-variable with $(k, T-2k)$ degrees of freedom. Chow (1960) proposed a test for equality of coefficients in two regression equations, namely assuming $\delta = 0$ in (3.1), i.e., when error variances are equal $(\rho = 1)$. In this case the test quantity $F(0)$ in Bayesian test is the same as that of Chow's test. The distribution of $F(0)$ has been studied by Chellah (1992), Conely and Manfield (1988) and Schmidt and Sickles (1977), when the error distribution is normal with unequal variance. In the next section we propose an extended version of $F(0)$ and derive its invariant distribution when $L(U) = S(I_n, \Sigma)$.

3.2 The Multivariate Extension

In this subsection we consider an extension of (2.1) with response matrices $Y_i : n_i \times p$ and coefficient matrices $\beta_i : k \times p$, p assumed to be known so that

$$Y_1 = X_1\beta_1 + U_1 \qquad\qquad Y_2 = X_2\beta_2 + U_2. \tag{3.2}$$

Parallel to (1.2) we can express (3.2), for $\beta_1 = \beta_2$, in the form

$$Y = X\beta + U \tag{3.3}$$

with

$$Y = \begin{bmatrix} Y_1 \\ Y_2 \end{bmatrix}, \; X = \begin{bmatrix} X_1 \; X_2 \end{bmatrix} \text{ and } U = \begin{bmatrix} U_1 \; U_2 \end{bmatrix}.$$

We assume $L(U) = S(I_T, \Sigma)$ with Σ a positive definite and non-diagonal matrix.

Theorem 3: Consider $S_1 = Y'(H^* - H)Y, S_2 = Y'(I - H^*)Y$, where $H = X(X'X)^{-1}X', H^* = \begin{bmatrix} H_1 & 0 \\ 0 & H_2 \end{bmatrix}$ and $H_i = X_i(X_i'X_i)^{-1}X_i'$, $i = 1, 2$. Also $L(U) = S(I_T, \Sigma)$ in model (3.3). Define matrix version of $F(0)$ in the form of $F^* = S_2^{-1/2}S_1(S_2^{-1/2})'$ where $S_2^{1/2}$ is a lower triangular square root of S_2 with positive diagonal such that $S_2^{1/2}(S_2^{1/2})' = S_2$. Then

a) the conditional distribution of F^* given τ is invariant and independent of Σ.

b) when $L(U) = N(I_T, \Sigma)$ (matrix normal), the conditional distribution is in (a) above as $F(k, T - 2k - p + 1;\ I_p)$ (matrix F-distribution).

Proof:

a) It is easy to show that $S_1 = U'(H^* - H)U$, $S_2 = U'(I - H)U$. Consider Theorem 1 and transformation $U \to UA, A \in G(p)$, the group of non-singular $p \times p$ matrices. The result is immediate.

b) Since $(H^* - H)'(I - H) = 0, S_1$ and S_2 are independent. Also $H^* - H, I - H$ are idempotent matrices. It follows from Eaton (1985, Proposition 8.5) that $L(S_1) = W(k, \Sigma), L(S_2) = W(T - 2k, \Sigma)$. The conditional distribution of S, given S, is $W(k, I)$. Hence the conditional distribution of $S_2^{-1/2}S_1(S_2^{-1/2})'$ given S is $W(k, S)$. Since $L(S) = IW(T - 2k - p + 1, I)$ we have the result from Dawid (1981),

$$L(S_2^{-1/2}S_1(S_2^{-1/2})') = F(k, T - 2k - p + 1;\ I_p).$$

Note that when $p = 1$ we have the univariate result of Kim (1991).

4. CONCLUDING REMARKS

This paper had suggested a matrix version of the Chow's test statistics based on Bayesian invariant approach. It have considered the problem of testing for stationary of regression parameters when the regression disturbances are spherically symmetric. It is observed that the new test is in a more general form than that of Kim (1991), Chelliah's (1992) F-test statistics and can be applied in the area of panel data modeling as that of Baltagi (1996).

ACKNOWLEDGEMENT

This research is partically supported by the office of the Research Administration of Kuwait University via project No. ss-050. I am thankful to Professor Ahsanullah for his constructive comments on an earlier version of this paper.

REFERENCES

Ahsanullah, M., Sinha, B., and Rukhin, A. (1995). *Applied change point problems in statistics*, Nova Science Publication Inc. New York, U.S.A.

Amemiya, T. (1985). *Advanced Econometrics, Cambridge*: Harvard University Press.

Ali, M. and Silver, J.L. (1985). Tests of equality between sets of coefficients in two linear regressions under hetroscedasticity, τ of the **Amer. Stat. Assoc. 80, 730-735.**

Ansari, M.J. (1991). An econometric model of structural-change: A case study of canada, **Applied Economics, 23, 791-798**.

Andrew, D.W.K. and Ploberger, W. (1992). Optimal tests when a nuisance parameter is present only under the alternative, Cowles Fundation Discussion Paper No. 1015, Yale University.

Andrew, D.W.K. (1993). Tests for parameter instability and structural change with unknown change point, **Econometrica, 61, 821-856**.

Baltagi, B.H. (1996). *Econometric Analysis of Panel Data*, Wiley, N.Y.

Bhatti, M.I., Khan, I.H. and C.J. Czerkawski (1998). Agricultural productivity in Shanghai region of China: An econometrics analysis, **Journal of Economic Sciences**, 1(1-2), 1-11.

Bhatti, M.I. (1993). Efficient estimation of random coefficient models based on survey data, **J. of Quantitative Economics, 9, 99-110**.

Bhatti, M.I. (1995). *Testing regression models based on sample survey data*, Aveburg, U.K.

Bhatti, M.I. and Barry, A.M. (1995). On Testing for equicorrelation in random coefficient models, **Biometrical Journal, 37(4), 383-391**.

Bollerslev, T., Chou, R.Y. and Kroner, K.F. (1992). ARCH modelling in finance: A rewiew of the theory and empirical evidence, **J. of. Econometrics, 52, 5-59**.

Broemeling, L.D. and Tsurumi, H. (1986). *Econometrics and structural change*, Marcel & Dekker, Inc. New York and Basel.

Chelliah, N. (1992). Invariant distributions of Chow statistics, **Statistics and Applied Prob. letters, 15, 401-402**.

Chow, G. (1960). Test of the equality between two sets of coefficients in two linear regression, **Econometrica, 28, 561-605**.

Conerly, M.D. and Mansfield, E.R. (1988). An approximate test for comparing heteroscedastic regression models, **J. Amer. Statist. Assoc, 33, 811-817**.

Cornerly, M.D. and Mansfield, E.R. (1989). An approximate test for comparing in dependent regression models with unequal error variances, **J. of Econometrics, 40, 239-259**.

Dawid, A.P. (1977). Spherical matrix distribution and a multivariate model, **J. Roy. Statist. Soc. Ser. B39, 254-261**.

Dawid, A.P. (1981). Some matrix variate distribution theory: Notational consideration and a bayesian application, **Biometrika, 68, 265-274**.

Dufour, J.M. (1980). Dummy variables and prediotive tests for structural change, **Economic letters, 6, 241-247**.

Dafour, J.M. (1982). Generalized Chow tests for structural change: A coordinate-free approach, **International Economic Review, 23, 565-575**.

Engle, R.F. and Bollerslev, T. (1986). Modelling the persistence of conditional variances, **Econometric Reviews**, 5, 1-50.

Fisher, F.M. (1970). Tests of equality between sets of coefficients in two linear regressions: An expository note, **Econometrica, 38, 361-366**.

Giles, D.E.A. and Scott, M. (1992). Some consequences of using the Chow test in the context of auto correlated disturbances. **Economic Letters, 38, 145-150**.

Hensen, B.E. (1993). Inference when a nuisance parameter is not identified under the null hypothesis, University of Rochester working paper.

Kim, D. (1991). A Bayesian significance test of the stationarity of regression parameters. **Biometrika, 78, 667-75**.

Schmidt, P. and Sickles, R. (1977). Some further evidence on the use of Chow test under heteroskedasticity. **Econometrica, 42, 601-608**.

Thursby, J.G. (1992). 4 comparison of several exact and approximate tests for structural shift under hetroscedosticity, **J. of Econometrics, 52, 363-386**.

Toyoda, T. (1974). Use of the Chow test under hetroscedosticity **Econometrica, 42, 601-608**.

Wu, X.P. and Bhatti, M.I. (1994). Testing for block effects and misspecification in regression models based on survey data, **J. of statistical computation and simulation**, 50(1-2), 75-90.

ON PROPERTIES OF STATISTICS CONNECTED WITH MINIMAL SPACING AND RECORD EXCEEDANCES

I. BAIRAMOV AND S. ERYILMAZ

Ankara University, Faculty of Sciences, Department of Statistics
06100, Tandoğan, Ankara, Turkey

SUMMARY

An additional insight to the distributional properties of statistics based on minimal spacing and record exceedances is provided. Behaviour of a sequence of independent identically distributed random variables with respect to a random threshold constructed by the help of minimal spacing and records is studied.

1. INTRODUCTION

This paper attempts to provide an additional insight into properties of exceedances and placement statistics based on minimal spacing and records. The importance of these concepts in construction of non-parametric tests of equality (identity) of distributions has been demonstrated by Katzenbeisser (1985), (1986), Matveychuk and Petunin (1990), Johnson and Kotz (1991), (1994), Wesolowski and Ahsanullah (1998), Bairamov, Gebizlioglu and Kaya (1999). Our study is also related to the theory of invariant confidence intervals which contain the main distributed mass or future observations. Invariant confidence intervals which involve future observations are similar to, but not identical with, the approach of tolerance limits. Tolerance limits are originally introduced by Shewart (1931). Tolerance limits and exceedance statistics are widely discussed by Wilks (1941), Robbins (1944), Gumbel and Shelling (1950), Epstein (1954), Sarkadi (1957), Siddiqui (1977) along with others. Further detailed discussion can be found in David (1981), Leadbetter at all (1983), Johnson at all (1992). A recent interest in the subject lies in the work of Wesolowski and Ahsanullah (1998), Bairamov and Kotz (2000), Bairamov and Eryılmaz (2000).

Invariant confidence intervals containing the main distributed mass were introduced in 1990 (c.f., Bairamov and Petunin (1991)). We mention here the definition and some results concerning invariant intervals. Let $X_1, X_2, ..., X_n$ be a random sample of size n with distribution function (d.f.) $F \in \Im$, $\Im$ is some class of d.f.'s. $f_1(.)$ and $f_2(.)$ are assumed to be two Borel functions satisfying $f_1(x_1, x_2, ..., x_n) \le f_2(x_1, x_2, ..., x_n)$, $\forall (x_1, x_2, ..., x_n) \in R^n$. The random interval $(f_1(X_1, X_2, ..., X_n),\ f_2(X_1,\ X_2, ..., X_n))$ is called an invariant confidence interval containing the main distributed mass (or invariant confidence interval containing the future observations) for class $\Im$, if $\exists\, \beta \in (0,1)$ such that $P\{X_{n+1} \in (f_1(X_1, X_2, ..., X_n),$ $f_2(X_1, X_2, ..., X_n))\} = \beta$, $\forall F \in \Im$. The quantity β is the same for all $F \in \Im$ and is called a confidence level of an invariant interval. Let $\Im = \Im_c$ be the class of all continuous d.f.'s. It is known that under some natural conditions any invariant confidence interval for the class $\Im_c$ can be generated only by the order statistics, so that $f_1(X_1, X_2, ..., X_n) = X_{(i)}$, $f_2(X_1, X_2, ..., X_n) = X_{(j)}, 1 \leq i < j \leq n$, where $X_{(1)} < X_{(2)} < ... < X_{(n)}$ are order statistics generated by X_1, X_2, ..., X_n , which is defined with probability 1 (c.f., Bairamov and Petunin (1991)), and $P\{X_{n+1} \in (X_{(i)}, X_{(j)})\} = (j-i)/(n+1)$.

Let $\{X_n\}_{n \geq 1}$ be a sequence of i.i.d. random variables with continuous d.f. $F(x)$. Let $X_{(1)} < X_{(2)} < ... < X_{(n)}$ be the almost sure (a.s.) defined order statistics generated by $X_1, X_2, ..., X_n$. Consider the spacings $X_{(1)} - X_{(0)}, X_{(2)} - X_{(1)}, X_{(3)} - X_{(2)}, ..., X_{(n)} - X_{(n-1)}$ with $X_{(0)} = \alpha(F) \equiv \inf\{x : F(x) > 0\}$. Define a random variable ν as follows: $\nu = k$ iff $X_{(k)} - X_{(k-1)} \leq X_{(i)} - X_{(i-1)}$, $i = 1, 2, .., n$. It is clear that ν is the index of a spacing having minimal length. Bairamov and Eryılmaz (2000) show that if $F(x) = 1 - \exp(-\lambda x)$, $x \geq 0, \lambda > 0$, then, for $s = 0, 1, ..., m$,

$$P\{X_{n+1}, X_{n+2}, ..., X_{n+s} \in (X_{(\nu-1)}, X_{(\nu)}), X_{n+s+1}, ..., X_{n+m} \notin (X_{(\nu-1)}, X_{(\nu)})\}$$

$$\begin{aligned} &= \frac{2}{n+1} \sum_{i=0}^{m-s} (-1)^i \binom{m-s}{i} \frac{n+s+i+1}{s+i+2} \left(\binom{\frac{n(n+1)}{2} + s + i}{s+i} \right)^{-1}, \qquad (1.1) \\ s &= 0, 1, 2, ..., m, \end{aligned}$$

where $X_{(0)} = 0$. Let S_m be the number of k, $k \in \{n+1, ..., n+m\}$, such that $X_k \in (X_{(\nu-1)}, X_{(\nu)})$. Denote by $A(m, n, s)$ the right hand side (r.h.s.) in (1.1). An almost straightforward consequence of (1.1) is that

$$P\{S_m = s\} = \binom{m}{s} A(m, n, s), \ s = 0, 1, 2, ..., m. \qquad (1.2)$$

2. ASYMPTOTIC DISTRIBUTION OF STATISTICS BASED ON MINIMAL SPACING

Let $X_1, X_2, ..., X_n$ be an independent copies of nonnegative random variable having continuous distribution with d.f. F. Also, let $X_{(1)} < X_{(2)} < ... < X_{(n)}$ be order statistics constructed on the basis of $X_1, X_2, ..., X_n$. Consider the spacings $X_{(1)} - X_{(0)}$, $X_{(2)} - X_{(1)}$, $X_{(3)} - X_{(2)}, ..., X_{(n)} - X_{(n-1)}$ with $X_{(0)} = 0$. Define a random variable ν as follows: $\nu = k$ iff $X_{(k)} - X_{(k-1)} \leq X_{(i)} - X_{(i-1)}$, $i = 1, 2, .., n$. It is clear that ν is the index of a spacing having minimal length. Let $Y_1, Y_2, ..., Y_m$ be another sample from nonnegative random variable Y with continuous d.f. G.

Let S_m be a number of observations $Y_1, Y_2, ..., Y_m$ falling above the random threshold $X_{(\nu-1)}$ and below the threshold $X_{(\nu)}$. It is clear that in case $F = G$ the second sample $Y_1, Y_2, ..., Y_m$ can be interpreted as a continuation $X_{n+1}, X_{n+2}, ..., X_{n+m}$ of the first sample $X_1, X_2, ..., X_n$.

Theorem 1. The asymptotic distribution of $\frac{S_m}{m}$ for large m is

$$\lim_{m\to\infty}\sup_{0\le x\le 1}\left|P\left\{\frac{S_m}{m}\le x\right\}-P\left\{G(X_{(\nu)})-G(X_{(\nu-1)})\le x\right\}\right|=0.$$

Proof. By definition

$$S_m=\sum_{i=1}^{m}\xi_i=\sum_{i=1}^{m}I_{(X_{(\nu-1)},X_{(\nu)})}(X_{n+i}), \tag{2.1}$$

where $I_B(x)=\begin{cases}1 \text{ if } x\in B\\ 0 \text{ if } x\notin B\end{cases}$. Using the representation (2.1) one has

$$P\left\{\frac{S_m}{m}\le x\right\}=P\left\{\frac{1}{m}\sum_{i=1}^{m}I_{(X_{(\nu-1)},X_{(\nu)})}(X_{n+i})\le x\right\}$$

$$=P\left\{\int_0^\infty I_{(X_{(\nu-1)},X_{(\nu)})}(u)dG_m^*(u)\le x\right\} \tag{2.2}$$

where $G_m^*(u)$ is the empirical distribution function constructed by the sample X_{n+1}, $X_{n+2},\ldots, X_{n+m}$ of size m. Note that G_m^* and $X_1, X_2,\ldots,X_n$ are independent random variables and $X_1(\omega), X_2(\omega),\ldots,X_n(\omega),\ X_{n+1}(\omega), X_{n+2}(\omega),\ldots,X_{n+m}(\omega),\ldots$ is considered as a sequence of i.i.d. random variables defined in probability space $\{\Omega,\Im,P\}$, where Ω is a set of points, $\Im$ is a $\sigma-$ field of subsets of Ω, and P is a probability measure given in $\Im$. Denote

$$H^*(G)=\int_0^\infty I_{(x,y)}(u)dG(u),\quad (x<y,\ x,y \text{ are fixed})$$

$$H(G)=\int_0^\infty I_{(X_{(k-1)},X_{(k)})}(u)dG(u).$$

$H(G)=H(G)(\omega)$ is a random variable defined in a probability space $\{\Omega,\Im,P\}$. By using Glivenko-Cantelli theorem one can observe that $H^*(G_m^*)\to H^*(G)$ almost surely. From (2.2) it follows that

$$P\left\{\frac{S_m}{m}\le x\right\}=\sum_{k=1}^{n}P\left\{\int_0^\infty I_{(X_{(k-1)},X_{(k)})}(u)dG_m^*(u)\le x,\nu=k\right\}$$

$$=\sum_{k=1}^{n}P\left\{\int_0^\infty I_{(X_{(k-1)},X_{(k)})}(u)dG_m^*(u)\le x, X_{(k)}-X_{(k-1)}\le X_{(i)}-X_{(i-1)},\right.$$

$$\left.i=1,2,..,n\right\}=\sum_{k=1}^{n}\int_0^\infty\int_0^y P\left\{\int_0^\infty I_{(X_{(k-1)},X_{(k)})}(u)dG_m^*(u)\le x, X_{(k)}-X_{(k-1)}\le\right.$$

$$\left.\le X_{(i)}-X_{(i-1)}, i=1,2,..,n \mid X_{(k-1)}=x, X_{(k)}=y\right\}dF_{k-1,k}(x,y), \tag{2.3}$$

where $F_{k-1,k}(x,y)$ is the joint distribution function of $(X_{(k-1)}, X_{(k)})$. After simplifying (2.3) it follows that

$$P\left\{\frac{S_m}{m} \leq x\right\} =$$

$$\sum_{k=1}^{n} \int_0^{\infty} \int_0^{y} P\left\{ \int_0^{\infty} I_{(x,y)}(u) dG_m^*(u) \leq x, x-y \leq X_{(i)} - X_{(i-1)}, i = 1,2,..,n, \right.$$

$$\left. X_{(k-1)} = x, X_{(k)} = y \right\} \frac{1}{P\{X_{(k-1)} = x, X_{(k)} = y\}} dF_{k-1,k}(x,y)$$

$$= \sum_{k=1}^{n} \int_0^{\infty} \int_0^{y} P\left\{ \int_0^{\infty} I_{(x,y)}(u) dG_m^*(u) \leq x \right\} P\{x-y \leq X_{(i)} - X_{(i-1)}, i = 1,2,..,n,$$

$$X_{(k-1)} = x, X_{(k)} = y\} \frac{1}{P\{X_{(k-1)} = x, X_{(k)} = y\}} dF_{k-1,k}(x,y) \overset{a.s.}{\rightarrow}$$

$$\overset{a.s.}{\rightarrow} \sum_{k=1}^{n} \int_0^{\infty} \int_0^{y} P\left\{ \int_0^{\infty} I_{(x,y)}(u) dG(u) \leq x \right\} P\{x-y \leq X_{(i)} - X_{(i-1)}, i = 1,2,..,n,$$

$$X_{(k-1)} = x, X_{(k)} = y\} \frac{1}{P\{X_{(k-1)} = x, X_{(k)} = y\}} dF_{k-1,k}(x,y)$$

$$= \sum_{k=1}^{n} \int_0^{\infty} \int_0^{y} P\left\{ \int_0^{\infty} I_{(x,y)}(u) dG(u) \leq x, x-y \leq X_{(i)} - X_{(i-1)}, i = 1,2,..,n, \right.$$

$$\left. X_{(k-1)} = x, X_{(k)} = y \right\} \frac{1}{P\{X_{(k-1)} = x, X_{(k)} = y\}} dF_{k-1,k}(x,y)$$

$$= \sum_{k=1}^{n} P\left\{ \int_0^{\infty} I_{(X_{(k-1)}, X_{(k)})}(u) dG(u) \leq x, X_{(k)} - X_{(k-1)} \leq X_{(i)} - X_{(i-1)}, i = 1,2,..,n \right\}$$

$$= P\left\{ \int_0^{\infty} I_{(X_{(\nu-1)}, X_{(\nu)})}(u) dG(u) \leq x \right\} = P\{G(X_{(\nu)}) - G(X_{(\nu-1)}) \leq x\}.$$

(Note that expressions of type $\frac{1}{P\{X_{(k-1)}=x, X_{(k)}=y\}}$ above should be understood as

$$\lim_{h\to 0} \frac{1}{P\{x \leq X_{(k-1)} < x+h, y \leq X_{(k)} < y+h\}}).$$

Theorem thus proved.

Remark 1. Let $F(x) = G(x) = 1 - \exp(-\lambda x)$, $x \geq 0$. Then

$$P\{F(X_{(\nu)}) - F(X_{(\nu-1)}) \leq x \mid \nu = 1\} = 1 - (1-x)^{\frac{n(n+1)}{2}}, 0 \leq x \leq 1.$$

Proof. In fact by definition of ν one has

$$P\left\{F(X_{(\nu)})-F(X_{(\nu-1)})\leq x \mid \nu=1\right\}=\frac{P\left\{F(X_{(\nu)})-F(X_{(\nu-1)})\leq x,\nu=1\right\}}{P\left\{\nu=1\right\}}.$$

Consider

$$P\left\{F(X_{(\nu)})-F(X_{(\nu-1)})\leq x,\nu=1\right\}=P\left\{F(X_{(1)})-F(X_{(0)})\leq x,\nu=1\right\}$$

$$=P\left\{F(X_{(1)})\leq x, X_{(1)}\leq X_{(2)}-X_{(1)},\ldots,X_{(1)}\leq X_{(n)}-X_{(n-1)}\right\}$$

$$=\int P\left\{F(X_{(1)})\leq x, X_{(1)}\leq X_{(2)}-X_{(1)},\ldots,X_{(1)}\leq X_{(n)}-X_{(n-1)} \mid X_{(1)}=t\right\}dF_1(t)$$

$$=n\lambda\int\limits_{F(t)\leq x}(e^{-\lambda t})^{\frac{n(n+1)}{2}}dt=n\lambda\int\limits_{0}^{-\frac{1}{\lambda}\ln(1-x)}(e^{-\lambda t})^{\frac{n(n+1)}{2}}dt$$

$$=\frac{2}{n+1}(1-(1-x)^{\frac{n(n+1)}{2}})\quad ,0\leq x\leq 1.$$

After some algebra, using $P\left\{\nu=1\right\}=\frac{2}{n+1}$ one obtains

$$P\left\{F(X_{(\nu)})-F(X_{(\nu-1)})\leq x \mid \nu=1\right\}=1-(1-x)^{\frac{n(n+1)}{2}}\quad ,0\leq x\leq 1.$$

Lemma 1. Let X_1,X_2,X_3 i.i.d. random variables with continuous d.f. F. Let $X_{(\nu)}-X_{(\nu-1)}\leq X_{(i)}-X_{(i-1)}$, $i=2,3$. The p.d.f. of random variable $F(X_{(\nu)})-F(X_{(\nu-1)})$ is

$$f^*(t)=6\int\limits_t^1\left[1-F(2F^{-1}(v)-F^{-1}(v-t))+F(2F^{-1}(v-t)-F^{-1}(v))\right]dv$$

$$0<t<1.$$

Proof. The d.f. of random variable $F(X_{(\nu)})-F(X_{(\nu-1)})$ can be found as follows :

$$F^*(t)=P\left\{F(X_{(\nu)})-F(X_{(\nu-1)})\leq t\right\}=\iint\limits_C f_{\nu-1,\nu}(x,y)dxdy \tag{2.4}$$

where $C=\{(x,y):x<y,F(y)-F(x)\leq t\}$ and $f_{\nu-1,\nu}(x,y)$ is the joint p.d.f. of $X_{(\nu-1)}$ and $X_{(\nu)}$. It is known that

$$f_{\nu-1,\nu}(x,y)=6f(x)f(y)\left[1-F(2y-x)+F(2x-y)\right]\quad ,x<y \tag{2.5}$$

(c.f., David (1981)).

Let us take in (2.4) $F(x)=u$ and $F(y)=v$. The Jacobian of this transformation is equal to $\left(f(F^{-1}(u))f(F^{-1}(v))\right)^{-1}$. Then from (2.4) using (2.5) one has

$$F^*(t)=6\int\limits_0^t\int\limits_0^v\left[1-F(2F^{-1}(v)-F^{-1}(u))+F(2F^{-1}(u)-F^{-1}(v))\right]dudv$$

$$+6\int_{t}^{1}\int_{v-t}^{v}\left[1-F(2F^{-1}(v)-F^{-1}(u))+F(2F^{-1}(u)-F^{-1}(v))\right]dudv.$$

The proof is completed by differentiating $F^{\star}(t)$ with respect to t.

Example. Let X_1, X_2, X_3 be a sample from uniform on $(0,1)$ distribution. Then

$$f^{\star}(t)=\begin{cases}6(2t-1)^2 & , \quad 0<t<\frac{1}{2}\\ 0 & , \quad otherwise\end{cases} \tag{2.6}$$

In fact, from Lemma 1 we have

$$f^{\star}(t)=6\int_{t}^{1}dv-6\int_{t}^{1}F(v+t)dv+6\int_{t}^{1}F(v-2t)dv. \tag{2.7}$$

It is evident that

$$F(v+t)=\begin{cases}0 & if \quad v\le -t\\ v+t & if \quad -t<v<1-t\\ 1 & if \quad v\ge 1-t\end{cases}, F(v-2t)=\begin{cases}0 & if \quad v\le 2t\\ v-2t & if \quad 2t<v<1+2t\\ 1 & if \quad v\ge 1+2t\end{cases}.$$

Let $0<t<\frac{1}{2}$. Then $t<1-t$ and one has

$$\begin{aligned}6\int_{t}^{1}F(v+t)dv &= 6\int_{t}^{1-t}(v+t)dv+6\int_{1-t}^{1}dv=3-12t^2+6t,\\ 6\int_{t}^{1}F(v-2t)dv &= 6\int_{2t}^{1}(v-2t)dv=3-12t+12t^2.\end{aligned} \tag{2.8}$$

Let $\frac{1}{2}<t<1$. Then $t>1-t$, $F(v+t)=1$, $F(v-2t)=0$ and one has

$$\begin{aligned}6\int_{t}^{1}F(v+t)dv &= 6\int_{t}^{1}dv=6(1-t),\\ 6\int_{t}^{1}F(v-2t)dv &= 0.\end{aligned} \tag{2.9}$$

Using (2.8) and (2.9) in (2.7) one obtains (2.6).

It is clear that in this case the d.f. of $F(X_{(\nu)})-F(X_{(\nu-1)})$ is

$$F^{\star}(t)=\begin{cases}8t^3-12t^2+6t, & 0<t<1/2\\ 1, & t\ge 1/2\end{cases}$$

Lemma 2. Let $X_1, X_2, ..., X_n$ i.i.d r.v.'s with d.f. $F(x)=1-\exp\{-\lambda x\}, x>0, \lambda>0$ and $(X_{(\nu-1)}, X_{(\nu)})$ is minimal spacing of this sample. In this case the d.f. of random variable $X_{(\nu)}-X_{(\nu-1)}$ is

$$P\left\{X_{(\nu)}-X_{(\nu-1)}\le x\right\}=1-\exp\left\{-\frac{n(n+1)}{2}\lambda x\right\} \quad , \quad x>0.$$

Remark 2. Let $X_1, X_2, ..., X_n$ i.i.d r.v.'s with d.f. F. If F is exponential d.f. with parameter λ then

$$\frac{n+1}{2}\left(X_{(\nu)} - X_{(\nu-1)}\right) \stackrel{d}{=} X_{(1)}.$$

Note that it is not known yet whether this property is characteristic property of exponential distribution or not.

3. RECORD STATISTICS

Let $\{X_n\}_{n\geq 1}$ be a sequence of i.i.d r.v.'s with continuous distribution function F. Define a record times of this sequence as follows: $u(1) = 1$, $u(n) = \min\{j : j > u(n-1),$ $X_j > X_{u(n-1)}, n > 1\}$. Let $X_{u(1)}, X_{u(2)}, ...$ be corresponding record values.

The d.f. and probability density function (p.d.f.) of record values can be expressed in terms of

$$R(x) = -\ln(1 - F(x)) \qquad \text{and} \qquad r(x) = \frac{d}{dx}R(x) = \frac{f(x)}{1 - F(x)}.$$

It is known that the distribution of n th record value is

$$F_n(x) = P\{X_{u(n)} \leq x\} = \int_{-\infty}^{x} \frac{R^{n-1}(u)}{(n-1)!} dF(u) \qquad , -\infty < x < \infty.$$

The joint p.d.f. of $X_{u(i)}$ and $X_{u(j)}$ is

$$f(x_i, x_j) = \frac{(R(x_i))^{i-1}}{(i-1)!} r(x_i) \frac{(R(x_j) - R(x_i))^{j-i-1}}{(j-i-1)!} f(x_j) \quad , -\infty < x_i < x_j < \infty$$

The details of record theory can be found in Galambos (1987), Nagaraja (1988), Nevzorov (1988), Ahsanullah (1995), Arnold, Balakrishnan and Nagaraja (1998) among others.

Let $X_{u(n)}$ be the n th record value of a sequence of i.i.d. r.v.'s.

Suppose that $X_{u(n)+1}, X_{u(n)+2}, ..., X_{u(n)+m}$ be the m observations that succeed $X_{u(n)}$. For $i = 1, 2, ..., m$ and $1 \leq r < s \leq n$, it is known that

$$P\{X_{u(n)+i} \in (X_{u(r)}, X_{u(s)})\} = \frac{1}{2^r} - \frac{1}{2^s}$$

(c.f., Bairamov (1997)).

Define the random variables $\xi_1, \xi_2, ..., \xi_m$ as follows:

$$\xi_i = \begin{cases} 1 & \text{if } X_{u(n)+i} \in (X_{u(r)}, X_{u(s)}) \\ 0 & \text{if } X_{u(n)+i} \notin (X_{u(r)}, X_{u(s)}) \end{cases} \quad , \quad i = 1, 2, ..., m \quad , \quad r < s.$$

Denote

$$\nu_m = \sum_{i=1}^{m} \xi_i$$

ν_m is the number of observations $X_{u(n)+1}, X_{u(n)+2}, ..., X_{u(n)+m}$ falling into interval $(X_{u(r)}, X_{u(s)})$. It is clear that the r.v.'s $\xi_1, \xi_2, ..., \xi_m$ are dependent. Bairamov and Eryılmaz (2000) show that for $r = k-1$ and $s = k$ ($k = 2, ..., n$)

$$P\{\nu_m = j\} = \binom{m}{j} \sum_{i=0}^{m-j} (-1)^i \binom{m-j}{i} \frac{1}{(j+i+1)^k}$$

$$j = 0, 1, ..., m.$$

Let $\{X_n\}_{n \geq 1}$ be a sequence of i.i.d. r.v.'s with continuous d.f. F , $X_{u(1)}, X_{u(2)}, ...$ is a sequence of record values. Let $X_1', X_2', ..., X_m'$ be i.i.d. random variables with continuous d.f. G. Define the following random variables

$$\xi_i^*(r,s) = \begin{cases} 1 & \text{if } X_i' \in (X_{u(r)}, X_{u(s)}) \\ 0 & \text{if } X_i' \notin (X_{u(r)}, X_{u(s)}) \end{cases} \quad , \quad i = 1, 2, ..., m \quad , \quad r < s.$$

and let

$$S_m(r,s) = \sum_{i=1}^{m} \xi_i^*(r,s).$$

It is clear that $S_m(r,s)$ denotes the number of X_i' 's falling above the threshold $X_{u(r)}$ and below the threshold $X_{u(s)}$. It is true that

$$\lim_{m \to \infty} \sup_{0 \leq x \leq 1} \left| P\left\{ \frac{S_m(r,s)}{m} \leq x \right\} - P\{G(X_{u(s)}) - G(X_{u(r)}) \leq x\} \right| = 0.$$

The case $F = G$ is of interest on its own and we have the following results for this case:

Let $F = G$. Then

$$\lim_{m \to \infty} \sup_{0 \leq x \leq 1} \left| P\left\{ \frac{S_m(r,s)}{m} \leq x \right\} - P\{F(X_{u(s)}) - F(X_{u(r)}) \leq x\} \right| = 0.$$

where $U_{rs} = F(X_{u(s)}) - F(X_{u(r)})$ has d.f.

$$\begin{aligned} P\{U_{rs} \leq x\} &= \frac{1}{(r-1)!(s-r-1)!} \int_0^x \int_0^{1-t_1} \left[\ln \frac{1}{1-t_2} \right]^{r-1} \frac{1}{1-t_2} \times \\ &\times \left[\ln \frac{1-t_1}{1-t_1-t_2} \right]^{s-r-1} dt_2 dt_1, \end{aligned}$$

and p.d.f.

$$\begin{aligned} f_{U_{rs}}(x) &= \frac{1}{(r-1)!(s-r-1)!} \int_0^{1-x} \left[\ln \frac{1}{1-t_2} \right]^{r-1} \frac{1}{1-t_2} \left[\ln \frac{1-x}{1-x-t_2} \right]^{s-r-1} dt_2, \\ 0 &< x < 1. \end{aligned}$$

Now let again $F = G$ and $r = k-1, s = k$ ($k = 2, ..., n$). Then the expected value and the variance of $S_m(r,s)$ is:

$$ES_m(k-1,k) = m/2^k,$$

$$VarS_m(k-1,k) = m(1/2^k - 1/3^k) + m^2(1/3^k - 1/2^{2k}).$$

Let $F = G$. For $r = k-1$ and $s = k$ $(k = 2, 3, ..., n)$. Then

$$\lim_{m\to\infty} \sup_{0\le x\le 1} \left| P\left\{ \frac{S_m(r,s)}{m} \le x \right\} - P\{U_{k-1,k} \le x\} \right| = 0,$$

where $U_{k-1,k} = F(X_{u(k)}) - F(X_{u(k-1)})$ has d.f.

$$D_k(x) \equiv P\{U_{k-1,k} \le x\} = \frac{1}{(k-1)!} \int_0^x \left[\ln \frac{1}{u}\right]^{k-1} du\ , \quad 0 < x < 1. \tag{3.1}$$

and p.d.f.

$$d_k(x) \equiv f_{U_{k-1,k}}(x) = \frac{1}{(k-1)!} \left[\ln \frac{1}{x}\right]^{k-1}, \quad 0 < x < 1.$$

Also it is true that

$$P\left\{ \frac{S_m(k-1,k) - ES_m(k-1,k)}{\sqrt{VarS_m(k-1,k)}} \le x \right\} \underset{m\to\infty}{\to} D_k(ax+b),$$

where

$$a = \sqrt{\frac{1}{3^k} - \frac{1}{2^{2k}}} \quad \text{and} \quad b = \frac{1}{2^k}.$$

Remark 3. Let $\{X_n\}_{n\ge1}$ be a sequence of i.i.d. r.v.'s with d.f. F. Let $L(1) = 1$, $L(n+1) = \min\{k : X_k < X_{L(n)}\}, n \ge 1$. $L(n)$ is called n th lower record time, $X_{L(n)}$ n th lower record of sequence $\{X_n\}_{n\ge1}$. The d.f. of $X_{L(n)}$ is $F_{L(n)}(x) = P\{X_{L(n)} \le x\} = \frac{1}{(n-1)!} \int_0^{F(x)} \left[\ln \frac{1}{u}\right]^{n-1} du$.

It is not difficult to observe that for the uniform distribution function $F(x) = x\,, 0 \le x \le 1\,, D_k(x)$ given in (3.1) coincides with $F_{L(k)}(x)$, i.e. $D_k(x) = P\{F(X_{u(k)}) - F(X_{u(k-1)}) \le x\} = \frac{1}{(k-1)!} \int_0^x \left[\ln \frac{1}{u}\right]^{k-1} du = F_{L(k)}(x)$. It is also interesting to observe that, for the uniform distribution over $[0,1]$.

$$X_{L(k)} \overset{d}{=} X_{u(k)} - X_{u(k-1)}.$$

References

Ahsanullah, M.(1995) *Record Statistics.* Nova Science publishers, Inc., Commack,NY.

Arnold, B.C. , Balakrishnan, N. and Nagaraja, H.N. (1998) *Records,* John Wiley & Sons, New York.

Bairamov, I.G. and Petunin, Yu.I. (1991) Structure of invariant confidence intervals containing the main distributed mass. Theor. Prob. Appl. 35, No_1, P.15-26.

Bairamov, I.G. and Gebizlioglu, O.L. (1998) Asymptotic distributions of the statistics based on order statistics, record values and invariant confidence intervals. Proceedings of International conference "Asymptotic Methods in Probability and Mathematical Statistics", St: Petersburg, June 24-28, 1998.

Bairamov, I.G., Gebizlioglu, O.L. and Kaya, M.(1999) Distributional properties of statistics based on invariant confidence intervals. İstatistik, Journal of the Turkish Statistical Association. 2, No 1, 15-33

Bairamov I. and Kotz S. (2000) On distributions of exceedances associated with order statistics and record values for arbitrary distributions. Statistical Papers. (to appear)

Bairamov, I.G. and Eryilmaz, S.N. (2000) Distributional properties of statistics based on minimal spacing and record exceedance statistics. Journal of Statistical Planning and Inference. *(to appear)*

David, H.A.(1981) *Order Statistics.* Wiley. New York.

Epstein, B. (1954) Tables for the distribution of the number of exceedances. *Ann. Math.. Statist.* 25, 762-768.

Galambos, J. (1987) *The Asymptotic Theory of Extreme Order Statistics.*(2nd ed). Klieger, Malabar (FL).

Gumbel, E.J. and Schelling, H, von (1950) The distribution of the number of exceedances. Ann. Math.. Statist. 21, 247-262.

Johnson, N.L., Kotz, S. and Kemp, A. (1992) *Univariate discrete distributions.* Wiley, New York.

Johnson, N. and Kotz, S. (1991) Some generalizations of Bernoulli and Polya-Eggenberger contagion models. Statist. Papers 32, P.1-17.

Johnson, N. and Kotz, S. (1994) Further comments on Matveychuk and Petunin's generalized Bernoulli model, and nonparametric tests of homogeneity. Journal of Statistical Planning and Inference. 41 , P. 61-72.

Katzenbeisser, W. (1985) The distribution of two sample location exceedance test statistics under Lehmann alternatives. Statist. Hefte (now *Papers*) 26,131-138.

Katzenbeisser, W. (1986) The exact power of two-sample location tests based on exceedances against shift alternatives. Math. Op. Forsch. (Statistics) 20, 47-54.

Leadbetter, M.L., Lindgren, G. and Rootzen, H. (1983) *Extremes and Related Properties of Random Sequences and Processes.* Springer-Verlag, New-York.

Matveychuk, S.A., Petunin Yu.I.(1990/1991) A generalization of the Bernoulli model arising in order statistics. I/II Ukr. Math. J. 42, P. 518-528/ 43, P. 779-785.

Nagaraja, H.N.(1988) Record values and related statistics. A Review.Commun in Statist.Theor Meth. 17, 2223-2238.

Nevzorov, V.B.(1988) Records. Theor.Prob.Apl. 32, 201-228.

Robbins, H. (1944) On distribution -free tolerance limits in random sampling. Ann. Math. Statist. 15, 214-216.

Siddiqui, M.M. (1970) Order statistics for a sample and extended sample . *Nonparametric Techniques in Statistical Inference.* (Ed. M.L.Puri), 417-423, Cambridge University Press, Cambridge.

Shewart W.A. (1931) *Economic control of quantity of manufactured product.* D. Van Nostrand, Princeton, N.J.

Sarkadi, K. (1957) On the distribution of the number of exceedances. Ann. Math. Stat., 28, 1021-1022.

Wesolowski, J. and Ahsanullah, M. (1998) Distributional properties of exceedance statistics. Ann. Inst. Statist. Math. 50, N_3, 543-565.

Estimation of Common Location Parameter of Two Exponential Distributions

K. Raghunandanan
Department of Mathematics
Temple University
Philadelphia, PA 19122

M. Ahsanullah
Department of Management Sciences
Rider University, Lawrenceville, NJ 08649-3099

ABSTRACT

In this paper we consider the estimation of the common location parameter of two exponential distributions with known coefficient of variation based on two independent samples. Three unbiased estimators of the location parameter are considered and their relative efficiencies are compared.

1. INTRODUCTION

Let $X_{11}, X_{12}, \ldots, X_{1n_1}$ be a random sample form an exponential distribution with probability density function (pdf)

$$f_1(x) = \sigma_1^{-1} \exp\{-(x-\mu_1)/\sigma_1\}, \quad x \geq \mu_1, \sigma_1 > 0 \qquad (1.1)$$

with $\sigma_1 = k\mu_1$.

Further let $X_{21}, X_{22}, \ldots, X_{2n_2}$ be a random sample from an exponential distribution with pdf

$$f_2(x) = \sigma_2^{-1} \exp\{-(x-\mu_2/\sigma_2)\} \;, x \geq \mu_2,\ \sigma_2 > 0 \qquad (1.2)$$

with $\sigma_2 = k\mu_2$.We will consider in this paper the case $\mu_1 = \mu_1 = \mu$.

Denote U = $\min\{X_{11}, X_{12}, \ldots, X_{1n_1}\}, V = \min\{X_{21}, X_{22}, \ldots, X_{2n_2}\}, Z = \min\{U, V\}$,
$T_i = n_i^{-1} \sum_{j=1}^{n_i} (X_{ij} - Z), i = 1,2$ and $a = \frac{n_1}{\sigma_1} + \frac{n_2}{\sigma_2}$. It can be easily shown that the MLE of μ, σ_1, and σ_2 are, respectively, $\hat{\mu} = Z,\ \widehat{\sigma_i} = T_i/n_i, i = 1,2$.

Ghosh and Razmpour [1984] showed that uniformly minimum variance unbiased estimator (UMVUE) $\hat{\mu}$ of μ for known σ_1 and σ_2 is

$$\hat{\mu} = Z - a^{-1}.$$

For the case when σ_1 and σ_2 are not known, they gave a modified MLE $\hat{\mu}^*$ of μ as

$$\hat{\mu}^* = Z - \hat{a}^{-1}, \text{where } \hat{a} = n_1/\hat{\sigma}_1 + n_2/\hat{\sigma}_2 .$$

This estimator is not unbiased. Pal and Sinha (1990) considered the estimation of the common means of the exponential distribution under the square loss function. See Jin and Cross (1998) for some related works.In this paper we will consider three unbiased estimator of μ based $\overline{X}_1, \overline{X}_2$, U, V and Z for known values of the coefficient of variations.

2. UNBIASED ESTIMATORS OF μ

We have

$$E(X_1) = \int_{\mu}^{\infty} x(k_1\mu)^{-1}\exp\{-(x-\mu)/k_1\mu\}dx$$

$$= \int_{0}^{\infty} \mu(1+k_1 t)e^{-t}dt = \mu(1+k_1)$$

$$E(X_1^2) = \int_{\mu}^{\infty} x^2(k_1\mu)^{-1}\exp\{-(x-\mu)/k_1\mu\}dx$$

$$= \int_{0}^{\infty} \mu^2(1+k_1 t)^2 e^{-t}dt = \mu^2(1+2k_1+2k_1^2)$$

Hence $\mathrm{Var}(X_1) = k_1^2\mu^2$, $\mathrm{Var}(\overline{X}_i) = \dfrac{k_i^2\mu^2}{n_i}, i = 1,2$, where $\overline{X}_i = \frac{1}{n_i}\sum_{j=1}^{n_i} X_{ij}$.
It can be shown that Z and $\dfrac{n_1}{k_1}\overline{X_1} + \dfrac{n_2}{k_2}\overline{X}_2$ are jointly sufficient for μ.
The pdf's of U, V and Z are, respectively,

$$f_u(u) = \frac{n_1}{k_1\mu}\exp\{-\tfrac{n_1}{k_1\mu}(u-\mu)\},\ u \geq \mu,\ k_1 > 0,$$

$$f_v(v) = \frac{n_2}{k_2\mu}\exp\{-\tfrac{n_2}{k_2\mu}(v-\mu)\},\ v \geq \mu,\ k_2 > 0 \text{ and}$$

$$f_z(z) = a\exp\{-a(z-\mu)\},\quad z \geq \mu,\ \text{where } a = \tfrac{n_1}{k_1\mu} + \tfrac{n_2}{k_2\mu}.$$

We will consider the following three estimators for μ. based on $\overline{X}_1, \overline{X}_2$, U, V and Z..

i) $\hat{\mu}_1 = c_1[\frac{n_1\overline{X_1}}{k_1} + \frac{n_2\overline{X_2}}{k_2}]$, where $c = \dfrac{k_1k_2}{n_1k_2(k_1+1)+n_2k_1(k_2+1)}$

ii) $\hat{\mu}_2 = \dfrac{n_1k_2+n_2k_1}{n_1k_2+n_2k_1+k_1k_2}Z$ and

iii) $\hat{\mu}_3 = a_1U + a_2V$, where constants a_1 and a_2 are so chosen so that $\hat{\mu}_3$ is unbiased and has minimum variance. Then the constants a_1 and a_2 are

$$a_1 = \frac{n_1k_2^2(n_1+k_1)}{k_1^2(n_2+k_2)^2+k_2^2(n_1+k_1)^2},\qquad a_2 = \frac{n_2k_1^2(n_2+k_2)}{k_1^2(n_2+k_2)^2+k_2^2(n_1+k_1)^2}$$

It can be easily shown that all these three estimators $\hat{\mu}_1, \hat{\mu}_2$ and $\hat{\mu}_3$ are unbiased and their corresponding variances are

$\text{Var}(\hat{\mu}_i) = d_i\mu^2$, where.

$d_1 = (n_1 + n_2)(\frac{1}{n_1+n_2+\frac{n_1}{k_1}+\frac{n_2}{k_2}})^2$, $d_2 = (\frac{1}{1+\frac{n_1}{k_1}+\frac{n_2}{k_2}})^2$ and

$d_3 = \frac{1}{(1+\frac{n_1}{k_1})^2+(1+\frac{n_2}{k_2})^2}$

For simplicity in comparison of these estimators we will take $n_1 = n_2 = n$ and $k_2 = k_1h$. Then the estimators and their variances are

$$\hat{\mu}_1 = \frac{1}{2k_1h + h + 1}(h\overline{X}_1 + \overline{X}_2), \qquad \hat{\mu}_2 = \frac{n(h+1)}{n(h+1)+k_1h}Z \text{ and}$$

$$\hat{\mu}_3 = \frac{1}{(n+k_1h)^2 + h^2(n+k_1)^2}[n(n+k_1)h^2U + n(n+hk_1)V].$$

The corresponding variances of the estimators are

$$Var(\hat{\mu}_1) = \tfrac{2}{n}(\frac{hk_1}{2hk_1 + h + 1})^2\mu^2, \qquad Var(\hat{\mu}_2) = (\frac{hk_1}{nh + n + k_1h})^2\mu^2 \text{ and}$$

$$Var(\hat{\mu}_3) = (\frac{h^2k_1^2}{h^2(n+k_1)^2 + (n+hk_1)^2})\mu^2$$

3. EFFICIENCIES OF THE ESTIMATORS

We will consider the efficiencies with respect to the variances of the estimators.

Let $e_{ij} = \frac{Var(\hat{\mu}_j)}{Var(\hat{\mu}_i)}$, i, j = 1, 2, 3. It can be shown that

$e_{21} > 1$ if $\tfrac{2}{n}\{\frac{h^2k_1^2}{(2hk_1 + h + 1)^2}\} > \frac{h^2k_1^2}{(nh + n + k_1h)^2}$

i.e., if $h[1 - k_1\sqrt{2/n}\,] \geq -1$. Since $h > 0$, $e_{21} \geq 1$, if $n \geq 2k_1^2$. If $n < 2k_1^2$, then

$e_{21} \geq 1, if\, h \leq \frac{\sqrt{n}}{k_1\sqrt{2} - n}$. Since k_1 is generally small 1, we will have $e_{21}>1$ for almost aa n.

For $e_{23} \geq 1$, we must have,

$$\frac{(hk_1)^2}{(nh + n + h + k)^2} \leq \frac{h^2k_1^2}{h^2(n+k_1)^2 + (n+hk_1)^2}, \text{ i.e., if } n \geq \frac{hk_1}{\sqrt{2}}.$$

Since $k_2\,(= hk_1)$ is small, we will have $e_{23} > 1$ almost all n.

For $e_{31} \geq 1$, we need

$$\frac{2}{n}(\frac{h^2k_1^2}{(2hk_1 + h + 1)^2}) \geq \frac{h^2k_1^2}{h^2(n + k_1^2) + (n + hk_1)^2}$$

i.e., if $k_1 \leq \frac{1}{2(2n-1)} \sqrt{\frac{n}{n-1}\{(2n-1)^2 - 1\}}$.

i.e.. if $k_1 \leq (1/2)\sqrt{n/(n-1)}$, *i.e.* if $k_1 \leq \sqrt{2}$.

We can say that for known coefficient of variations the estimator $\hat{\mu}_2$ has minimum variance than the estimators $\hat{\mu}_1$ and $\hat{\mu}_3$ if the coefficient of the variations are not too large.

4. SIMULATION RESULTS

We did a simulation study to compare the variances of the different estimators. Using IMSL program, we generated n random samples from uniform distribution in the interval (0,1). These values were used to generate random samples from the exponential distributions. By changing the values of a, h and k the three estimators $\hat{\mu}_1, \hat{\mu}_2$ and $\hat{\mu}_3$ were computed. Variances of these three estimators were then calculated using thousand simulations. We did this for n=10, 20, 25 and 50; a = 0.5, 1, 10; h = 1, 2,3 and 5; k = 0.1,0.25 and 1. The following tables give the values of the variances of the estimators for selected values of, a, h and k. The variances given are variances divided by μ^2. The notation, for example, $E-4$ indicates the variance is multiplied by 10^{-4}.

We found that estimator $\hat{\mu}_2$ has smaller variances for all the cases. Since the same conclusion was reached for all values of a, table is given only for a = 1. We also noticed that the variances of the estimators decreases as the value of h increases and they increased as k increased. Values of the estimators and their variances based on simulation are given in Tables 1, 2 and 3. In the tables, E is the estimator and V is the variance.

REFERENCES

Ghosh, M. and Razmpour,A. (1984). Estimation of the common location parameter of several exponentials.Sankhya, 48,A, 383-394.

Jin, C. and Crouse, R. H. (1998). A note on the common location parameter of several exponential populations.(1998). Commun.Statist.-Theory Meth. 27(11), 2777-2789.

Pal, N. and Sinha, B.K. (1990),. Estimation of a common location of several exponentials. Statistics & Decisions. 8, 27-36.

Table 1

		k = 0.1 E	k = 0.1 V	k = 0.25 E	k = 0.25 V	k = 1 E	k = 1 V
$n = 10$	$\hat{\mu}_1$	1.0000	$3.8849E-4$	1.0026	$2.0159E-3$	1.0015	$1.2110E-2$
$h = 1$	$\hat{\mu}_2$	1.0000	$2.5247E-5$	1.0002	$1.6934E-4$	0.9995	$2.2406E-3$
	$\hat{\mu}_3$	0.9999	$4.9465E-5$	1.0002	$2.9224E-4$	0.9993	$3.8716E-3$

		k = 0.1		k = 0.25		k = 1	
$n = 10$	$\hat{\mu}_1$	0.9999	$7.1356E-4$	0.9974	$3.2904E-3$	0.9964	$1.7106E-2$
$h = 2$	$\hat{\mu}_2$	0.9998	$3.9783E-5$	1.0007	$2.8121E-4$	0.9972	$3.7160E-3$
	$\hat{\mu}_3$	1.0000	$7.6143E-5$	1.0008	$5.2934E-4$	0.9948	$5.6023E-3$

		k = 0.1		k = 0.25		k = 1	
$n = 10$	$\hat{\mu}_1$	0.9978	$8.2810E-4$	1.0019	$3.9286E-3$	0.9995	$1.7752E-2$
$h = 3$	$\hat{\mu}_2$	0.9999	$6.0778E-5$	1.0001	$3.4603E-4$	1.0017	$4.8633E-3$
	$\hat{\mu}_3$	0.9998	$8.9551E-5$	0.9995	$4.7156E-4$	1.0040	$8.2194E-3$

Table 2

		k = 0.1		k = 0.25		k = 1	
$n = 25$	$\hat{\mu}_1$	1.0001	$1.7857E-4$	0.9998	$8.6727E-4$	0.9978	$4.7703E-3$
$h = 1$	$\hat{\mu}_2$	0.9999	$3.7082E-6$	0.9997	$2.2463E-5$	0.9996	$3.6582E-4$
	$\hat{\mu}_3$	0.9999	$7.9919E-6$	0.9996	$4.6433E-5$	0.9998	$6.7398E-4$

		k = 0.1		k = 0.25		k = 1	
$n = 25$	$\hat{\mu}_1$	1.0006	$2.7625E-4$	0.9997	$1.2511E-3$	1.0007	$6.9438E-3$
$h = 2$	$\hat{\mu}_2$	1.0000	$6.5712E-6$	0.9999	$4.4495E-5$	1.0007	$7.6272E-4$
	$\hat{\mu}_3$	1.0000	$1.2822E-5$	0.9997	$7.4715E-5$	1.0021	$1.2799E-3$

$k = 0.1$ | k=0.25 | k=1

$n = 25$	$\hat{\mu}_1$	0.9999	$3.4572E-4$	0.9993	$1.4129E-3$	1.0009	$7.4745E-3$
$h = 3$	$\hat{\mu}_2$	0.9999	$8.8804E-6$	1.0000	$5.8685E-5$	1.0004	$8.0660E-4$
	$\hat{\mu}_3$	0.9999	$1.3263E-5$	1.0002	$8.8298E-5$	0.9991	$1.2744E-3$

Table 3

		k = 0.1		k = 0.25		k = 1	
$n = 50$	$\hat{\mu}_1$	0.9993	$7.9344E-5$	0.9999	$3.7248E-4$	0.9999	$2.6991E-3$
$h = 1$	$\hat{\mu}_2$	0.9999	$8.7587E-7$	1.0000	$6.3205E-6$	1.0004	$9.6881E-5$
	$\hat{\mu}_3$	0.9999	$1.7983E-6$	0.9999	$1.2120E-5$	1.0006	$2.0694E-4$

		k = 0.1		k = 0.25		k = 1	
$n = 50$	$\hat{\mu}_1$	0.9995	$1.3522E-4$	1.0016	$6.4303E-4$	0.9996	$3.1116E-3$
$h = 2$	$\hat{\mu}_2$	0.9999	$1.6196E-6$	1.0001	$1.1469E-5$	1.0006	$1.9852E-4$
	$\hat{\mu}_3$	0.9999	$3.2428E-6$	1.0002	$2.2694E-5$	1.0004	$3.3004E-4$

		k = 0.1		k = 0.25		k = 1	
$n = 50$	$\hat{\mu}_1$	1.0005	$1.7247E-4$	0.9995	$7.4591E-4$	1.0022	$3.6167E-3$
$h = 3$	$\hat{\mu}_2$	0.9999	$2.3601E-6$	1.0000	$1.5608E-5$	1.0004	$2.2148E-4$
	$\hat{\mu}_3$	0.9999	$3.7797E-6$	1.0000	$2.3234E-5$	1.0005	$3.3991E-4$

THE STOCK MARKET: DETERMINISTIC OR RANDOM?

G.R. DARGAHI-NOUBARY

Department of Mathematics

Bloomsburg University, Bloomsburg, PA 17815

SUMMARY

In recent years selection of stocks in an almost random fashion has become popular among inventors on the ground that it works as well as expert advice. The individuals who are familiar with probability theory rely on the fact that random walk, Brownian motion or models of this type provide reasonable description of the stock market. While there is a general agreement that such models are reasonable, why they do work is not entirely understood or explained. This article analyzes the general characteristics of the such models in an attempt to explain their connection to the behavior of typical investors. It also discusses both deterministic and probabilistic analysis and makes some comparison.

1. INTRODUCTION

Paul Cootner was one of the first economists to recognize the importance of what was then largely statistical analyses of stock price movements and to begin to provide a theoretical foundation for the apparently random behavior of stock price changes. In his first paper on stock market prices (1962) he provided an operational definition of market efficiency and its relation to random walk. The ideas later were discussed in more detail in a collection of essays that he edited --- the Random Characters of Stock Market Prices (1964).

Is stock market random? As is pointed out by many experts, it would be a negation of economic law if competitive prices truly moved in a classical random walk. This is because given enough time prices could wander anywhere. Still, an economist would be ill advised to ignore the probabilistic jitter inherent in speculative prices. This observation has a long history and goes back to Maurice Kendall's (1953) survey of the statistics of speculative price. For spot and futures prices of commodities, for single stocks and stock indices numbers of stocks, the noneconomist Kendall found white spectra-almost zero autocorrelations. As he said, to the initial outrage of the economists in his audience at the Royal Statistical Society, it was as if the devil drew the price changes out of an urn. Kendall's observations accord with the classical Wiener definition of a random walk in terms of a process with independent increments. The random walk was discussed as early as 1900 in Sorbonne thesis of L. Bachelier and also in his popular 1914 books. See also Fischer (1987), Malkiel (1985),

Stewart (1989) and Zweig (1990). Later several corrections were made to his strict model of the arithmetic random walk. Paul Samuelson in the middle 1950s repaired the Bachelier long run anomaly by stipulating the geometric Brownian motion based on proportionate changes. This led, as the physicist M.F.M. Osborne observed empirically at much the same time, to the Log-normal Wiener process. Samuelson argued that the Bachelier anomalies could be avoided once the absolute or arithmetic Brownian motion of physics is replaced by the geometric Brownian motion of economics. This, of course, is not very surprising since most of the modern option theory operates in terms of Log-normal distribution of prices which has a longer (fatter) tail than normal distribution. Mandelbrot (1963) has pointed out that even fat-tailed distributions of the Pareto type may occur in speculative stochastic processes. This is because meandering chance could lead the real world away from economic law even faster and further than the Wiener process had suggested. Cootner (1964) has argued that, for stock prices, the correct model is not a Gaussian process, nor even a Log-normal random walk, but a martingale. There are, also, some arguments in favor of Markov process, ergodic probability distribution, and, more recently, chaos, see e.g. Chorafas (1994), Peters (1991), Stewart (1989) and Williams (1995).

2. WHY PROBABILISTIC ANALYSIS

Consider a single stock and the collection of individuals who are interested in it. In order to study the price changes let us pick a starting point and look at the situation as a system. A priori, the choice of initial conditions presents a very considerable problem. To see this, imagine only one popular stock and examine the possibility of representing investors behavior through some deterministic theory.

1. We would need to know the initial investment and initial direction (investment plan, motive, etc.) for every investor. This makes the initial data, enormous.
2. To perform the calculation we would need to solve a very large system of equations featuring initial conditions.

   The result would consist of very many numbers, representing the state of the system at some given later time.
3. Finally, we would need to calculate the readily measurable macroscopic performance variables as sums over these output values.

Given our knowledge of 20th-century, we can see that attempts to implement such a deterministic program must run into several insurmountable obstacles. First, there is the practical impossibility of knowing simultaneously the investment and the direction (the plan) taken by individual investors. Next, even if we had this information, such calculations would require a very large memory and would be so lengthy as to be impracticable. Finally, there is a methodological objection which one can formulate as follows: since the aim is to calculate "macroscopically" measurable quantities from microscopic data, might one not do this more easily in some other fashion than a deterministic analysis?

The story is similar to what happened in physics when Maxwell in 1859 could not but realize that in practice deterministic calculation related to molecules of even one mole of gas is impossible. Here the initial investment and direction taken by an individual is similar to initial position and velocity of a molecule. It was in this context that Maxwell made the conceptual leap of introducing into the theory the notion of unpredictable. His breathtaking intuition was developed further by Boltzman and was confirmed half a century later by the work of Albert Einstein (1905) and of Jean Perrin (1908) on Brownian motion. It is important to note that Maxwell does not reject determinism as such, but he mitigates our ignorance of the initial conditions by introducing assumptions of probabilistic nature. His approach can be placed within a framework of the so-called "probabilities through ignorance". The introduction of this approach in 1872 led to the birth of statistical physics.

In a similar fashion when modeling the stock market one can fruitfully exploit probabilities that represent our ignorance. Basically, we admit the deterministic nature of actions taken by investors, but try to find a way to treat them probabilistically. In other words, in order to construct the economic theory governing a deterministic system of large number of investors it is possible to exploit probabilities through ignorance, and have a complete success.

The reason for this claim deserves discussion and explanation. The Soviet physicist Lev Landau has shown that a classical system requiring infinitely many parameters would behave in a totally random fashion: in other words, it would be random unavoidably, and not merely by reason of our ignorance. A system of several million investors depends on very large number of parameters which though, theoretically not infinite, can be considered as infinite for practical purposes. In this way one is led from chance by reason of ignorance to chance unavoidable, which is governed by the laws of probability. That is why we have struck so lucky with our appeal to the random walk and Brownian motion and other models, and why the use of probabilities through ignorance, far from being a mere second-best reflecting the limits of our numerical capabilities, in fact takes us very close to the laws of market.

3. HOW ABOUT DETERMINISTIC CHAOS?

In the preceding section we saw, using Maxwell's idea, how chance could generate determinism. The present section briefly looks at a situation called chaos where determinism can generate chance and points out the way the resulting theory may be considered appropriate for stock market analysis. Chaos is a particularly unfortunate name because unlike what we understand from the word, it actually refers to a higher degree of order. To appreciate its importance, we can refer to the fact that the heart has to be largely regular or you die. But the brain has to be largely irregular; if not, you have epilepsy. This shows that irregularity, chaos, leads to complex systems. It is not all disorder. The chaos theory was pioneered by Henri Poincarè near the end of the last century, but only the advent of fast computers in the early 1960s has made its development possible. Today it is a very active field of research, and has induced a far-reaching revolution in our concepts. In order to classify it, we make a semantic distinction between random and chaotic processes.

A dust particle suspended in water moves around randomly, executing what is called Brownian motion. This stems from molecular agitation, through the impacts of water molecules on the dust particle. Every molecule, much like an investor, is a direct or indirect cause of the motion, and we can say that the Brownian motion of the dust particle is governed by very many variables. In such cases one speaks of a *random process*; to treat it mathematically we use the calculus of probabilities.

A compass needle acted on simultaneously by a fixed and by a rotating field constitutes a very simple physical system pending on only three variables. However, one can choose experimental conditions under which the motion of the magnetized needle is so unsystematic that prediction seems totally impossible. In such very simple cases whose evolution is nevertheless unpredictable one speaks of *chaos* and of *chaotic processes*; these are the terms one uses whenever the variables characterizing the system are few.

Thus, when the system such as the one we based on behavior of individual investors is studied, one looks at a very large number of variables, and hence, random process is more appropriate. However, when only a few variables such as interest rate, currency rate, etc. can characterize the system, the chaos may be used.

The relevance of chaos to market behavior may also be explained by the following example. A pendulum swings to and from with a regular back-and-forth motion, but if it is struck by the ball of a second pendulum before reaching its zenith, both pendulums may begin swinging in wildly erratic patters. In the financial market, a trend is enhanced or undermined by surprises in governmental announcements or economic actions by one or more influential nations.

4. CONCLUDING REMARKS

Based on the proceeding discussion, the possible models may be described in the following way:

1. Stochastic models use randomness as a basic concept, so that market is governed, at least partially, by chance and associated laws of probability.
2. Chaos offers the fascinating possibility of describing randomness as the <u>result</u> of a known deterministic market. Randomness may lie on the choice of initial conditions and

the subsequent mechanism of the market change. Note that the term chaos is usually reserved for dynamical systems whose state can be described with differential equations in continuous time or difference equations in discrete time. This means that it is possible to represent the market using these models.

The theory of chaos is fascinating if for no other reason than the blurring of the long-held distinction between random and deterministic phenomena. It is potentially capable of "explaining" very complex processes with simple, parsimonious models and essentially without error. The theory has the potential to offer traders entirely new perspectives on the movements of market

There are many important questions about the new theory that are currently unresolved, and it is likely that some of these issues will never be resolved. One such a question in present context is the question of choosing between deterministic versus stochastic modeling of market. One reason for this is appropriateness of the random walk model. To clarify this we note that random walk is stochastic version of first order difference equation and structurally has similarly to logistic equation. On the other hand it is an autoregressive model with unit root. To clarify this we note that there is a connection between random subdivision and self-similarity which is the basis of chaos.

Kolmogorov was probably the first to invoke the connection between random subdivision and self-similarity. His argument ran as follows: if we have some large unit (say, a large rock) which is broken up into n_i smaller pieces, then fragmented again and again so that a fragment belonging to the jth generation gives rise to n_j smaller fragments, then the number of rocks at the ith stage of fragmentation is $n_1 n_2 ... n_i$. Assuming that the process of fragmentation is homogeneous in space and in time (i.e., that the mechanism of breakage is always the same), the numbers n_i are identically distributed random variables.

By a straightforward reasoning, Komogorov found that the distribution of sizes of fragments must be lognormal. the property of lognormality has, indeed, been known for many years to describe the granulometry of sands and natural aggregates. There are many other examples of lognormality in variables resulting from processes of infinite random fragmentation.

REFERENCES

Chorafas, D. N.(1994) *Chaos Theory in the Financial Markets*. Chicago, Probus Publ. Co.

Cootner, P., (1060) Returns to Speculators: Telser vs. Keynes, *Journal of Political Economy 68*, 396-404; and "Rejoinder," ibid., 415-18.

_____, (1962) Stock Prices: Random vs. Systematic Changes, *Industrial Management Review 3*, 24-45, reproduced in The Random Character of Stock Market Prices, P. Cootner, ed. Cambridge, Mass.: M.I.T. Press, 1964.

Cootner, P. ed.,(1964 and 1967), *The Random Character of Stock Market Prices.* Cambridge, Mass.: M.I.T.
Press, hereafter referred to as *Cootner Anthology.* For Cootner's own contributions, see pp. 1-6, 79-83, 189-197, 231-252, 333-337, 373-375.

Fischer, K. L. (1987). *The Wall Street Waltz.* Chicago: Contemporary Books.

Kendall, M. G.,(1953). The Analysis of Economic Time Series, Part I: Prices, *Journal of the Royal Statistical*
Society 116, 11-25; reproduced in *Cootner Anthology*, pp. 85-99.

Malkiel, B. (1963). *A Random Walk Down Wall Street,* 4th ed. New York: W. W. Norton, 1985.

Mandelbrot,
B., "Variation of Speculative Prices," *Journal of Business 36*, 394-419; reproduced in *Cootner Anthology*, pp. 307-32.

Markowitz, H.M.(1959), Portfolio Selection: Efficient Diversification of Investment. Cowles Foundation
Monograph 16 (New Haven, CT: Tale University Press).

Osborne, M.F.M.(1959). Brownian Motion in the Stock Market, *Operations Research 7*, 145-73 (reproduced
in *Cootner Anthology*, pp. 100-128).

Peters, E. (1991). *Fractal Market Analysis.* New York: John Wiley & Sons.

Samuelson, P.A.,(1965). Intertemporal Price Equilibrium: A Prologue to the Theory of Speculation,
Weltwirtschaftliches Archiv, (1957); reprinted in *CSP 2.*

_______, (1965a) Proof that Properly Anticipated Prices Fluctuate Randomly, *Industrial Management*
Review 6, 41-49; reprinted in *CSP 3,* 1972.

________, (1965b) Rational Theory of Warrant Pricing and Appendix: A Free Boundary Problem for the
Heat Equation Arising from a Problem in Mathematical Economics, by H.P. McKean, Jr., *Industrial Management Review 6* , 13-39; reprinted in *CSP 3* 1972 and in *Cootner Anthology*, pp. 506-22.

_______, (1971). Stochastic Speculative Price, *Proceedings of the National Academy of Sciences 68*,
335-37; reprinted in *CSP 4,* 1972.

_______, (1973). Proof that Properly Discounted Present Values of Assets Vibrate Randomly, *The Bell*
Journal of Economics 4, 369-74; reprinted in *CSP 4*, 1977.

_______, (1976). Is Real-World Price a Tale Told by the Idiot of Chance? *The Review of Economics*
and Statistics 58, 120-23; reprinted in *CSP 4*, 1977.

_______, (1980). Stochastic Land Valuation Total Yield as Martingale Implying Price a Negatively
Correlated Walk, unpublished manuscript.

Stewart, I. *Does God Play Dice? The Mathematics of Chaos.* Cambridge, MA:Blackwell, (1989).

Williams,
B. *Trading Chaos.* New York: Wiley, 1995.

Zweig, M.(1990). *Winning on Wall Street.* New York: Warner Books.

EVALUATION OF THE RELIABILITY OF SERUM CHOLESTEROL MEASUREMENTS

by

CATHERINE R. ALLAN, R. CHOUDARY HANUMARA
Department of Computer Science and Statistics
University of Rhode Island, Kingston, R.I.

and

KATE LAPANE
Pawtucket Heart Health Program, Memorial Hospital
Pawtucket, R.I.

SUMMARY

The reliability of the inexpensive Reflotron desk-top analyzer used at workplace in the determination of cholesterol measurements, choice of lot of reagent strips, and sources of variation have been studied in this paper. The measurements are negatively biased in reference to measurements made in the laboratory. The shelf life of certain lot of reagent strips is a contributing factor to this and those lots were not used in the experiment. The mixed model analysis of variance pointed out the largest source of variation was due to technicians. A two-step adjustment procedure of Reflotron measurements was suggested to reflect the true cholesterol values of individuals.

Key Words and Phrases: Accuracy, Precision, Mixed Model, Tolerance Intervals, Maximum Percent Error.

I. INTRODUCTION

Coronary heart disease (CHD) is the leading cause of death in the United States (AHA, 1989). Heightened awareness of the importance of cholesterol as a primary modifiable risk factor for heart disease has increased the number of individuals having their cholesterol levels measured. For each 1% decrease in total blood cholesterol, the rate of coronary heart disease is reduced by 2% (NIH, 1984). An estimated 36% of the population has hypercholesterolemia as defined by the National Cholesterol Education Program (NCEP). An estimated 41% of American adults have cholesterol levels which justify further testing and possible treatment (Sempos, 1989). The NCEP recommends that all persons over the age of twenty have their cholesterol levels measured at least once every five years, and those with high cholesterol (>240mg/dl) receive treatment to lower their cholesterol levels. In response, "simple" cholesterol testing procedures have become common with the development of portable desk-top analyzers which allow for inexpensive, fast, accurate measurements of serum cholesterol.

The Worksites, Occupational Nurses, and Cholesterol Change Program is a study affiliated with the Memorial Hospital in Pawtucket, Rhode Island, designed to address many specific aims. One of the most important of these aims is to determine whether occupational nurses within a worksite *(Internal)* are more effective in lowering total blood cholesterol than technicians employed by an outside screening organization *(External)* to direct cholesterol screening, educate individuals about cholesterol, and conduct referral programs for individuals at risk. Another important goal is to determine if automated cholesterol assessment and feedback of suggested behavior changes *(RT)* is as effective in reducing CHD modifiable risk factors as a face-to-face delivery of these recommendations in a brief 10-12 minute counseling session *(FF)*.

Twenty-five worksites in Southern Massachusetts and Rhode Island with full-time occupational nurses were recruited and then assigned randomly to one of five experimental units, each condition consisting of five worksites (Table 1). A random sample of the employees within each worksite were tested at six-month and twelve-month follow-up periods. The cholesterol measurements were taken utilizing a Reflotron portable desk-top analyzer.

Table 1

Experimental Conditions

	Face to Face Feedback (FF)	Automated Feedback (RT)
Occupational Nurses (Internal)	5 Worksites	5 Worksites
Project Staff (External)	5 Worksites	5 Worksites
Usual Care (Minimal education about cholesterol such as usual public screener)	5 Worksites	

The objective of this work is to determine how accurate these cholesterol measurements are, and how they should be adjusted in the final stage so that they actually represent the true cholesterol values of individuals in the study. Approximately a 30- L aliquot of human blood or sera is pipetted onto an application pad on a reagent strip. The reagent system is impregnated into the membranous layer on the strip (Sedor, 1988). When the regent strip is inserted into the Reflotron machine, contact occurs between the reagent and sample layers which starts a chemical reaction. The individual's cholesterol values are displayed in 175 seconds.

It has been previously documented that variability exists between manufactured lots of reagent strips (Powers, 1992). Since each lot of reagent strips is only "good" for so long, each lot having a specific expiration date, testing is necessary for baseline, 6-month follow-up, and 12-month follow-up worksite wide cholesterol screening. The analysis of the data was done for lots labeled 941, 942 (twice in Feb. '93 and once in Aug. '93) and 564 (Aug. '93) for comparison purposes and also to understand the components of variation in measurements. However, the results for only August 1993, lot 941 are presented here for brevity sake. The confidence intervals on variance components are obtained but are not presented here.

II. ACCURACY TESTING

The initial accuracy testing was performed in order to determine which lot of reagent strips give the best, most accurate results. This was done by determining what percentage of Reflotron-obtained values for each lot of reagent strips fell within specific guidelines set by the NCEP. These guidelines are set at 14.2% and 8.9% of the target or true value.

In this initial lot testing, approximately 20-30 volunteers are utilized. Each individual has his/her total blood cholesterol measured three times, twice by Reflotron and once by venous blood sample. To account for machine-to-machine variability, approximately the first half of the volunteers' first finger-stick values are obtained with reagent strip A in machine 1 and the second half of the volunteers' first finger-stick values are obtained with reagent strip A in machine 2. Likewise, the second finger-stick values are obtained for the first half of the volunteers with reagent strip B in machine 2 and for the second half of the volunteers with reagent strip B in machine 1. Finally, to obtain the third cholesterol value, a venous blood sample is drawn and analyzed at the Center for Disease Control standardized laboratory at Miriam Hospital. These cholesterol values are taken as the true cholesterol value or "gold standard" for each person since the reliability of these estimates is high.

Percent Differences

The first lot testing in February, 1993 consisted of 19 volunteers. The two initial lots of reagent strips compared were lot 941 and lot 942. Let Y_{ij} ($i=1,2$; $j=1,2,\ldots,n$) denote the ith fingerstick measurement (i=1 if reagent strip from lot 941 and i=2 if regent strip from lot 942) on the jth volunteer, and let X_j denote the gold standard value for the jth volunteer.

To calculate the percent difference between the Reflotron obtained cholesterol values and the gold standard values, standardization is required as follows:

$$\nabla\% = \frac{Y_{ij}-X_j}{X_j} * 100$$

In this study, 100% of the Reflotron observations fall within ±14.2% of the gold standard values regardless of the lot of regent strip utilized. However, lot 941 fared slightly better with 84.2% of its fingerstick values falling within ±8.9% as opposed to lot 942 with 78.9% of its fingerstick values falling within ±8.9%. Both of these lots of reagent strips are fairly accurate and warrant further testing. The reagent strips from lot 941 were used to collect baseline worksite-wide cholesterol values.

In August 1993, a new lot of regent strips is needed. The shelf-life of lot 941 is quickly approaching. The second lot testing experiment was run with reagent strips from lot 941 and lot 943 and consisted of 32 volunteers. Of those measured with reagent strips from lot 941, 60.6% fell within ±8.9% and 75.8% fell within ±14.2% of the true value. While for lot 943, only 27.3% were within ±8.9% and 72.7% within ±14.2% of the true value. Thus, lot 943 is not acceptable, and no further testing of it is planned.

Still searching for a new lot of reagent strips, the third initial lot testing is conducted with lot 941 and lot 564. For lot 941, only 43.3% of the fingerstick observations are within ±8.9% and 73.3% within ±14.2%. The lot 564 appears promising with 70.0% within ±8.9% and 100% within ±14.2%. For the next round of testing, lot 564 is the lot chosen to collect 6 month followup worksite-wide cholesterol values.

The overall mean percent differences between the desktop results and the reference lab results were not all within the established NCEP guidelines for accuracy (5.0%). Only the February 1993 lot testing of 941 and 942 meets the guidelines with -3.6% and -5-1% respectively. Lot testing of 564 resulted in an overall mean percent difference of -6.5% which is outside the guidelines, yet it still appears to be a viable lot of reagent strips.

Finally, it is concluded that all lots of reagent strips result in negatively biased cholesterol values, estimating lower than the gold standard values. Thus, Reflotron obtained cholesterol measurements underestimate the true cholesterol value of an individual.

III. PRECISION TESTING

Design of Experiment

In the second stage of lot testing reagent strips, precision of the process is to be assessed. The experiments are run in order to set tolerance limits for the purpose of quality control checks when performing mass cholesterol screening at worksites. There are several sources of variability associated with cholesterol measurement via the Reflotron desk-top analyzer. In these experiments, frozen samples of sera are utilized here as opposed to an individual's whole blood.

Each vial of frozen sera is actually a pool of approximately 30 different whole blood samples. Hence the exact cholesterol concentration is not known. However, the range in which each vial's cholesterol value should fall is known. The ranges are defined as high (>240 mg/dl), medium (200-239 mg/dl), and low (<200 mg/dl). The assigned technician extracts the sera sample from a specific tube with a glass capillary tube and transfers it onto a reagent strip from a specific vial of strips. The strip is then inserted into a designated Reflotron machine to obtain the cholesterol measurement.

Let the pool the sera belongs to be factor P, the round of experimentation factor R, and the sample of sera from a specific tube factor S. Since each round is different for each pool of sera, factor R is nested within factor P. Also, factor S is nested within factor R and P since each sample of sera is unique for each round within each pool. Let the machine analyzing the sera sample be factor M and the technician conducting the experiment be factor T. Let the vial of reagent strips used to test the sera sample be factor V. Since there should be no significant interactions between factors, no crossed effects are in the model.

The nested, unbalanced, mixed model (Montgomery, 1991) for the February 1993 experiments is:

$$Y_{ijkmnq} = P_i + R_{j(i)} + S_{k(ij)} + M_m + T_n + \varepsilon_{q(ijkmn)}$$

And, the model for the August 1993 and subsequent experiments is given by:

$$Y_{ijkmnpq} = P_i + R_{j(i)} + S_{k(ij)} + M_m + T_n + V_p + \varepsilon_{q(ijkmnp)}$$

where P_i are the fixed pool effects, $R_{j(i)}$ are the random round (within pool) effects, $S_{k(ij)}$ are the random sera samples effects (within round and pool), M_m are the random machine effects, T_n are the random technician effects, V_p are the random strip vial effects, and $\varepsilon_{q(ijkmn)}$ are the random error effects. We assume $R_{j(i)}$, $S_{k(ij)}$, M_m, T_n, V_p, and $\varepsilon_{q(ijkmn)}$ are normally distributed with expectations zero and variances σ_R^2, σ_S^2, σ_M^2, σ_T^2, σ_V^2, and σ_ε^2 , respectively. The variances associated with these random effects are termed variance components and are analyzed in later evaluation in the calculation of tolerance limits, maximum percent error, and confidence intervals for not only the variance components themselves but for the total variability of the model (observation).

Generation of Experimental Design

The SAS (1990) *Proc plan* generates designs by generating a selection of factor levels (for a given factor) for all combinations of levels for factors which precede it. The technicians then run the experiment as designed following strict protocol. First the Reflotrons' optical eyes are checked to insure that the machines are reporting cholesterol values in the "correct" range. With a reagent strip from a lot of previously analyzed strips, a quality control check is performed. A sample of high sera and a sample of low sera are measured separately. If both initial sera samples are within their respective 95% tolerance limits previously set, the machine is accepted and the optical eye cleaned. If either sample fails by falling outside of known 95% tolerance limits, another sample from the failed pool is measured. If the resulting value is within the limits, the experiment continues as mentioned above. Otherwise, the malfunctioning Reflotron machine is replaced with a back-up. Once all machines are calibrated, testing begins with a specified reagent strip and is recorded by the selected technician. At the termination of the experiment, the optical eye in each machine is cleaned once again.

Computation of Tolerance Intervals

The ANOVA results are presented in Table 2. The estimates of variance components from four methods, ANOVA Estimation (TYPE 1), Minimum Variance Unbiased Estimation (MINQUE or MIVQUE(0), Maximum Likelihood Estimation (ML), Restricted Maximum Likelihood Estimation (REML) are given in Table 3.

Table 2

Analysis of Variance with Type III Unweighted Sums of Squares

for Unbalanced Two-Fold Nested Mixed Model

August, 1993 Lot 941

Source of Variation	DF	SS	MS	EMS
Pool	2	64700.33	32350.16	$\theta_{0U} = \sigma_{\varepsilon}^{2} + 2.2116\sigma_{S}^{2} + 11.058\sigma_{R}^{2} + Q_{P}$
Round3	11.91	$3.97\theta_{1U} = \sigma_{\varepsilon}^{2} + 2.6237\sigma_{S}^{2} + 13.119\sigma_{R}^{2}$		
Sample	24	292.06	12.17	$\theta_{2U} = \sigma_{\varepsilon}^{2} + 2.3423\sigma_{S}^{2}$
Machine	14	270.88	19.35	$\theta_{3U} = \sigma_{\varepsilon}^{2} + 3.8464\sigma_{M}^{2}$
Technician	2	25.87	12.93	$\theta_{4U} = \sigma_{\varepsilon}^{2} + 30\sigma_{T}^{2}$
Strip Vial	4	1364.99	341.25	$\theta_{5U} = \sigma_{\varepsilon}^{2} + 8.5411\sigma_{V}^{2}$
Error	40	602.05	15.05	$\theta_{6} = \sigma_{\varepsilon}^{2}$
Total	89	106640.10		

Table 3

Estimation of Variance Components by Different Methods

August, 1993 Lot 941

Variance Component	Method of Estimation			
	TYPE I	MIVQUE	ML	REML
Round σ_R^2	-0.1289	0.1082	0	0
Sample σ_S^2	-0.3428	-0.5242	0	0
Machine σ_M^2	3.8312	4.7897	1.9901	2.0027
Technician σ_T^2	-0.0706	0.0447	0	0
Strip Vial σ_V^2	38.1914	41.2412	32.4448	40.6123
Error σ_ε^2	15.0513	11.5927	13.5874	13.9373

Tolerance limits are derived using REML estimates of the variance components. For each pool of sera, the (1-α)% tolerance limits (De Vor, 1992) are constructed for the February lot testing as

$$\overline{P_i} \pm Z_{\alpha/2}\sqrt{\sigma_R^2 + \sigma_S^2 + \sigma_M^2 + \sigma_T^2 + \sigma_\varepsilon^2}$$

and for the August (Table 4) and subsequent lot testing as:

$$\overline{P_i} \pm Z_{\alpha/2}\sqrt{\sigma_R^2 + \sigma_S^2 + \sigma_M^2 + \sigma_T^2 + \sigma_V^2 + \sigma_\varepsilon^2}$$

where $\overline{P_i}$ is the overall mean cholesterol value for pool i,i=High, Medium, Low.

These tolerance limits are used in the field experiment when using reagent strips from lot 941 to conduct the worksite-wide cholesterol screening.

Table 4

95% Tolerance Limits for Pool i of Sera

August, 1993 Lot 941

POOL	LOWER BOUND	UPPER BOUND
High	211	241
Medium	192	223
Low	133	164

When the random effect for vial of strips is added as a source of variation in the August 1993 precision testing, this variance component is significantly large. Thus, the tolerance intervals computed from this experiment are wider than those from the February 1993 experiment. Action should be taken if precision continues to decrease. Since this effect accounts for most of the variability in the model and is not a controllable factor, the overall experiment seems to run well. The results on precision testing of lot 564 are similar and it is the chosen lot for the 6 month follow-up field experiment.

IV. MAXIMUM PERCENT ERROR

One standardized procedure for reporting and monitoring the accuracy of cholesterol measurements is *maximum percent error* (McShane, 1991, 1992). The maximum percent error is an easily understandable measure of laboratory precision which incorporates all sources of variability in the model.

Since the primary concern in this part of the overall investigation is how well the cholesterol reading obtained by the Reflotron reflects the true cholesterol value, the variance of a particular cholesterol value is modeled. For a particular pool of sera:

$$\overline{Y}_i = \sum_j \sum_k \sum_n Y_{i(jkmnpq)} / JKN$$

Where JKN denotes partially the replication scheme in which J denotes the number of rounds of testing, K denotes the number of samples of sera per round, and N denotes the number of technicians involved in the testing.

For the February 1993 experiments, for a particular pool i=1,

$$Y_{1(jkmnq)} = \sum_{j=1}^{J} \sum_{k=1}^{K} \sum_{n=1}^{N} (P_1 + R_{j(1)} + S_{k(1j)} + M_m + T_n + \varepsilon_{1(jkmnq)}$$

and for the August 1993 and subsequent experiments, for particular pool i=1,

$$Y_{1(jkmnpq)} = \sum_{j=1}^{J} \sum_{k=1}^{K} \sum_{n=1}^{N} (P_1 + R_{j(1)} + S_{k(1j)} + M_m + T_n + V_p + \varepsilon_{1(jkmnpq)}.$$

If we assume that $\overline{Y}_i$ is approximately normally distributed, a measure of accuracy termed the maximum percent error is given by (for the February 1993 lot testing):

$$P(\alpha) = z_{\alpha/2} \left[\frac{\hat{\sigma}_R^2}{c_1} + \frac{\hat{\sigma}_S^2}{c_2} + \frac{\hat{\sigma}_M^2}{c_3} + \frac{\hat{\sigma}_T^2}{c_4} + \frac{\hat{\sigma}_\varepsilon^2}{c_5} \right]^{\frac{1}{2}} \Big/ \overline{P}_i$$

and (for the August 1993 lot testing):

$$P(\alpha) = z_{\alpha/2} \left[\frac{\hat{\sigma}_R^2}{c_1} + \frac{\hat{\sigma}_S^2}{c_2} + \frac{\hat{\sigma}_M^2}{c_3} + \frac{\hat{\sigma}_T^2}{c_4} + \frac{\hat{\sigma}_V^2}{c_5} + \frac{\hat{\sigma}_\varepsilon^2}{c_6} \right]^{\frac{1}{2}} \Big/ \overline{P}_i$$

$$P(\alpha) = z_{\alpha/2} \left[\frac{\hat{\sigma}_R^2}{c_1} + \frac{\hat{\sigma}_S^2}{c_2} + \frac{\hat{\sigma}_M^2}{c_3} + \frac{\hat{\sigma}_T^2}{c_4} + \frac{\hat{\sigma}_V^2}{c_5} + \frac{\hat{\sigma}_\varepsilon^2}{c_6} \right]^{\frac{1}{2}} \Big/ \overline{P}_j$$

where c_1, c_2, c_3, c_4, c_5, c_6 are constants dependent on the number of times each particular factor level is employed in the experiment. They are derived separately for each experiment since the factor levels of each effect in the model are subject to change. This implies that the cholesterol measurement $\overline{Y_i}$ will be within 100*P(α)% of the true value of the sample mean with probability 1-α (McShane, 1992). From this we can also find the coefficient of variation of $\overline{Y_i}$ as: CV($\overline{Y_i}$)=P(α)/$Z_{\alpha/2}$

For example, in the August 1993 precision experiment using lot 941, the three different pools of sera were measured in two rounds (J=2) with five samples per round (K=5) on a randomly assigned machine (M=15) by a particular technician (N=3) with a reagent strip from a randomly assigned vial of strips (P=5). The variance of each effect can be derived.

$$V\left(\sum_{j=1}^{J}\sum_{k=1}^{K}\sum_{n=1}^{N} R_{j(1)}\right) = V\left(\sum_{j=1}^{J} 15R_{j(1)}\right) = 225 * V\left(\sum_{j=1}^{J} R_{j(1)}\right) = 225 * 2\sigma_R^2 = 450\sigma_R^2$$

$$V\left(\sum_{j=1}^{J}\sum_{k=1}^{K}\sum_{n=1}^{N} S_{k(1j)}\right) = V\left(\sum_{j=1}^{J}\sum_{k=1}^{K} 3S_{k(1j)}\right) = 9 * V\left(\sum_{j=1}^{J}\sum_{k=1}^{K} S_{k(1j)}\right) = 9 * (5 * 2)\sigma_S^2 = 90\sigma_S^2$$

$$V\left(\sum_{j=1}^{J}\sum_{k=1}^{K}\sum_{n=1}^{N} M_m\right) = V(a_1M_1 + a_2M_2 + \ldots + a_mM_m) = a_1^2\sigma_M^2 + a_2^2\sigma_M^2 + \ldots + a_m^2\sigma_M^2 = 84\sigma_M^2$$

where $a_1 + a_2 + \ldots + a_m = 30$, a_w= the number of times machine w is utilized in the experiment, and m=1,2, ... , 15.

$$V\left(\sum_{j=1}^{J}\sum_{k=1}^{K}\sum_{n=1}^{N} T_n\right) = V\left(\sum_{n=1}^{N} 5 * 2T_n\right) = 100 * V\left(\sum_{n=1}^{N} T_n\right) = 100 * 3\sigma_T^2 = 300\sigma_T^2$$

$$V\left(\sum_{j=1}^{J}\sum_{k=1}^{K}\sum_{n=1}^{N} V_p\right) = V(b_1V_1 + b_2V_2 + \ldots + b_pV_p) = b_1^2\sigma_v^2 + b_2^2\sigma_v^2 + \ldots + b_p^2\sigma_v^2 = 180\sigma_v^2$$

where b_z=6=the number of times vial of reagent strips z is utilized in the experiment and p=1, 2, ... , 5.

$$V\left(\sum_{j=1}^{J}\sum_{k=1}^{K}\sum_{n=1}^{N}\varepsilon_{1jkmnpq}\right)=30\sigma_{\varepsilon}^{2}$$

Thus, this implies that the variance of $\overline{Y}_i$ is given by:

$$V(\overline{Y}_i)=\frac{450\sigma_R^2+90\sigma_S^2+84\sigma_M^2+300\sigma_T^2+180\sigma_V^2+30\sigma_\varepsilon^2}{30^2}=\frac{\sigma_R^2}{2}+\frac{\sigma_S^2}{10}+\frac{7\sigma_M^2}{75}+\frac{\sigma_T^2}{3}+\frac{\sigma_V^2}{5}+\frac{\sigma_\varepsilon^2}{30}$$

which in turn gives the following equation for the maximum percent error:

$$P(\alpha)=z_{\alpha/2}\left[\frac{\sigma_R^2}{2}+\frac{\sigma_S^2}{10}+\frac{7\sigma_M^2}{75}+\frac{\sigma_T^2}{3}+\frac{\sigma_V^2}{5}+\frac{\sigma_\varepsilon^2}{30}\right]^{\frac{1}{2}}\Big/\overline{P}_i$$

The maximum percent error of the HIGH pool in the August 1993 lot testing is:

$$P(.05)=1.96\left[\frac{0}{2}+\frac{0}{10}+\frac{7(2.00268)}{75}+\frac{0}{3}+\frac{40.61229}{5}+\frac{13.93725}{30}\right]^{\frac{1}{2}}/226.04042=0.0257\infty 2.57\%$$

With 95% probability, the summary measurement for a pool will be within 2.57% of the true value. And the coefficient of variation is:

$$CV=P(.05)\Big/z_{\alpha/2}=0.0257/1.96=0.0131\infty 1.31\%$$

The maximum percent errors and coefficients of variation are given in Table 5. For each experiment run, the maximum percent error for each pool of sera is small (less than 4%). The results from the testing were well within the allowable ranges for coefficient of variation (cv<5%) set by the NCEP.

Table 5

Maximum Percent Error & Coefficient of Variation

August, 1993 Lot 941

POOL	MAXIMUM PERCENT ERROR	COEFFICIENT OF VARIATION
High	2.57%	1.31%
Medium	2.79%	1.43%
Low	3.91%	1.99%

V. THE FIELD EXPERIMENT

The types of worksites recruited for this study are primarily manufacturing and distribution companies with 250 to 800 employees that employ at least one full-time occupational nurse. All worksites chosen are within sixty miles of Memorial Hospital in Pawtucket, Rhode Island.

The volunteers participating in this experiment are required to meet certain criteria such as they must be at least 20 years old and not pregnant, must not have suffered a stroke, heart attack, or surgery in the previous 2 months.

The proposed daily quality control protocol employed in the field is the following. At the beginning and end of every day that a Reflotron machine is used the technician must:

1. Test the first sample of sera (HIGH or LOW).
2. If the first sample falls within the 95% tolerance range (for the specific lot of reagent strips being used), document the value and proceed with the scheduled screenings.
3. If the first sample is outside of the 95% tolerance range, test another sera sample (from the same pool).
4. If the second sample is outside of the range in the same direction, reject the machine. Begin procedure again with another Reflotron desktop analyzer. The rejected machine is eliminated from further use until it has been repaired.

This type of quality control check is suggested (CSCS, 1990; Morgan, 1986) to be made periodically throughout the day while cholesterol testing is in progress (after every 20th measurement).

The Two-Step Adjustment

These Reflotron obtained cholesterol values must be adjusted to reflect the actual or true cholesterol value. The object is to transform the fingerstick value to a gold standard value.

The first step in the adjustment of cholesterol measurements is to make a daily calibration as follows. For each individual cholesterol measurement taken, certain information is recorded about the machine and technician involved. For this particular machine and technician on a specific day are recorded quality control check values. The median values for the high and low sera QC checks ($\overline{P_{mi}}$) are calculated. Now, these median values are linked to the overall mean cholesterol values for pool high and pool low ($\overline{P_i}$) in the precision lot testing for the particular lot of reagent strips used. Regressing the daily median high and low QC checks on the overall high and low mean cholesterol values gives the following equation: $Z=\alpha_0+\alpha_1 Y$ where Z is the observed cholesterol value measured on a particular day at a specified worksite, and Y is the observed cholesterol measurement adjusted for time. Let z_0 be the observed of Z. A point estimate (Ott, 1993) of the corresponding value of Y is given by: $y_0 = (z_0 - \alpha_0)/\alpha_1$

Now, an adjustment must be made to transform a Reflotron obtained cholesterol value into a true cholesterol value. The relationship between fingerstick obtained cholesterol values measured by a specific lot of reagent strips and gold standard values was determined as:

$$Y = \beta_0 + \beta_1 X$$

Since y_0 is the observed value of Y, a point estimate of the corresponding value of X is given by:

$$x_0 = \frac{y_0 - \beta_0}{\beta_1}$$

The following example illustrates this two-step adjustment procedure. Suppose "Joe Smith" had his cholesterol level measured at his worksite on a particular day. The technician in charge used Reflotron machine "C" with a reagent strip from lot 941. The resulting fingerstick value obtained is 230mg/dl.

On that day, the median values for the high and low sera quality control checks were

Low Sera	148	174	157	156	$\overline{P_{ml}} = 156.5$
High Sera	240	234		235	$\overline{P_{mh}} = 235$

Notice that there is one more low sera check than high sera check. This is because the low sera check value of 174mg/dl fell outside of the 95% tolerance interval for the low pool. The technician was following protocol by checking another low sera sample to determine if the machine should be rejected. However, the next sera sample resulted in a cholesterol value within the specified range so testing continued with machine "C".

Next, Z ($\overline{P_{ml}}$ and $\overline{P_{mh}}$) are regressed on Y ($\overline{P_l}$ and $\overline{P_h}$) and the overall mean cholesterol values per pool from the precision testing are $\overline{P_l} = 157.28$ and $\overline{P_h} = 239.75$. The resulting equation is $Z = 6.791+0.952Y$ and Joe Smith's Reflotron obtained cholesterol value $z_0 = 230$ after adjustment for day-to-day variation is: $y_0 = \frac{z_0 - \alpha_0}{\alpha_1} = \frac{230-6.791}{0.952} = 234.50$

The second step of the adjustment procedure will transform Joe Smith's adjusted Reflotron obtained cholesterol value to a gold standard value. It was determined from the initial lot testing of lot 941 that the Reflotron value (Y) is related to the true cholesterol value (X) by the following regression equation: $Y = 5.624 + 0.936X$

Taking $y_0 = 234.50$, Joe Smith's estimated true cholesterol value is:

$$x_0 = \frac{y_0 - \beta_0}{\beta_1} = \frac{234.50-5.624}{0.936} = 244.43$$

Hence Joe Smith, who obtained a cholesterol value of 230mg/dl from the worksite screening, is placed in the moderate risk category. However, his estimated true cholesterol level is 14mg/dl higher. Joe Smith actually falls into the high cholesterol risk category. The 95% prediction interval (Ott, 1993) for x_0 is computed to be (223.0, 266.8). Joe Smith should be concerned with his cholesterol level.

VI. CONCLUSION

The Reflotron obtained cholesterol measurements are negatively biased, consistently underestimating the true cholesterol values. The maximum percent errors are small for each experiment conducted (P(.05)<=4%). The precision testing resulted in selecting lot 941 for the baseline field experiment and lot 564 for the 6-month follow-up. The major component to the variability of the model appears to be the technician-to-technician variability in the February experiments and the vial-to-vial variation in reagent strips in the August experiments.

The field experiment yields cholesterol measurements on participants at worksites. These values underestimate the true cholesterol levels of the participants. The two-step adjustment procedure utilizes the results of the accuracy and precision experiments to determine the estimated true cholesterol values. These predicted values are to be utilized in the analysis of the field experiment. The construction of prediction intervals on individual predicted cholesterol estimates help minimize the misclassification of individuals into the cholesterol risk categories.

The most important advantage of the Reflotron machine is its portability to where people are thus increasing participation rates and decreasing expenses. However, some precautions are necessary in using these machines in the field. Adequate cleaning of the Reflotron machine is critical to the accuracy of the test. Protocol must be followed to maintain reliable results. Deviations from protocol can reduce the accuracy, as can distractions in the testing area. Even subjecting the Reflotron machine and materials to excessive temperature changes can reduce the accuracy.

ACKNOWLEDGMENTS

We thank Dr. Tom Lasater of the Pawtucket Heart Health Program for permission for the use of the data and to his associates for valuable suggestions.

REFERENCES

American Heart Association (1989). *1989 Heart Facts*. Dallas Texas: American Heart Association.

Canadian Society of Clinical Chemists (1990). Out-of-Laboratory Measurement of the Blood Cholesterol Level. *Canadian Medical Association Journal*. 143, 721-723.

DeVor, R., et al. (1992). *Statistical Quality Design and Control*. Macmillan Publishing Co., NY, NY.

McShane, L., et al. (1991). Reporting the Accuracy of Biochemical Measurements for Epidemiologic and Nutrition Studies. *American Journal of Clinical Nutrition*. 53, 1354-1360.

McShane, L., et al. (1992). Application of Variance Components Methods to Laboratory Quality Control for Biochemical Measurements. *American Statistical Association Proceedings of the Biopharmaceutical Section.* 250-256.

Montgomery, D. (1991). *Design and Analysis of Experiments, Third Edition.* John Wiley and Sons, Inc., NY, NY.

Morgan, J., MRFIT (1986). Quality Control of Technical Procedures and Data Acquisition. *Controlled Clinical Trials: Design, Methods, and Analysis*. 7, IS-202S.

National Institute of Health (1984). Treatment of Hypertriclyceridemia. NIH Consensus Development Conference Summary. *Arteriosclerosis*. 4, 296-301.

Ott, R. (1993). *An Introduction to Statistical Methods and Data Analysis, Fourth Edition.* Duxbury Press, Belmont, CA.

Powers, D. (1992). Establishing and Maintaining Performance Claims. *Arch Pathol Lab Med.* 116, 718-725.

Sedor, F., et al. (1988). Reflotron Cholesterol Measurement Evaluated as a Screening Technique. *Clinical Chemistry*. 34, 2542-2545.

Sempos, C., et al. (1989). The Prevalence of High Blood Cholesterol Levels Among Adults in the United States. *Journal of the American Medical Association.* 262, 45-52.

SAS Institute. (1990). *SAS/STAT User's Guide, Version 6, Fourth Edition,* SAS Institute, Inc., Cary, NJ.

INDEX

L

M

O

P

Q

R

S

T

U

V